Ulrich Langer, Olaf Steinbach (Eds.)
Space-Time Methods

Radon Series on Computational and Applied Mathematics

Volume 25

Space-Time Methods

Applications to Partial Differential Equations

Edited by
Ulrich Langer and Olaf Steinbach

DE GRUYTER

Mathematics Subject Classification 2010
Primary: 65M38, 65M60, 65M50; Secondary: 35F16, 35K05, 35L05

Editors
Prof. Dr. Ulrich Langer
Institute for Computational Mathematics
Johannes Kepler University Linz
Altenberger Straße 69
4040 Linz
Austria
ulanger@numa.uni-linz.ac.at

Prof. Dr. Olaf Steinbach
Institute of Applied Mathematics
Graz University of Technology
Steyrergasse 30
8010 Graz
Austria
o.steinbach@tugraz.at

ISBN 978-3-11-054787-0
e-ISBN (PDF) 978-3-11-054848-8
e-ISBN (EPUB) 978-3-11-054799-3
ISSN 1865-3707

Library of Congress Control Number: 2019947134

Bibliographic information published by the Deutsche Nationalbibliothek
The Deutsche Nationalbibliothek lists this publication in the Deutsche Nationalbibliografie; detailed bibliographic data are available on the Internet at http://dnb.dnb.de.

Typesetting: VTeX UAB, Lithuania
Printing and binding: CPI books GmbH, Leck

www.degruyter.com

Preface

This volume contains seven chapters that provide some recent developments in the formulation, analysis, and implementation of space-time discretization methods for the numerical solution of time-dependent partial differential equations. The contributions result from the workshop on "Space-time methods for partial differential equations" held at the Radon Institute for Computational and Applied Mathematics (RICAM) in Linz, Austria, November 7–11, 2016. This workshop was the second workshop within the special semester on "Computational methods in science and engineering", which took place in Linz, October 10–December 16, 2016, see also the website

https://www.ricam.oeaw.ac.at/specsem/specsem2016/

Almost 70 scientists participated in the workshop, see Figure 1. Within 36 presentations, an overwiew on space-time discretization methods was given. Although the analysis of time-dependent partial differential equations is well established since a long time, global simulation techniques for the numerical solution in the space-time domain are rather new, which is mainly due to today's availability of high-performing computing facilities. However, first applications in elastodynamics were already presented in the 1980s.

Figure 1: Participants of the second workshop of the special semester 2016 at RICAM.

https://doi.org/10.1515/9783110548488-201

The main characteristics of space-time methods is that the discretization is done with respect to the space-time domain, interpreting time as additional dimension, and hence requesting four-dimensional finite element meshes in the most general case. Whereas the solution of the resulting global system of linear or nonlinear algebraic equations requires the use of parallel computing architectures to handle the larger amount of data due to the space-time discretization, it also allows the use of time-parallel solution methods, and the use of adaptive resolutions simultaneously in space and time.

Applications of space-time discretization methods are numerous, and still in its infancy. Probably the most prominent examples are optimal control problems subject to time-dependent partial differential equations, where the optimality conditions couple the primal problem with the adjoint problem, which is backward in time. Therefore, it requires the solution of a coupled system in the space-time domain. Other application areas cover moving geometries and moving interfaces, and acoustic and electromagnetic scattering problems.

This collection of selected contributions contains original and review papers that are arranged in an alphabetical order. We are now going to give short summaries of these contributions.

The contribution of Dohr, Niino, and Steinbach describes the use of space-time boundary element methods for the numerical solution of the heat equation as model problem. Whereas the analysis of boundary integral equations for the heat equation and related time-stepping boundary element methods is well established, the focus of this paper is on approximation properties of rather general space-time boundary element spaces.

Dörfler, Findeisen, Wieners, and Ziegler use a goal-oriented dual-weighted error estimator to drive an adaptive discontinuous Galerkin method in space and time for the solution of acoustic and elastic waves. The resulting linear system is solved in parallel, using a multilevel preconditioner.

The contribution of Ernesti and Wieners discusses a space-time discontinuous Petrov–Galerkin scheme for the simulation of acoustic waves. This approach is based on first-order Friedrichs systems, and it utilizes results for operators and semigroups for hyperbolic systems. A simplified DPG method with ansatz functions, which are discontinuous on the faces of the space-time skeleton, is then used to compute a diffraction pattern.

Gopalakrishnan and Sepúlveda present a space-time discontinuous Petrov–Galerkin method for the linear wave equation. Error estimates and numerical results are given for different meshes of the space-time domain, and a built-in error estimator is used to drive an adaptive refinement strategy.

The contribution of Langer, Matculevich, and Repin combines the space-time approach for parabolic evolution problems with an isogeometric analysis and a posteriori error estimators of the functional type to drive an adaptive scheme. They provide a rigorous numerical analysis, and the numerical results illustrate an improved convergence of the computed global approximations.

Neumüller and Karabelas describe the generation of admissible space-time simplex finite element meshes for moving objects in three space dimensions. These meshes are then used to solve the transient Stokes equations in moving domains, using an interior penalty Galerkin approach in space, and an upwind scheme in time.

The last contribution by Steinbach and Yang gives a detailed review on space-time finite element methods for the numerical solution of parabolic evolution equations in the Bochner space-setting. Besides an overview on different discretization strategies, the particular focus is on different a posteriori error estimation technique, the use of resulting adaptive refinement strategies, and the solution by means of geometric and algebraic multigrid problems.

The careful reviewing process was only possible with the help of the anonymous referees who did an invaluable work that helped the authors to improve their contributions. Furthermore, we would like to thank the administrative and technical staff of RICAM for their support during the special semester. Last but not least, we express our thanks to Apostolos Damialis and Nadja Schedensack from de Gruyter Berlin for continuing support and patience while preparing this volume.

Linz, Graz
March 2019

Ulrich Langer
Olaf Steinbach

Contents

Stefan Dohr, Kazuki Niino, and Olaf Steinbach

1 Space-time boundary element methods for the heat equation

Abstract: In this chapter, we describe a space-time boundary element method for the numerical solution of the time-dependent heat equation. As model problem, we consider the initial Dirichlet boundary value problem, where the solution can be expressed in terms of given Dirichlet and initial data, and the unknown Neumann datum, which is determined by the solution of an appropriate boundary integral equation. For its numerical approximation, we consider a discretization, which is done with respect to a space-time decomposition of the boundary of the space-time domain. This space-time discretization technique allows us to parallelize the computation of the global solution of the whole space-time system. Besides the widely-used tensor product approach, we also consider an arbitrary decomposition of the space-time boundary into boundary elements, allowing us to apply adaptive refinement in space and time simultaneously. In addition to the analysis of the boundary integral operators and the formulation of boundary element methods for the initial Dirichlet boundary value problem, we state a priori error estimates of the approximations. Moreover, we present numerical experiments to confirm the theoretical findings.

Keywords: Heat equation, space-time boundary element methods, a priori error estimates

MSC 2010: 65M38, 65R20, 65M50, 65M12

1 Introduction

Let $\Omega \subset \mathbb{R}^n$ $(n = 1,2,3)$ be a bounded domain with, for $n = 2,3$, Lipschitz boundary $\Gamma := \partial\Omega$, $T \in \mathbb{R}$ with $T > 0$, and $\alpha \in \mathbb{R}$ a fixed heat capacity constant with $\alpha > 0$. As model problem, we consider the initial Dirichlet boundary value problem

Acknowledgement: This work was supported by the International Research Training Group 1754, funded by the German Research Foundation (DFG) and the Austrian Science Fund (FWF). K. Niino was supported by "The John Mung Program", the Kyoto University Global Frontier Project for Young Professionals.

Stefan Dohr, Olaf Steinbach, Institute of Applied Mathematics, TU Graz, Steyrergasse 30, 8010 Graz, Austria, e-mails: stefan.dohr@tugraz.at, o.steinbach@tugraz.at
Kazuki Niino, Department of Advanced Mathematical Sciences, Kyoto University, 606-8501 Kyoto, Japan, e-mail: niino@acs.i.kyoto-u.ac.jp

https://doi.org/10.1515/9783110548488-001

$$\begin{aligned} \alpha\partial_t u(x,t) - \Delta_x u(x,t) &= f(x,t) && \text{for } (x,t) \in Q := \Omega \times (0,T), \\ u(x,t) &= g(x,t) && \text{for } (x,t) \in \Sigma := \Gamma \times (0,T), \\ u(x,0) &= u_0(x) && \text{for } x \in \Omega \end{aligned} \tag{1.1}$$

with given source term f, Dirichlet datum g, and initial datum u_0. Unique solvability of problem (1.1) in the setting of anisotropic Sobolev spaces [22] was shown in, for example, [5, 13, 36]. An explicit formula describing the solution of problem (1.1) is given by the so-called representation formula for the heat equation, see, for example, [2]. For $(x,t) \in Q$ we have

$$\begin{aligned} u(x,t) = &\int_\Omega U^\star(x-y,t)u_0(y)\,dy + \frac{1}{\alpha}\int_Q U^\star(x-y,t-\tau)f(y,\tau)\,dy\,d\tau \\ &+ \frac{1}{\alpha}\int_\Sigma U^\star(x-y,t-\tau)\partial_{n_y}u(y,\tau)\,ds_y\,d\tau \\ &- \frac{1}{\alpha}\int_\Sigma \partial_{n_y}U^\star(x-y,t-\tau)g(y,\tau)\,ds_y\,d\tau, \end{aligned} \tag{1.2}$$

where

$$U^\star(x-y,t-\tau) = \begin{cases} \left(\frac{\alpha}{4\pi(t-\tau)}\right)^{n/2} \exp\left(\frac{-\alpha|x-y|^2}{4(t-\tau)}\right), & x,y \in \mathbb{R}^n, 0 \le \tau < t, \\ 0, & \text{else} \end{cases}$$

denotes the fundamental solution of the heat equation [12]. Due to this representation of the solution, it suffices to determine the Neumann datum $\partial_n u_{|\Sigma}$ to compute the solution u of problem (1.1). Hence, the problem is reduced to the lateral boundary Σ of the space-time domain Q. We can determine the unknown Neumann datum $\partial_n u_{|\Sigma}$ by applying the Dirichlet trace operator to the representation formula (1.2), and solving the resulting space-time boundary integral equation. The approximation of the solution only requires a decomposition of the space-time boundary Σ into boundary elements. Thus, in the case of space-time boundary element methods, the dimension of the problem is reduced to n compared to $n+1$ for space-time finite elements methods discussed in, for example, [34, 36].

Boundary integral equations and corresponding boundary element methods for the approximation of the solution of the initial Dirichlet boundary value problem for the heat equation (1.1) have been studied for a long time [2, 3, 5, 18]. Besides well-known time-stepping methods [4], the convolution quadrature method [23] or the Nyström method [38, 39], one can use the Galerkin approach [5, 16, 25, 26, 27, 29] for the discretization of the global space-time integral equation. Space-time discretization methods in general are gaining in popularity due to their ability to drive adaptivity in space and time simultaneously, and to use parallel iterative solution strategies for time-dependent problems [6, 14, 28]. The global space-time nature of the system matrices leads to improved parallel scalability in distributed memory systems in contrast to time-stepping methods, where the parallelization is limited to the spatial dimension. However, to get a competitive space-time solver compared to, for example, time-stepping schemes, an efficient iterative solution technique for the global space-time system is necessary. That is, the solution requires an application of suitable preconditioners. In [6, 7, 8], a robust preconditioning strategy for space-time integral equations for the heat equation based on boundary integral operators of opposite order [17, 35] is discussed. A parallel solver for space-time boundary element methods for the heat equation was introduced in [10] and extended to the preconditioned system in [9].

In this paper, we analyze the heat potentials in (1.2) and the arising boundary integral operators, and the solvability of the space-time boundary integral equations. The analysis of the boundary integral operators and equations is mainly based on [2, 3, 5]. We start with a discussion of the domain variational formulation of (1.1) (see [36]), and derive the mapping properties of the related boundary integral operators, and the ellipticity of the single layer and hypersingular boundary integral operators. Moreover, we discuss two different space-time discretization methods to compute an approximation of the unknown Neumann datum $\partial_n u_{|\Sigma}$. The first one is the so-called tensor product approach [26, 29], originating from a separate decomposition of the boundary Γ and the time interval $(0, T)$. In this case, we use space-time tensor product spaces for the discretization of the boundary integral equation. The second approach is using boundary element spaces, which are defined with respect to a shape-regular triangulation of the whole space-time boundary $\Sigma = \Gamma \times (0, T)$ into boundary elements. This approach additionally allows us to apply adaptive refinement in space and time simultaneously, while maintaining the regularity of the boundary element mesh. We also present some numerical experiments to confirm the theoretical results.

The structure of the paper is as follows: In Section 2, we give a short overview of the functional framework for the numerical analysis of problem (1.1), that is, introducing anisotropic Sobolev spaces on the space-time domain Q and anisotropic Sobolev spaces on the space-time boundary Σ [21, 22]. In Section 3, we recall existence and uniqueness results [13, 20, 36] for the domain variational formulation of (1.1). This domain variational formulation is later used to prove the ellipticity of the single layer and hypersingular boundary integral operators. Sections 4 and 5 are devoted to the analysis of the arising boundary integral operators and boundary integral equations. In Section 6, we introduce the already mentioned space-time discretization techniques, define suitable boundary element spaces, and derive approximation properties of related L^2 projection operators. The space-time trial and test spaces are then used for the discretization of the boundary integral equations in Section 7, where we also derive a priori error estimates for the Galerkin approximation of the unknown Neumann datum. In Section 8, we provide results of numerical experiments validating the introduced discretization techniques, and we conclude with a brief outlook in Section 9.

2 Functional framework

The analysis of problem (1.1) is done in anisotropic Sobolev spaces, which are introduced and discussed in this section. Under certain conditions, we can define trace operators acting on those spaces, and therefore provide conditions for the given Dirichlet datum g and the unknown Neumann datum $\partial_n u_{|\Sigma}$ of the solution, resulting in existence and uniqueness theorems for solutions of the model problem (1.1). The definitions and results in this section are mainly based on [21, 22, 36]. We start with the definition of anisotropic Sobolev spaces on the space-time domain Q in Section 2.1, and extend this to the space-time boundary Σ in Sections 2.2 and 2.3.

2.1 Anisotropic Sobolev spaces on Q

The anisotropic Sobolev space $H^{1,1/2}(Q)$ is defined as

$$H^{1,1/2}(Q) := L^2(0,T;H^1(\Omega)) \cap H^{1/2}(0,T;L^2(\Omega)),$$

where

$$H^{1/2}(0,T;L^2(\Omega)) := \{u \in L^2(Q) : |u|_{H^{1/2}(0,T;L^2(\Omega))} < \infty\}$$

with

$$|u|^2_{H^{1/2}(0,T;L^2(\Omega))} := \int_0^T\int_0^T \frac{\|u(\cdot,t)-u(\cdot,\tau)\|^2_{L^2(\Omega)}}{|t-\tau|^2}\, d\tau\, dt.$$

The space $L^2(0,T;H^1(\Omega))$ denotes the Bochner space as introduced in, for example, [40, Section 23.2]. The norm of a function $u \in H^{1,1/2}(Q)$ is given by

$$\|u\|^2_{H^{1,1/2}(Q)} := \|u\|^2_{L^2(Q)} + \|\nabla_x u\|^2_{L^2(Q)} + |u|^2_{H^{1/2}(0,T,L^2(\Omega))}.$$

Moreover, we define the space of functions in $H^{1,1/2}(Q)$ with zero initial conditions:

$$H^{1,1/2}_{;0,}(Q) := \{u \in H^{1,1/2}(Q) : \|u\|_{H^{1,1/2}_{;0,}(Q)} < \infty\},$$

where

$$\|u\|^2_{H^{1,1/2}_{;0,}(Q)} := \|u\|^2_{H^{1,1/2}(Q)} + |u|^2_{H^{1/2}_{0,}(0,T;L^2(\Omega))}$$

with

$$|u|^2_{H^{1/2}_{0,}(0,T;L^2(\Omega))} := \int_0^T \frac{\|u(\cdot,t)\|^2_{L^2(\Omega)}}{t}\, dt, \tag{2.1}$$

and write $H^{1,1/2}_{;0,}(Q) = L^2(0,T;H^1(\Omega)) \cap H^{1/2}_{0,}(0,T;L^2(\Omega))$, where

$$H^{1/2}_{0,}(0,T;L^2(\Omega)) = \{u = \tilde{u}_{|(0,T)} : \tilde{u} \in H^{1/2}(-\infty,T;L^2(\Omega)) : \tilde{u}(t) = 0,\ t < 0\}.$$

The space of functions in $H^{1,1/2}_{;0,}(Q)$ having zero boundary conditions is defined as

$$H^{1,1/2}_{0;0,}(Q) := L^2(0,T;H^1_0(\Omega)) \cap H^{1/2}_{0,}(0,T;L^2(\Omega)),$$

and is equipped with the norm

$$\|u\|^2_{H^{1,1/2}_{0;0,}(Q)} := \|\nabla_x u\|^2_{L^2(Q)} + |u|^2_{H^{1/2}(0,T,L^2(\Omega))} + |u|^2_{H^{1/2}_{0,}(0,T;L^2(\Omega))}.$$

In the same way, we introduce the space of functions in $H^{1,1/2}(Q)$ vanishing at the final time T, that is,

$$H^{1,1/2}_{;,0}(Q) := L^2(0,T;H^1(\Omega)) \cap H^{1/2}_{,0}(0,T;L^2(\Omega)),$$

and

$$H^{1,1/2}_{0;,0}(Q) := L^2(0,T;H^1_0(\Omega)) \cap H^{1/2}_{,0}(0,T;L^2(\Omega)).$$

In this case, the seminorm (2.1) is replaced by

$$|u|^2_{H^{1/2}_{,0}(0,T;L^2(\Omega))} := \int_0^T \frac{\|u(\cdot,t)\|^2_{L^2(\Omega)}}{T-t}\,dt. \tag{2.2}$$

Moreover, we define the space

$$H^{1,1/2}_{;0,}(Q,\mathcal{L}) := \{u \in H^{1,1/2}_{;0,}(Q) : \mathcal{L}u \in L^2(Q)\},$$

where $\mathcal{L} := \alpha\partial_t - \Delta_x$ denotes the differential operator of the heat equation. The norm of a function $u \in H^{1,1/2}_{;0,}(Q,\mathcal{L})$ is then given by

$$\|u\|^2_{H^{1,1/2}_{;0,}(Q,\mathcal{L})} := \|u\|^2_{H^{1,1/2}_{;0,}(Q)} + \|\mathcal{L}u\|^2_{L^2(Q)}.$$

The definition of the space $H^{1,1/2}_{;,0}(Q,\mathcal{L}')$, where $\mathcal{L}' := -\alpha\partial_t - \Delta_x$ denotes the operator of the adjoint heat equation, follows the same path.

2.2 Anisotropic Sobolev spaces on Σ

The spaces $H^{r,s}(\Sigma)$ for $r, s \geq 0$ are defined in a similar way. We set

$$H^{r,s}(\Sigma) := L^2(0,T;H^r(\Gamma)) \cap H^s(0,T;L^2(\Gamma)).$$

For a smooth spatial boundary Γ, these spaces are defined for arbitrary $r,s \geq 0$. However, for a general Lipschitz boundary Γ, the spaces $H^{r,s}(\Sigma)$ are only defined for $0 \leq r \leq 1$ and $s \geq 0$. For $r,s \in (0,1)$, a norm is given by

$$\|u\|^2_{H^{r,s}(\Sigma)} := \|u\|^2_{L^2(\Sigma)} + |u|^2_{L^2(0,T;H^r(\Gamma))} + |u|^2_{H^s(0,T;L^2(\Gamma))}$$

with

$$|u|^2_{L^2(0,T;H^r(\Gamma))} := \int\limits_\Gamma \int\limits_\Gamma \frac{\|u(x,\cdot) - u(y,\cdot)\|^2_{L^2(0,T)}}{|x-y|^{n-1+2r}} \, ds_y \, ds_x,$$

and

$$|u|^2_{H^s(0,T;L^2(\Gamma))} := \int\limits_0^T \int\limits_0^T \frac{\|u(\cdot,t) - u(\cdot,\tau)\|^2_{L^2(\Gamma)}}{|t-\tau|^{1+2s}} \, d\tau \, dt.$$

The following lemma is essential for the numerical analysis of the approximation properties of L^2 projections on boundary element spaces, which are defined with respect to an arbitrary triangulation of the space-time boundary Σ. Since we will work with shape-regular elements only, the lemma basically implies that we can use the approximation properties in standard Sobolev spaces $H^s(\Sigma)$ for $s \geq 0$ (see, for example, [24, 33]), to obtain the convergence results in anisotropic spaces.

Lemma 2.1. *For $r,s \in [0,1]$, the continuous embeddings*

$$H^{\max(r,s)}(\Sigma) \hookrightarrow H^{r,s}(\Sigma) \hookrightarrow H^{\min(r,s)}(\Sigma) \tag{2.3}$$

hold.

Proof. Let $u \in H^{r,s}(\Sigma)$ for $r,s \in [0,1]$, and define $m := \min(r,s)$ and $M := \max(r,s)$. Since $H^r(\Gamma) \hookrightarrow H^m(\Gamma)$ [19, Theorem 4.2.2], and $H^s((0,T)) \hookrightarrow H^m((0,T))$, we have

$$\begin{aligned}\|u\|^2_{H^{m,m}(\Sigma)} &\cong \|u\|^2_{H^{m,0}(\Sigma)} + \|u\|^2_{H^{0,m}(\Sigma)} \\ &\leq c\big(\|u\|^2_{H^{r,0}(\Sigma)} + \|u\|^2_{H^{0,s}(\Sigma)}\big) \leq c\|u\|^2_{H^{r,s}(\Sigma)};\end{aligned} \tag{2.4}$$

therefore, $H^{r,s}(\Sigma) \hookrightarrow H^{m,m}(\Sigma)$. According to [24, Theorem B.11 ff.] and [21, 22], and since $H^1(\Sigma) \cong H^{1,1}(\Sigma)$, we have

$$H^{m,m}(\Sigma) = [L^2(\Sigma); H^{1,1}(\Sigma)]_m \cong [L^2(\Sigma); H^1(\Sigma)]_m = H^m(\Sigma). \tag{2.5}$$

Hence, $\|u\|_{H^{m,m}(\Sigma)} \cong \|u\|_{H^m(\Sigma)}$. Therefore, $H^{r,s}(\Sigma) \hookrightarrow H^m(\Sigma)$ follows. The proof of the first equality in (2.5) follows the same path as described in [22, Proposition 2.1] in the case of anisotropic Sobolev spaces on Q.

To prove the continuous embedding $H^M(\Sigma) \hookrightarrow H^{r,s}(\Sigma)$, we use $H^M(\Gamma) \hookrightarrow H^r(\Gamma)$ and $H^M((0,T)) \hookrightarrow H^s((0,T))$. Analogously, to estimate (2.4) and relation (2.5), we obtain $H^M(\Sigma) \cong H^{M,M}(\Sigma)$ and $\|u\|^2_{H^{r,s}(\Sigma)} \leq c\|u\|^2_{H^{M,M}(\Sigma)}$. Therefore, we conclude $H^M(\Sigma) \hookrightarrow H^{r,s}(\Sigma)$. □

Let us now introduce the subspace

$$H^{r,s}_{;0,0}(\Sigma) := L^2(0,T;H^r(\Gamma)) \cap H^s_{0,0}(0,T;L^2(\Gamma)),$$

which is the closure in $H^{r,s}(\Sigma)$ of the subspace of functions vanishing in a neighborhood of $t = 0$ and $t = T$. Anisotropic Sobolev spaces on Σ with negative order $r, s < 0$ are defined as

$$H^{r,s}(\Sigma) := [H^{-r,-s}_{;0,0}(\Sigma)]', \quad \widetilde{H}^{r,s}(\Sigma) := [H^{-r,-s}(\Sigma)]'.$$

Remark 2.1. For $r \geq 0$ and $0 \leq s < \frac{1}{2}$, we have $H^{r,s}_{;0,0}(\Sigma) = H^{r,s}(\Sigma)$; therefore, $H^{-r,-s}(\Sigma) = \widetilde{H}^{-r,-s}(\Sigma)$.

For a function $u \in C(\overline{Q})$, we define the interior Dirichlet trace

$$\gamma_0^{\text{int}} u(x,t) := \lim_{\Omega \ni \tilde{x} \to x \in \Gamma} u(\tilde{x}, t) \quad \text{for } (x,t) \in \Sigma.$$

Hence, $\gamma_0^{\text{int}} u$ coincides with the restriction of u to the space-time boundary Σ. That is, we have $\gamma_0^{\text{int}} u = u_{|\Sigma}$. This operator can be extended to the anisotropic Sobolev space $H^{1,1/2}(Q)$.

Theorem 2.1 (Trace theorem, [22, Theorem 2.1]). *The interior Dirichlet trace operator*

$$\gamma_0^{int} : H^{1,1/2}(Q) \to H^{1/2,1/4}(\Sigma)$$

is linear and bounded, satisfying

$$\|\gamma_0^{int} u\|_{H^{1/2,1/4}(\Sigma)} \leq c_T \|u\|_{H^{1,1/2}(Q)} \quad \textit{for all } u \in H^{1,1/2}(Q).$$

Lemma 2.2 ([5, Lemma 2.4]). *The interior Dirichlet trace operator γ_0^{int} is bounded and surjective from $H^{1,1/2}_{;0,}(Q)$ to $H^{1/2,1/4}(\Sigma)$.*

Theorem 2.2 (Inverse trace theorem). *The interior Dirichlet trace operator γ_0^{int} : $H^{1,1/2}_{;0,}(Q) \to H^{1/2,1/4}(\Sigma)$ has a continuous right inverse operator*

$$\mathcal{E}_0 : H^{1/2,1/4}(\Sigma) \to H^{1,1/2}_{;0,}(Q),$$

satisfying $\gamma_0^{int} \mathcal{E}_0 v = v$ for all $v \in H^{1/2,1/4}(\Sigma)$ and

$$\|\mathcal{E}_0 v\|_{H^{1,1/2}_{;0,}(Q)} \leq c_{IT} \|v\|_{H^{1/2,1/4}(\Sigma)} \quad \textit{for all } v \in H^{1/2,1/4}(\Sigma).$$

Proof. The proof is similar to [13, Theorem 4.9], see also [5]. □

2.3 Piecewise smooth functions on Σ

For a closed, piecewise smooth boundary $\Gamma = \bigcup_{j=1}^{J} \overline{\Gamma}_j$ with $\Gamma_i \cap \Gamma_j = \emptyset$ for $i \neq j$, where Γ_j are open parts of the boundary Γ, we set $\Sigma_j := \Gamma_j \times (0,T)$ for $j = 1,\dots,J$. We then have $\overline{\Sigma} = \bigcup_{j=1}^{J} \overline{\Sigma}_j$. For $r \geq 0$ and $s \geq 0$, we define the anisotropic Sobolev space on the open part Σ_j of the space-time boundary Σ,

$$H^{r,s}(\Sigma_j) := \{v = \widetilde{v}_{|\Sigma_j} : \widetilde{v} \in H^{r,s}(\Sigma)\},$$

and the space of piecewise smooth functions on Σ,

$$H^{r,s}_{\mathrm{pw}}(\Sigma) := \{v \in L^2(\Sigma) : v_{|\Sigma_j} \in H^{r,s}(\Sigma_j) \text{ for } j = 1,\dots,J\},$$

with the norm

$$\|v\|_{H^{r,s}_{\mathrm{pw}}(\Sigma)} := \left(\sum_{j=1}^{J} \|v_{|\Sigma_j}\|^2_{H^{r,s}(\Sigma_j)} \right)^{1/2}.$$

For $r, s < 0$, the anisotropic Sobolev space on Σ_j is defined as the corresponding dual space:

$$\widetilde{H}^{r,s}(\Sigma_j) := [H^{-r,-s}(\Sigma_j)]'.$$

The space of piecewise smooth functions on Σ with negative order is then given by

$$H^{r,s}_{\mathrm{pw}}(\Sigma) := \prod_{j=1}^{J} \widetilde{H}^{r,s}(\Sigma_j),$$

with the norm

$$\|w\|_{H^{r,s}_{\mathrm{pw}}(\Sigma)} := \sum_{j=1}^{J} \|w_{|\Sigma_j}\|_{\widetilde{H}^{r,s}(\Sigma_j)}.$$

Lemma 2.3. *For $r, s < 0$ and $w \in H^{r,s}_{pw}(\Sigma)$, there holds*

$$\|w\|_{\widetilde{H}^{r,s}(\Sigma)} \leq \|w\|_{H^{r,s}_{pw}(\Sigma)}.$$

Proof. Let $w \in H^{r,s}_{\mathrm{pw}}(\Sigma)$. By duality, we conclude

$$\begin{aligned}
\|w\|_{\widetilde{H}^{r,s}(\Sigma)} &= \sup_{0 \neq v \in H^{-r,-s}(\Sigma)} \frac{|\langle w, v\rangle_\Sigma|}{\|v\|_{H^{-r,-s}(\Sigma)}} \leq \sup_{0 \neq v \in H^{-r,-s}(\Sigma)} \sum_{j=1}^{J} \frac{|\langle w, v\rangle_{\Sigma_j}|}{\|v\|_{H^{-r,-s}(\Sigma)}} \\
&\leq \sup_{0 \neq v \in H^{-r,-s}(\Sigma)} \sum_{j=1}^{J} \frac{|\langle w_{|\Sigma_j}, v_{|\Sigma_j}\rangle_{\Sigma_j}|}{\|v_{|\Sigma_j}\|_{H^{-r,-s}(\Sigma_j)}} \\
&\leq \sum_{j=1}^{J} \sup_{0 \neq v_j \in H^{-r,-s}(\Sigma_j)} \frac{|\langle w_{|\Sigma_j}, v_j\rangle_{\Sigma_j}|}{\|v_j\|_{H^{-r,-s}(\Sigma_j)}} = \|w\|_{H^{r,s}_{\mathrm{pw}}(\Sigma)}.
\end{aligned}$$

□

Note that for a Lipschitz boundary Γ, we have to assume $|r| \leq 1$ to keep the validity of the above statements.

3 Domain variational formulation

In this section, we introduce and analyze the domain variational formulation of problem (1.1) in the setting of anisotropic Sobolev spaces. We derive Green's formulae for the heat equation in Subsection 3.1. In Subsection 3.2, we recall existence and uniqueness results for the solution of the variational formulation of the model problem with zero initial conditions. The unique solvability of problem (1.1) with a nonhomogeneous initial datum is discussed in Subsection 3.3. The presented results are based on [5, 13, 36, 40]. In Subsection 3.4, we analyze the Neumann trace of solutions of the model problem (1.1).

3.1 Green's formulae

This subsection is devoted to the derivation of Green's first and second formula for the heat equation with respect to the previously introduced setting of anisotropic Sobolev spaces. These formulae are later on used to derive the representation formula for the heat equation and for the analysis of related boundary integral operators, see Section 4. Recall that $\Omega \subset \mathbb{R}^n$ is assumed to be a bounded domain with, for $n = 2,3$, Lipschitz boundary $\Gamma := \partial\Omega$.

Theorem 3.1 ([1, Corollary 7.8]). *Let $u \in C^2(\Omega) \cap C^1(\overline{\Omega})$. Then there holds the classical Green's formula, that is,*

$$\int\limits_{\Omega} [\Delta u(x)v(x) + \nabla u(x) \cdot \nabla v(x)]\,\mathrm{d}x = \int\limits_{\Gamma} \partial_n u(x)v(x)\,\mathrm{d}s_x$$

for all $v \in C^1(\Omega) \cap C(\overline{\Omega})$.

Now consider $u \in C^2(\overline{Q})$. By applying Theorem 3.1, we get

$$\begin{aligned}\int\limits_0^T\int\limits_{\Omega} &[\alpha\partial_t u(x,t) - \Delta_x u(x,t)]v(x,t)\,\mathrm{d}x\,\mathrm{d}t \\ &= \int\limits_0^T\int\limits_{\Omega} [\alpha\partial_t u(x,t)v(x,t) + \nabla_x u(x,t) \cdot \nabla_x v(x,t)]\,\mathrm{d}x\,\mathrm{d}t \\ &\quad - \int\limits_0^T\int\limits_{\Gamma} \partial_{n_x} u(x,t)v(x,t)\,\mathrm{d}s_x\,\mathrm{d}t.\end{aligned} \tag{3.1}$$

This equation is the so-called Green's first formula for the heat equation. Using integration by parts on the first term of the right hand side and rearranging the terms yields

$$
\begin{aligned}
\alpha \int_\Omega u(x,T)v(x,T)\,\mathrm{d}x
&= \alpha \int_\Omega u(x,0)v(x,0)\,\mathrm{d}x \\
&\quad + \int_0^T \int_\Omega [\alpha\partial_t u(x,t) - \Delta_x u(x,t)]v(x,t)\,\mathrm{d}x\,\mathrm{d}t + \int_0^T \int_\Gamma \partial_{n_x} u(x,t)v(x,t)\,\mathrm{d}s_x\,\mathrm{d}t \\
&\quad + \int_0^T \int_\Omega [\alpha u(x,t)\partial_t v(x,t) - \nabla_x u(x,t)\cdot\nabla_x v(x,t)]\,\mathrm{d}x\,\mathrm{d}t.
\end{aligned}
$$

Again, by applying Theorem 3.1, we get

$$
\begin{aligned}
\alpha \int_\Omega u(x,T)v(x,T)\,\mathrm{d}x = \alpha \int_\Omega u(x,0)v(x,0)\,\mathrm{d}y
&+ \int_0^T \int_\Omega [\alpha\partial_t u(x,t) - \Delta_x u(x,t)]v(x,t)\,\mathrm{d}x\,\mathrm{d}t \\
&- \int_0^T \int_\Omega [-\alpha\partial_t v(x,t) - \Delta_x v(x,t)]u(x,t)\,\mathrm{d}x\,\mathrm{d}t \\
&+ \int_0^T \int_\Gamma \partial_{n_x} u(x,t)v(x,t)\,\mathrm{d}s_x\,\mathrm{d}t - \int_0^T \int_\Gamma \partial_{n_x} v(x,t)u(x,t)\,\mathrm{d}s_x\,\mathrm{d}t.
\end{aligned}
\tag{3.2}
$$

This equation is the so-called Green's second formula for the heat equation. Our aim is to extend these formulae to the more general case of functions in $H^{1,1/2}(Q)$. To do so, we use the following density results:

Lemma 3.1 ([5, Lemma 2.22]). *Let $C_0^\infty(\overline{\Omega} \times (0,T])$ be the space of restrictions to $\overline{Q}$ of functions in $C_0^\infty(\mathbb{R}^n \times (0,\infty))$. Then $C_0^\infty(\overline{\Omega} \times (0,T])$ is dense in $H^{1,1/2}_{;0,}(Q,\mathcal{L})$.*

Analogously, we obtain the following result, where $C_0^\infty(\overline{\Omega} \times [0,T))$ is the space of restrictions to $\overline{Q}$ of functions in $C_0^\infty(\mathbb{R}^n \times (-\infty,T))$.

Corollary 3.1. *The space $C_0^\infty(\overline{\Omega} \times [0,T))$ is dense in $H^{1,1/2}_{;,0}(Q,\mathcal{L}')$.*

Before we introduce Green's formulae for functions in anisotropic Sobolev spaces, we have to ensure that the bilinear form $\langle \partial_t u, v\rangle_Q$ is well defined. In [36], it was shown that the bilinear form $\langle \partial_t u, v\rangle_Q$ can be extended to functions $u \in H^{1,1/2}_{;0,}(Q)$, $v \in H^{1,1/2}_{;,0}(Q)$, and that there exists a constant $c > 0$, such that

$$
\langle \partial_t u, v\rangle_Q \le c\|u\|_{H^{1,1/2}_{;0,}(Q)}\|v\|_{H^{1,1/2}_{;,0}(Q)} \tag{3.3}
$$

for all $u \in H^{1,1/2}_{;0,}(Q)$ and $v \in H^{1,1/2}_{;,0}(Q)$. Here, and in the following, $\langle\cdot,\cdot\rangle_Q$ denotes the duality pairing as extension of the inner product in $L^2(Q)$.

For a function $u \in C^1(\overline{Q})$, we define the interior Neumann trace

$$\gamma_1^{\text{int}} u(x,t) := \lim_{\Omega \ni \tilde{x} \to x \in \Gamma} n_x \cdot \nabla_{\tilde{x}} u(\tilde{x}, t) \quad \text{for } (x,t) \in \Sigma,$$

which coincides with the conormal derivative of u, that is, we have $\gamma_1^{\text{int}} u = \partial_{n_x} u_{|\Sigma}$. The definition of the Neumann trace operator γ_1^{int} can be extended to the anisotropic Sobolev space $H^{1,1/2}(Q, \mathcal{L})$.

Lemma 3.2. *The interior Neumann trace operator*

$$\gamma_1^{int} : H^{1,1/2}(Q, \mathcal{L}) \to H^{-1/2,-1/4}(\Sigma)$$

is linear and bounded, satisfying

$$\|\gamma_1^{int} v\|_{H^{-1/2,-1/4}(\Sigma)} \le c_{NT} \|v\|_{H^{1,1/2}(Q,\mathcal{L})} \quad \textit{for all } v \in H^{1,1/2}(Q, \mathcal{L}).$$

For $u \in C^2(\overline{Q})$, we have $\gamma_1^{int} u = \partial_{n_x} u_{|\Sigma}$ in the distributional sense.

Proof. Follows the lines of [5, Proposition 2.18]. □

Theorem 3.2 (Green's first formula). *For $u \in H^{1,1/2}_{;0,}(Q, \mathcal{L})$ and $v \in H^{1,1/2}_{;,0}(Q)$, there holds*

$$\alpha \langle \partial_t u, v \rangle_Q + \langle \nabla_x u, \nabla_x v \rangle_{L^2(Q)} = \langle \gamma_1^{int} u, \gamma_0^{int} v \rangle_\Sigma + \langle \mathcal{L} u, v \rangle_Q, \tag{3.4}$$

where $\langle \cdot, \cdot \rangle_\Sigma$ denotes the duality pairing on $H^{-1/2,-1/4}(\Sigma) \times H^{1/2,1/4}(\Sigma)$.

Proof. Let $u \in C_0^\infty(\overline{\Omega} \times (0,T])$. According to (3.1), there holds

$$\langle \mathcal{L} u, v \rangle_{L^2(Q)} = \alpha \langle \partial_t u, v \rangle_{L^2(Q)} + \langle \nabla_x u, \nabla_x v \rangle_{L^2(Q)} - \langle \partial_{n_x} u, v \rangle_{L^2(\Sigma)} \tag{3.5}$$

for all $v \in C_0^\infty(\overline{\Omega} \times [0,T))$. All the terms are continuous with respect to v in the $H^{1,1/2}_{;,0}(Q)$ norm. Hence, we can extend (3.5) by continuity to $v \in H^{1,1/2}_{;,0}(Q)$. Whereas, for fixed $v \in H^{1,1/2}_{;,0}(Q)$, all the terms are continuous with respect to u in the $H^{1,1/2}_{;0,}(Q, \mathcal{L})$ norm. Hence, by applying Lemma 3.1, we can extend (3.5) to $u \in H^{1,1/2}_{;0,}(Q, \mathcal{L})$, which concludes the proof. □

Theorem 3.3 (Green's second formula). *For $u \in H^{1,1/2}_{;0,}(Q, \mathcal{L})$ and $v \in H^{1,1/2}_{;,0}(Q, \mathcal{L}')$, there holds*

$$\langle \mathcal{L} u, v \rangle_Q - \langle u, \mathcal{L}' v \rangle_Q = -\langle \gamma_1^{int} u, \gamma_0^{int} v \rangle_\Sigma + \langle \gamma_0^{int} u, \gamma_1^{int} v \rangle_\Sigma. \tag{3.6}$$

Proof. For $u \in C_0^\infty(\overline{\Omega} \times (0,T])$ and $v \in C_0^\infty(\overline{\Omega} \times [0,T))$, there holds

$$\langle \mathcal{L} u, v \rangle_{L^2(Q)} - \langle u, \mathcal{L}' v \rangle_{L^2(Q)} = -\langle \gamma_1^{\text{int}} u, \gamma_0^{\text{int}} v \rangle_{L^2(\Sigma)} + \langle \gamma_0^{\text{int}} u, \gamma_1^{\text{int}} v \rangle_{L^2(\Sigma)}.$$

As in the proof of Theorem 3.2, we can extend this formula to $u \in H^{1,1/2}_{;0,}(Q, \mathcal{L})$ and $v \in H^{1,1/2}_{;,0}(Q, \mathcal{L}')$ by applying Lemma 3.1 and Corollary 3.1. □

3.2 Homogeneous initial datum

In the following subsection, we discuss, based on [36], the unique solvability of problem (1.1) with zero initial conditions. Let $f \in [H^{1,1/2}_{0;,0}(Q)]'$ and $g \in H^{1/2,1/4}(\Sigma)$ be given. We consider the initial Dirichlet boundary value problem

$$\begin{aligned} \alpha\partial_t u(x,t) - \Delta_x u(x,t) &= f(x,t) && \text{for } (x,t) \in Q, \\ u(x,t) &= g(x,t) && \text{for } (x,t) \in \Sigma, \\ u(x,0) &= 0 && \text{for } x \in \Omega. \end{aligned} \tag{3.7}$$

The variational formulation of problem (3.7) is to find $u \in H^{1,1/2}_{;0,}(Q)$, $u = g$ on Σ, such that

$$a(u,v) = \langle f, v\rangle_Q \quad \text{for all } v \in H^{1,1/2}_{0;,0}(Q) \tag{3.8}$$

with the bilinear form

$$a(u,v) := \alpha\langle \partial_t u, v\rangle_Q + \langle \nabla_x u, \nabla_x v\rangle_{L^2(Q)}$$

for $u \in H^{1,1/2}_{;0,}(Q)$ and $v \in H^{1,1/2}_{;,0}(Q)$. The bilinear form

$$a(\cdot,\cdot) : H^{1,1/2}_{;0,}(Q) \times H^{1,1/2}_{;,0}(Q) \to \mathbb{R}$$

is bounded. That is, there exists a constant $c_2^A > 0$, such that

$$|a(u,v)| \le c_2^A \|u\|_{H^{1,1/2}_{;0,}(Q)} \|v\|_{H^{1,1/2}_{;,0}(Q)}$$

for all $u \in H^{1,1/2}_{;0,}(Q)$, $v \in H^{1,1/2}_{;,0}(Q)$. For the given Dirichlet datum $g \in H^{1/2,1/4}(\Sigma)$, we consider the decomposition $u := \bar{u} + \tilde{u}_g$, where $\tilde{u}_g := \mathcal{E}_0 g$ is an extension of g to the space-time domain Q, satisfying $\gamma_0^{\text{int}}\tilde{u}_g = g$. The boundedness of the inverse trace operator $\mathcal{E}_0 : H^{1/2,1/4}(\Sigma) \to H^{1,1/2}_{;0,}(Q)$ then implies

$$\|\tilde{u}_g\|_{H^{1,1/2}_{;0,}(Q)} \le c_{IT} \|g\|_{H^{1/2,1/4}(\Sigma)}. \tag{3.9}$$

Hence, the variational formulation (3.8) changes to: Find $\bar{u} \in H^{1,1/2}_{0;0,}(Q)$ such that

$$a(\bar{u},v) = \langle f, v\rangle_Q - a(\tilde{u}_g, v) \quad \text{for all } v \in H^{1,1/2}_{0;,0}(Q). \tag{3.10}$$

Theorem 3.4 (Existence and uniqueness [36]). *The variational formulation* (3.10) *implies an isomorphism*

$$\mathcal{L} : H^{1,1/2}_{0;0,}(Q) \to [H^{1,1/2}_{0;,0}(Q)]',$$

satisfying

$$\|\bar{u}\|_{H^{1,1/2}_{0;0,}(Q)} \le 2\|\mathcal{L}\bar{u}\|_{[H^{1,1/2}_{0;,0}(Q)]'} \quad \textit{for all } \bar{u} \in H^{1,1/2}_{0;0,}(Q).$$

Note that $\mathcal{L} : H^{1,1/2}_{0;0,}(Q) \to [H^{1,1/2}_{0;,0}(Q)]'$ is the continuous extension of the heat operator as considered in Section 2.1.

Hence, we conclude that the variational problem (3.10) is uniquely solvable; therefore, $u = \bar{u} + \tilde{u}_g$ is the unique solution of the variational problem (3.8). A direct consequence of Theorem 3.4 is the stability estimate

$$\frac{1}{2}\|\bar{u}\|_{H^{1,1/2}_{0;0,}(Q)} \le \sup_{0\neq v\in H^{1,1/2}_{0;,0}(Q)} \frac{a(\bar{u},v)}{\|v\|_{H^{1,1/2}_{0;,0}(Q)}} \quad \text{for all } \bar{u} \in H^{1,1/2}_{0;0,}(Q). \tag{3.11}$$

Theorem 3.5. *For $f \in [H^{1,1/2}_{0;,0}(Q)]'$ and $g \in H^{1/2,1/4}(\Sigma)$, there exists a unique solution $u \in H^{1,1/2}_{0;0,}(Q)$ of the variational problem* (3.8)*, satisfying*

$$\|u\|_{H^{1,1/2}_{;0,}(Q)} \le c_R\|f\|_{[H^{1,1/2}_{0;,0}(Q)]'} + c_B\|g\|_{H^{1/2,1/4}(\Sigma)}. \tag{3.12}$$

Proof. Unique solvability is a result of Theorem 3.4. The stability condition (3.11) and the boundedness of the bilinear form $a(\cdot,\cdot)$ imply

$$\begin{aligned}\frac{1}{2}\|\bar{u}\|_{H^{1,1/2}_{0;0,}(Q)} &\le \sup_{0\neq v\in H^{1,1/2}_{0;,0}(Q)} \frac{a(\bar{u},v)}{\|v\|_{H^{1,1/2}_{0;,0}(Q)}} = \sup_{0\neq v\in H^{1,1/2}_{0;,0}(Q)} \frac{\langle f,v\rangle_Q - a(\tilde{u}_g,v)}{\|v\|_{H^{1,1/2}_{0;,0}(Q)}}\\ &\le \|f\|_{[H^{1,1/2}_{0;,0}(Q)]'} + c\|\tilde{u}_g\|_{H^{1,1/2}_{;0,}(Q)}.\end{aligned}$$

The assertion follows by using the triangle inequality for $u = \bar{u} + \tilde{u}_g$, the Poincaré inequality, and the stability (3.9) of the inverse trace operator. □

3.3 Nonhomogeneous initial datum

The following analysis, in the case of a given initial datum and zero boundary conditions, is mainly based on [40, Chapter 23] and [34]. In this subsection, we only recall the main results. Let $u_0 \in L^2(\Omega)$ be given. We consider the initial Dirichlet boundary value problem

$$\begin{aligned}\alpha\partial_t u(x,t) - \Delta_x u(x,t) &= 0 &&\text{for } (x,t) \in Q,\\ u(x,t) &= 0 &&\text{for } (x,t) \in \Sigma,\\ u(x,0) &= u_0(x) &&\text{for } x \in \Omega.\end{aligned} \tag{3.13}$$

The analysis of problem (3.13) is done in the space $\mathcal{V}_0(Q)$, defined as

$$\mathcal{V}_0(Q) := L^2(0,T;H^1_0(\Omega)) \cap H^1(0,T;H^{-1}(\Omega)).$$

The norm of a function $u \in \mathcal{V}_0(Q)$ is given by

$$\|u\|^2_{\mathcal{V}_0(Q)} := \|u\|^2_{L^2(0,T;H^1_0(\Omega))} + \|\alpha\partial_t u\|^2_{L^2(0,T;H^{-1}(\Omega))},$$

where

$$\|u\|_{L^2(0,T;H_0^1(\Omega))} := \|\nabla_x u\|_{L^2(Q)},$$

and

$$\|\alpha\partial_t u\|_{L^2(0,T;H^{-1}(\Omega))} := \sup_{0\neq v\in L^2(0,T;H_0^1(\Omega))} \frac{\langle \alpha\partial_t u, v\rangle_Q}{\|v\|_{L^2(0,T;H_0^1(\Omega))}}.$$

Analogously, we define the space $\mathcal{V}(Q)$ of functions with nonhomogeneous boundary conditions, that is,

$$\mathcal{V}(Q) := L^2(0,T;H^1(\Omega)) \cap H^1(0,T;H^{-1}(\Omega))$$

with norm

$$\|u\|_{\mathcal{V}(Q)}^2 := \|u\|_{L^2(Q)}^2 + \|u\|_{\mathcal{V}_0(Q)}^2.$$

Theorem 3.6 ([40, Theorem 23.A]). *For $u_0 \in L^2(\Omega)$, there exists a unique solution $u \in \mathcal{V}_0(Q)$ of problem* (3.13), *satisfying the stability estimate*

$$\|u\|_{\mathcal{V}_0(Q)} \leq c_I \|u_0\|_{L^2(\Omega)}.$$

The spaces $\mathcal{V}(Q)$ and $\mathcal{V}_0(Q)$ are dense subspaces of $H^{1,1/2}(Q)$ and $H^{1,1/2}_{0;,}(Q)$, respectively [5, 21]. Moreover, the following norm equivalence holds:

Lemma 3.3. *For $u \in \mathcal{V}(Q)$ with $\mathcal{L}u = 0$ in Q, the norms of $\mathcal{V}(Q)$ and $H^{1,1/2}(Q)$ are equivalent. That is, there exist constants $c_1, c_2 > 0$, such that*

$$\|u\|_{\mathcal{V}(Q)} \leq c_1 \|u\|_{H^{1,1/2}(Q)} \leq c_2 \|u\|_{\mathcal{V}(Q)}.$$

Proof. Follows the lines of [5, Lemma 2.15]. □

Additionally, if $u \in \mathcal{V}_0(Q)$, that is, u vanishes on the boundary Σ, we immediately conclude that there exist constants $\tilde{c}_1, \tilde{c}_2 > 0$, such that

$$\|u\|_{\mathcal{V}_0(Q)} \leq \tilde{c}_1 \|u\|_{H^{1,1/2}_{0;,}(Q)} \leq \tilde{c}_2 \|u\|_{\mathcal{V}_0(Q)}. \tag{3.14}$$

This follows by using the Poincaré inequality and Lemma 3.3.

An important property of functions $u \in \mathcal{V}(Q)$ is the continuity in time, that is, we have

$$u \in C([0,T]; L^2(\Omega)). \tag{3.15}$$

Hence, the initial trace $\tau_0 u := u_{|t=0} \in L^2(\Omega)$ of the solution u of problem (3.13) is well defined.

The unique solution $u \in H^{1,1/2}(Q)$ of the fully nonhomogeneous initial Dirichlet boundary value problem is then given as $u = \bar{u}_g + \bar{u}_0$, where $\bar{u}_g \in H^{1,1/2}_{;0,}(Q)$ is the unique solution of problem (3.7) with zero initial conditions, and $\bar{u}_0 \in \mathcal{V}_0(Q)$ is the unique solution of problem (3.13). By applying the stability estimate of Theorem 3.5, the Poincaré inequality, estimate (3.14), and Theorem 3.6, we obtain the following stability estimate for the solution $u \in H^{1,1/2}(Q)$:

$$\|u\|_{H^{1,1/2}(Q)} \leq c_R \|f\|_{[H^{1,1/2}_{0;,0}(Q)]'} + c_B \|g\|_{H^{1/2,1/4}(\Sigma)} + c\|u_0\|_{L^2(\Omega)}.$$

The initial trace of solutions $u \in H^{1,1/2}(Q)$ of problem (1.1) is well defined due to $u = \bar{u}_g + \bar{u}_0$ with $\bar{u}_g \in H^{1,1/2}_{;0,}(Q)$ and $\bar{u}_0 \in \mathcal{V}_0(Q)$. We set $\tau_0 u := \tau_0 \bar{u}_0 \in L^2(\Omega)$ according to (3.15).

3.4 Neumann trace operator

For the solution $u \in H^{1,1/2}_{;0,}(Q, \mathcal{L})$ of (3.7) with $f \in L^2(Q)$, we can determine the associated conormal derivative $\gamma_1^{\text{int}} u \in H^{-1/2,-1/4}(\Sigma)$ as the unique solution of the variational problem

$$\langle \gamma_1^{\text{int}} u, z \rangle_\Sigma = a(u, \mathcal{E}_T z) - \langle f, \mathcal{E}_T z \rangle_Q \quad \text{for all } z \in H^{1/2,1/4}(\Sigma), \tag{3.16}$$

where $\mathcal{E}_T := \mathcal{H}_T \mathcal{E}_0 : H^{1/2,1/4}(\Sigma) \to H^{1,1/2}_{;,0}(Q)$. Note that Green's first formula for the heat equation is, as in (3.16), considered for test functions $v \in H^{1,1/2}_{;,0}(Q)$. To end up with a Galerkin–Bubnov setting, we are interested in an operator $\mathcal{H}_T : H^{1,1/2}_{;0,}(Q) \to H^{1,1/2}_{;,0}(Q)$, which is defined as follows: For $u \in L^2(Q)$, we consider the series representation

$$u(x,t) = \sum_{i=1}^{\infty} U_i(t)\phi_i(x), \quad U_i(t) = \sum_{k=0}^{\infty} u_{i,k} v_k(t) \quad \text{for } (x,t) \in Q, \tag{3.17}$$

where $\phi_i \in H^1_0(\Omega)$ are the eigenfunctions of the Dirichlet eigenvalue problem

$$-\Delta\phi = \mu\phi \quad \text{in } \Omega, \quad \phi = 0 \quad \text{on } \Gamma,$$

and $v_k \in H^1_{0,}(0,T)$ are given by

$$v_k(t) = \sin\left(\left(\frac{\pi}{2} + k\pi\right)\frac{t}{T}\right), \quad k \in \mathbb{N}_0. \tag{3.18}$$

The coefficients $u_{i,k}$ in (3.17) are given by

$$u_{i,k} = \frac{2}{T}\int_0^T \int_\Omega u(x,t) v_k(t)\phi_i(x)\,dx\,dt.$$

Then $\mathcal{H}_T u$ is defined as

$$(\mathcal{H}_T u)(x,t) = \sum_{i=1}^{\infty} (\mathcal{H}_T U_i)(t)\phi_i(x) \quad \text{for } (x,t) \in Q,$$

where

$$(\mathcal{H}_T U_i)(t) := \sum_{k=0}^{\infty} u_{i,k} \cos\left(\left(\frac{\pi}{2} + k\pi\right)\frac{t}{T}\right).$$

The operator $\mathcal{H}_T : L^2(Q) \to L^2(Q)$ and its restriction $\mathcal{H}_T : H^{1,1/2}_{;0,}(Q) \to H^{1,1/2}_{;,0}(Q)$ define isometric isomorphisms. That is, we have

$$\|\mathcal{H}_T u\|_{L^2(Q)} = \|u\|_{L^2(Q)} \quad \text{for all } u \in L^2(Q),$$

and

$$\|\mathcal{H}_T u\|_{H^{1,1/2}_{;,0}(Q)} = \|u\|_{H^{1,1/2}_{;0,}(Q)} \quad \text{for all } u \in H^{1,1/2}_{;0,}(Q).$$

A detailed analysis of the transformation operator $\mathcal{H}_T$ is given in [36].

We now get the following stability estimate for the Neumann trace of solutions u of problem (3.7):

Theorem 3.7. *Let $u \in H^{1,1/2}_{;0,}(Q)$ be the unique solution of problem (3.7) with $f \in L^2(Q)$ and $g \in H^{1/2,1/4}(\Sigma)$. Then the Neumann trace $\gamma_1^{int} u \in H^{-1/2,-1/4}(\Sigma)$ satisfies the stability estimate*

$$\|\gamma_1^{int} u\|_{H^{-1/2,-1/4}(\Sigma)} \le c_{IT}\left(\|f\|_{[H^{1,1/2}_{;,0}(Q)]'} + c_2^A \|u\|_{H^{1,1/2}_{;0,}(Q)}\right).$$

Proof. Using (3.16), the boundedness of the bilinear form $a(\cdot,\cdot)$ and of the operator $\mathcal{E}_T$ yields

$$\begin{aligned} \|\gamma_1^{\text{int}} u\|_{H^{-1/2,-1/4}(\Sigma)} &= \sup_{0 \neq z \in H^{1/2,1/4}(\Sigma)} \frac{\langle \gamma_1^{\text{int}} u, z\rangle_\Sigma}{\|z\|_{H^{1/2,1/4}(\Sigma)}} \\ &= \sup_{0 \neq z \in H^{1/2,1/4}(\Sigma)} \frac{a(u, \mathcal{E}_T z) - \langle f, \mathcal{E}_T z\rangle_Q}{\|z\|_{H^{1/2,1/4}(\Sigma)}} \\ &\le c_{IT}\left(c_2^A \|u\|_{H^{1,1/2}_{;0,}(Q)} + \|f\|_{[H^{1,1/2}_{;,0}(Q)]'}\right). \end{aligned}$$

□

In particular, for the solution u of the initial Dirichlet boundary value problem (1.1) with homogeneous right hand side and initial datum, i. e., $f \equiv 0$ and $u_0 \equiv 0$, we get

$$\|\gamma_1^{\text{int}} u\|_{H^{-1/2,-1/4}(\Sigma)} \le c_{IT} c_2^A \|u\|_{H^{1,1/2}_{;0,}(Q)}.$$

The following lemma is essential for the derivation of the jump conditions of the boundary integral operators in Section 4:

Lemma 3.4 ([5, Lemma 2.23]). *The combined trace map*

$$(\gamma_0^{int}, \gamma_1^{int}) : u \mapsto (\gamma_0^{int} u, \gamma_1^{int} u)$$

maps $C_0^\infty(\overline{\Omega} \times (0,T])$ *onto a dense subspace of* $H^{1/2,1/4}(\Sigma) \times H^{-1/2,-1/4}(\Sigma)$.

Remark 3.1. Lemma 3.4 is also valid if we replace the space $C_0^\infty(\overline{\Omega} \times (0,T])$ by $C_0^\infty(\overline{\Omega} \times [0,T))$.

4 Boundary integral operators

To express the solution of the initial Dirichlet boundary value problem (1.1) by means of heat potentials as in (1.2), the existence of a fundamental solution is essential. In Subsection 4.1, we derive the fundamental solution of the heat equation and the related representation formula. In Subsections 4.2–4.7, we introduce and analyze the heat potentials and the resulting boundary integral operators.

4.1 Representation formula for the heat equation

In this subsection, we derive the representation formula (1.2) for the heat equation. Therefore, we consider Green's second formula (3.2) for $u \in C^2(\overline{Q})$. We want the third integral on the right hand side to be zero. That is, we search for a function v, which is a solution of the adjoint homogeneous heat equation

$$-\alpha \partial_\tau v(y,\tau) - \Delta_y v(y,\tau) = 0 \quad \text{for } (y,\tau) \in Q.$$

Since we want to find a representation of the solution $u = u(x,t)$ of the model problem (1.1), we define v as

$$v(y,\tau) := U(y-x, t-\tau),$$

where $(x,t) \in Q$ is fixed. In this case, we have

$$\partial_\tau v(y,\tau) = \partial_\tau U(y-x,t-\tau) = -\partial_\theta U(y-x,\theta),$$

where $\theta = t - \tau$. Thus,

$$\alpha \partial_\theta U(y-x,\theta) - \Delta_y U(y-x,\theta) = 0 \quad \text{for } (y,\theta) \in Q.$$

We assume the function U to be spherically symmetric. That is, $U(y-x,\theta) = \widetilde{U}(r,\theta)$, where $r = |y-x|$. For $r \neq 0$, we get

$$\alpha \partial_\theta \widetilde{U}(r,\theta) - \partial_{rr} \widetilde{U}(r,\theta) - (n-1)\frac{1}{r}\partial_r \widetilde{U}(r,\theta) = 0. \tag{4.1}$$

With

$$\widetilde{U}(r,\theta) = \theta^\gamma g(z), \quad z = \frac{r}{\sqrt{\theta}}, \quad \gamma \in \mathbb{R}, \quad \theta = t - \tau > 0, \quad \tau < t,$$

we get

$$\begin{aligned}
\partial_\theta \widetilde{U}(r,\theta) &= \gamma\theta^{\gamma-1} g(z) - \frac{1}{2}\theta^{\gamma-1} z g'(z), \\
\partial_r \widetilde{U}(r,\theta) &= g'(z)\theta^{\gamma-\frac{1}{2}}, \\
\partial_{rr} \widetilde{U}(r,\theta) &= g''(z)\theta^{\gamma-1};
\end{aligned}$$

therefore, equation (4.1) becomes

$$\alpha\left[\gamma\theta^{\gamma-1} g(z) - \frac{1}{2}\theta^{\gamma-1} z g'(z)\right] - g''(z)\theta^{\gamma-1} - (n-1)\frac{1}{r} g'(z)\theta^{\gamma-\frac{1}{2}} = 0,$$

which is equivalent to

$$\alpha\left[\gamma g(z) - \frac{1}{2} z g'(z)\right] - g''(z) - (n-1)\frac{1}{z} g'(z) = 0. \tag{4.2}$$

It remains to solve this ordinary differential equation. First, we consider the one-dimensional case $n = 1$. That is, we have

$$\alpha\gamma g(z) - \alpha\frac{1}{2} z g'(z) - g''(z) = 0,$$

which can be written as

$$\alpha\left[\gamma + \frac{1}{2}\right] g(z) - \frac{\mathrm{d}}{\mathrm{d}z}\left[\alpha\frac{1}{2} z g(z) + g'(z)\right] = 0.$$

By choosing $\gamma = -\frac{1}{2}$, we get

$$\frac{\mathrm{d}}{\mathrm{d}z}\left[\alpha\frac{1}{2} z g(z) + g'(z)\right] = 0,$$

and hence

$$\alpha\frac{1}{2} z g(z) + g'(z) = c_0 \in \mathbb{R}$$

follows. In particular, for $c_0 = 0$, and using separation of variables, we get

$$\ln g(z) = -\alpha\frac{1}{4} z^2 + c_1, \quad c_1 \in \mathbb{R},$$

and for $c_1 = 0$, we conclude

$$g(z) = \exp\left(-\frac{\alpha}{4} z^2\right), \tag{4.3}$$

which is a solution of the differential equation (4.2) for $n = 1$. When inserting (4.3) into (4.2) for general n, we get

$$\begin{aligned} 0 &= \alpha\Big[\gamma \exp\Big(-\frac{\alpha}{4}z^2\Big) + \frac{\alpha}{4}z^2 \exp\Big(-\frac{\alpha}{4}z^2\Big)\Big] + \frac{\alpha}{2}\exp\Big(-\frac{\alpha}{4}z^2\Big) \\ &\quad - \frac{\alpha^2}{4}z^2 \exp\Big(-\frac{\alpha}{4}z^2\Big) + (n-1)\frac{\alpha}{2}\exp\Big(-\frac{\alpha}{4}z^2\Big) \\ &= \exp\Big(-\frac{\alpha}{4}z^2\Big)\alpha\Big[\gamma + \frac{n}{2}\Big]. \end{aligned}$$

Thus, (4.3) is also a solution in the two- and three-dimensional case if $\gamma = -\frac{n}{2}$. Recalling the definition of the functions U and $\widetilde{U}$, we therefore have

$$U(y-x, t-\tau) = (t-\tau)^{-n/2}\exp\Big(-\frac{\alpha|y-x|^2}{4(t-\tau)}\Big) \quad \text{for } \tau < t.$$

Due to the singularity of the function U at $(x,t) = (y,\tau)$, we consider the space-time cylinder $Q_{t-\varepsilon} := \Omega \times (0, t-\varepsilon)$, where $0 < \varepsilon < t$. Analogously to (3.2), we get

$$\begin{aligned} \alpha\int_\Omega u(y,t-\varepsilon)v(y,t-\varepsilon)\,dy &= \alpha\int_\Omega u(y,0)v(y,0)\,dy \\ &\quad + \int_0^{t-\varepsilon}\int_\Omega [\alpha\partial_\tau u(y,\tau) - \Delta_y u(y,\tau)]v(y,\tau)\,dy\,d\tau \\ &\quad - \int_0^{t-\varepsilon}\int_\Omega [-\alpha\partial_\tau v(y,\tau) - \Delta_y v(y,\tau)]u(y,\tau)\,dy\,d\tau \\ &\quad + \int_0^{t-\varepsilon}\int_\Gamma \partial_{n_y}u(y,\tau)v(y,\tau)\,ds_y\,d\tau - \int_0^{t-\varepsilon}\int_\Gamma \partial_{n_y}v(y,\tau)u(y,\tau)\,ds_y\,d\tau. \end{aligned}$$

With $v(y,\tau) = U(y-x, t-\tau)$, we now obtain

$$\begin{aligned} \alpha\int_\Omega u(y,t-\varepsilon)U(y-x,\varepsilon)\,dy &= \alpha\int_\Omega u(y,0)U(y-x,t)\,dy \\ &\quad + \int_0^{t-\varepsilon}\int_\Omega [\alpha\partial_\tau u(y,\tau) - \Delta_y u(y,\tau)]U(y-x,t-\tau)\,dy\,d\tau \\ &\quad + \int_0^{t-\varepsilon}\int_\Gamma \partial_{n_y}u(y,\tau)U(y-x,t-\tau)\,ds_y\,d\tau \\ &\quad - \int_0^{t-\varepsilon}\int_\Gamma \partial_{n_y}U(y-x,t-\tau)u(y,\tau)\,ds_y\,d\tau. \end{aligned} \tag{4.4}$$

Let us consider the integral on the left hand side. That is,

$$\alpha \int_{\Omega} u(y,t-\varepsilon)U(y-x,\varepsilon)\,dy = \alpha \int_{\Omega} \varepsilon^{-n/2} u(y,t-\varepsilon)\exp\left(-\frac{\alpha|y-x|^2}{4\varepsilon}\right)dy.$$

By using the Taylor expansion

$$u(y,t-\varepsilon) = u(x,t) + (y-x)^{\top}\nabla_x u(\xi_x,\xi_t) - \varepsilon\partial_t u(\xi_x,\xi_t)$$

with

$$\begin{pmatrix}\xi_x\\ \xi_t\end{pmatrix} = \begin{pmatrix} x+\sigma(y-x)\\ t-\sigma\varepsilon\end{pmatrix}, \qquad \sigma \in (0,1),$$

we get

$$\begin{aligned}
&\frac{\alpha}{\varepsilon^{n/2}}\int_{\Omega} u(y,t-\varepsilon)\exp\left(-\frac{\alpha|y-x|^2}{4\varepsilon}\right)dy\\
&= u(x,t)\frac{\alpha}{\varepsilon^{n/2}}\int_{\Omega}\exp\left(-\frac{\alpha|y-x|^2}{4\varepsilon}\right)dy\\
&\quad+\frac{\alpha}{\varepsilon^{n/2}}\int_{\Omega}(y-x)^{\top}\nabla_x u(\xi_x,\xi_t)\exp\left(-\frac{\alpha|y-x|^2}{4\varepsilon}\right)dy \qquad (4.5)\\
&\quad-\frac{\alpha}{\varepsilon^{n/2-1}}\int_{\Omega}\partial_t u(\xi_x,\xi_t)\exp\left(-\frac{\alpha|y-x|^2}{4\varepsilon}\right)dy.
\end{aligned}$$

Next, we are going to show the convergence of the first integral on the right hand side. First, we consider the spatially one-dimensional case $n = 1$. That is, $\Omega = (a,b)$ with $a,b \in \mathbb{R}$ and $x \in (a,b)$. We have

$$\begin{aligned}
A := {}&\frac{\alpha}{\varepsilon^{1/2}}\int_a^b \exp\left(-\frac{\alpha(y-x)^2}{4\varepsilon}\right)dy\\
= {}&\frac{\alpha}{\varepsilon^{1/2}}\int_a^x \exp\left(-\frac{\alpha(y-x)^2}{4\varepsilon}\right)dy + \frac{\alpha}{\varepsilon^{1/2}}\int_x^b \exp\left(-\frac{\alpha(y-x)^2}{4\varepsilon}\right)dy.
\end{aligned}$$

By using the substitution $z = \frac{x-y}{x-a}$ for the first integral and $z = \frac{y-x}{b-x}$ for the second one we get

$$\begin{aligned}
A = {}&\frac{\alpha}{\varepsilon^{1/2}}(x-a)\int_0^1 \exp\left(-\frac{\alpha(x-a)^2z^2}{4\varepsilon}\right)dz\\
&+\frac{\alpha}{\varepsilon^{1/2}}(b-x)\int_0^1 \exp\left(-\frac{\alpha(b-x)^2z^2}{4\varepsilon}\right)dz.
\end{aligned}$$

The substitution $\frac{\alpha(x-a)^2 z^2}{4\varepsilon} = \eta^2$ for the first and $\frac{\alpha(b-x)^2 z^2}{4\varepsilon} = \eta^2$ for the second integral leads to

$$A = 2\sqrt{\alpha} \int\limits_0^{\frac{(x-a)}{2}\sqrt{\frac{\alpha}{\varepsilon}}} \exp(-\eta^2)\, \mathrm{d}\eta + 2\sqrt{\alpha} \int\limits_0^{\frac{(b-x)}{2}\sqrt{\frac{\alpha}{\varepsilon}}} \exp(-\eta^2)\, \mathrm{d}\eta,$$

and we finally obtain

$$\lim_{\varepsilon \to 0} A = 4\sqrt{\alpha} \int\limits_0^{\infty} \exp(-\eta^2)\, \mathrm{d}\eta = 2\sqrt{\alpha\pi}.$$

In the two-dimensional case, we choose $R > 0$, such that $B_R(x) \subset \Omega$, and consider

$$A := \frac{\alpha}{\varepsilon} \int\limits_{B_R(x)} \exp\left(-\frac{\alpha|y-x|^2}{4\varepsilon}\right) \mathrm{d}y.$$

The integral over $\Omega \setminus B_R(x)$ converges to 0, since $\frac{\alpha}{\varepsilon} \exp(-\frac{\alpha|y-x|^2}{4\varepsilon}) \to 0$ for $y \neq x$ as $\varepsilon \to 0$. By using polar coordinates, we get

$$\begin{aligned} A &= \frac{\alpha}{\varepsilon} \int\limits_0^R \int\limits_0^{2\pi} \exp\left(-\frac{\alpha r^2}{4\varepsilon}\right) r\, \mathrm{d}\varphi\, \mathrm{d}r = \frac{2\pi\alpha}{\varepsilon} \int\limits_0^R \exp\left(-\frac{\alpha r^2}{4\varepsilon}\right) r\, \mathrm{d}r \\ &= 4\pi\left[1 - \exp\left(-\frac{\alpha R^2}{4\varepsilon}\right)\right] \xrightarrow{\varepsilon \to 0} 4\pi. \end{aligned}$$

In the three-dimensional case, we also choose $R > 0$, such that $B_R(x) \subset \Omega$, and consider

$$A := \frac{\alpha}{\varepsilon^{3/2}} \int\limits_{B_R(x)} \exp\left(-\frac{\alpha|y-x|^2}{4\varepsilon}\right) \mathrm{d}y.$$

As in the two-dimensional case, the integral over $\Omega \setminus B_R(x)$ vanishes. By using spherical coordinates, we obtain

$$\begin{aligned} A &= \frac{\alpha}{\varepsilon^{3/2}} \int\limits_0^R \int\limits_0^{2\pi} \int\limits_0^{\pi} \exp\left(-\frac{\alpha r^2}{4\varepsilon}\right) r^2 \sin\theta\, \mathrm{d}\theta\, \mathrm{d}\varphi\, \mathrm{d}r \\ &= \frac{4\pi\alpha}{\varepsilon^{3/2}} \int\limits_0^R \exp\left(-\frac{\alpha r^2}{4\varepsilon}\right) r^2\, \mathrm{d}r. \end{aligned}$$

The substitution $\eta^2 = \frac{\alpha r^2}{4\varepsilon}$ leads to

$$A = \frac{32\pi}{\sqrt{\alpha}} \int\limits_0^{\sqrt{\frac{\alpha}{4\varepsilon}} R} \exp(-\eta^2)\eta^2\, \mathrm{d}\eta \xrightarrow{\varepsilon \to 0} \frac{32\pi}{\sqrt{\alpha}} \int\limits_0^{\infty} \exp(-\eta^2)\eta^2\, \mathrm{d}\eta = \frac{8\pi^{3/2}}{\sqrt{\alpha}}.$$

The other two integrals in (4.5) vanish as $\varepsilon \to 0$ due to the boundedness of $\nabla_x u$ and $\partial_t u$. We finally get the representation formula by taking the limit $\varepsilon \to 0$ in (4.4). That is, we have

$$\begin{aligned} u(x,t) &= \int_\Omega u(y,0)U^\star(x-y,t)\,dy + \frac{1}{\alpha}\int_Q \mathcal{L}u(y,\tau)U^\star(x-y,t-\tau)\,dy\,d\tau \\ &\quad + \frac{1}{\alpha}\int_\Sigma \partial_{n_y} u(y,\tau)U^\star(x-y,t-\tau)\,ds_y\,d\tau \\ &\quad - \frac{1}{\alpha}\int_\Sigma u(y,\tau)\partial_{n_y} U^\star(x-y,t-\tau)\,ds_y\,d\tau, \end{aligned} \tag{4.6}$$

where

$$U^\star(x-y,t-\tau) = \left(\frac{\alpha}{4\pi(t-\tau)}\right)^{n/2} \exp\left(\frac{-\alpha|x-y|^2}{4(t-\tau)}\right) \quad \text{for } \tau < t.$$

The function

$$U^\star(x,t) = \begin{cases} \left(\frac{\alpha}{4\pi t}\right)^{n/2} \exp\left(\frac{-\alpha|x|^2}{4t}\right), & (x,t) \in \mathbb{R}^n \times (0,\infty), \\ 0, & \text{else}, \end{cases} \tag{4.7}$$

is called the fundamental solution of the heat equation, and due to construction, $U^\star$ is a solution of the homogeneous heat equation on $\mathbb{R}^n \times (0,\infty)$, see, for example, [12]. That is,

$$[\alpha\partial_t - \Delta_x]U^\star(x,t) = 0 \quad \text{for } (x,t) \in \mathbb{R}^n \times (0,\infty).$$

Additionally, the fundamental solution has the following properties:

Lemma 4.1. *For $t > 0$, there holds*

$$\int_{\mathbb{R}^n} U^\star(x,t)\,dx = 1.$$

Proof. Let $t > 0$. We have

$$\begin{aligned} \int_{\mathbb{R}^n} U^\star(x,t)\,dx &= \left(\frac{\alpha}{4\pi t}\right)^{n/2} \int_{\mathbb{R}^n} \exp\left(\frac{-\alpha|x|^2}{4t}\right) dx = \pi^{-n/2} \int_{\mathbb{R}^n} \exp(-|z|^2)\,dz \\ &= \pi^{-n/2} \prod_{i=1}^n \int_{\mathbb{R}} \exp(-z_i^2)\,dz_i = 1. \end{aligned}$$

□

Lemma 4.2. *Let $u \in C(\Omega) \cap L^\infty(\Omega)$. For $x \in \Omega$, there holds*

$$\lim_{t\to 0} \int_\Omega U^\star(x-y,t)u(y)\,dy = u(x). \tag{4.8}$$

Proof. Let $\varepsilon > 0$ and $u \in C(\Omega) \cap L^\infty(\Omega)$. We define the function $\tilde{u}$ as

$$\tilde{u}(x) = \begin{cases} u(x) & \text{for } x \in \Omega, \\ 0 & \text{else.} \end{cases}$$

Moreover, let $(x,t) \in \Omega \times (0,\infty)$. Due to Lemma 4.1, and since $U^\star > 0$ on $\mathbb{R}^n \times (0,\infty)$, we have

$$\left| \int_\Omega U^\star(x-y,t)u(y)\,dy - u(x) \right| = \left| \int_{\mathbb{R}^n} U^\star(x-y,t)[\tilde{u}(y) - \tilde{u}(x)]\,dy \right|$$
$$\leq \int_{\mathbb{R}^n} U^\star(x-y,t)|\tilde{u}(y) - \tilde{u}(x)|\,dy.$$

Since u is continuous on Ω and $x \in \Omega$, there exists a constant $\delta > 0$, such that

$$|\tilde{u}(y) - \tilde{u}(x)| < \varepsilon/2 \quad \text{if } |y-x| < \delta.$$

Thus, we write the last integral as

$$\int_{\mathbb{R}^n} U^\star(x-y,t)|\tilde{u}(y) - \tilde{u}(x)|\,dy = \int_{\mathbb{R}^n \backslash B_\delta(x)} U^\star(x-y,t)|\tilde{u}(y) - \tilde{u}(x)|\,dy$$
$$+ \int_{B_\delta(x)} U^\star(x-y,t)|\tilde{u}(y) - \tilde{u}(x)|\,dy.$$

The second integral can be estimated by

$$\int_{B_\delta(x)} U^\star(x-y,t) \underbrace{|\tilde{u}(y) - \tilde{u}(x)|}_{<\varepsilon/2}\,dy < \frac{\varepsilon}{2} \int_{\mathbb{R}^n} U^\star(x-y,t)\,dy = \frac{\varepsilon}{2}.$$

For the first integral, we obtain, due to $u \in L^\infty(\Omega)$,

$$\int_{\mathbb{R}^n \backslash B_\delta(x)} U^\star(x-y,t)|\tilde{u}(y) - \tilde{u}(x)|\,dy \leq 2\|u\|_{L^\infty(\Omega)} \int_{\mathbb{R}^n \backslash B_\delta(x)} U^\star(x-y,t)\,dy.$$

The substitution $z = x - y$ yields

$$\int_{\mathbb{R}^n \backslash B_\delta(x)} U^\star(x-y,t)\,dy = \int_{\mathbb{R}^n \backslash B_\delta(0)} U^\star(z,t)\,dz$$
$$= \left(\frac{\alpha}{4\pi t}\right)^{n/2} \int_{\mathbb{R}^n \backslash B_\delta(0)} \exp\left(\frac{-|z|^2\alpha}{4t}\right) dz,$$

and by using polar coordinates, we get

$$\int_{\mathbb{R}^n \backslash B_\delta(x)} U^\star(x-y,t)\,dy \leq Ct^{-n/2} \int_\delta^\infty r^{n-1} \exp\left(\frac{-r^2\alpha}{4t}\right) dr$$
$$= C' \int_{at^{-1/2}}^\infty \rho^{n-1} \exp(-\rho^2)\,d\rho$$

with suitable constants $C, C' > 0$ and $a = \delta(\frac{\alpha}{4})^{1/2}$. The last integral vanishes as $t \to 0$. That is, for t small enough, there holds

$$\int_{\mathbb{R}^n \setminus B_\delta(x)} U^\star(x-y,t)\big|\tilde{u}(y)-\tilde{u}(x)\big|\,\mathrm{d}y < \varepsilon/2.$$

Altogether, we obtain

$$\left|\int_\Omega U^\star(x-y,t)u(y)\,\mathrm{d}y - u(x)\right| < \varepsilon$$

for t small enough, which concludes the proof. □

One can show that for sufficiently regular input data f, g, and u_0, the solution of the initial Dirichlet boundary value problem (1.1) is given by the representation formula. For $(x,t) \in Q$ we have

$$\begin{aligned} u(x,t) = &\int_\Omega U^\star(x-y,t)u_0(y)\,\mathrm{d}y + \frac{1}{\alpha}\int_Q U^\star(x-y,t-\tau)f(y,\tau)\,\mathrm{d}y\,\mathrm{d}\tau \\ &+ \frac{1}{\alpha}\int_\Sigma U^\star(x-y,t-\tau)\partial_{n_y}u(y,\tau)\,\mathrm{d}s_y\,\mathrm{d}\tau \\ &- \frac{1}{\alpha}\int_\Sigma \partial_{n_y}U^\star(x-y,t-\tau)g(y,\tau)\,\mathrm{d}s_y\,\mathrm{d}\tau. \end{aligned} \tag{4.9}$$

Due to the given representation (4.9) for the solution of problem (1.1), it suffices to determine the yet unknown Cauchy datum $\partial_n u_{|\Sigma}$ to compute the solution in the space-time domain Q. This can be done by applying the Dirichlet trace operator to (4.9), and solving the resulting boundary integral equation on the space-time boundary Σ. The following subsections are devoted to the analysis of the heat potentials in (4.9) and the resulting boundary integral operators.

4.2 Initial potential

Let $u_0 \in L^2(\Omega)$. The function

$$(\widetilde{M}_0 u_0)(x,t) := \int_\Omega U^\star(x-y,t)u_0(y)\,\mathrm{d}y \quad \text{for } (x,t) \in \mathbb{R}^n \times (0,T) \tag{4.10}$$

is called initial potential of the heat equation with initial condition u_0.

Lemma 4.3. *For $u_0 \in L^2(\Omega)$, the initial potential $\widetilde{M}_0 u_0$ satisfies the homogeneous heat equation. That is,*

$$[\alpha\partial_t - \Delta_x](\widetilde{M}_0 u_0)(x,t) = 0 \quad \textit{for all } (x,t) \in \mathbb{R}^n \times (0,T).$$

Proof. For $(x,t) \in \mathbb{R}^n \times (0,T)$, there exists a compact neighborhood O of (x,t), such that $O \subset \mathbb{R}^n \times (0,T)$. The restriction of $U^\star(x-y,t)$ to $(x,t) \in O$ and $y \in \Omega$ is bounded and differentiable on O for $y \in \Omega$. Moreover, $U^\star(x-\cdot,t)$ is integrable over Ω. The Leibniz integral rule then implies that we can interchange differentiation and integration, and we obtain

$$[\alpha\partial_t - \Delta_x](\widetilde{M}_0 u_0)(x,t) = \int_\Omega [\alpha\partial_t - \Delta_x] U^\star(x-y,t) u_0(y)\,\mathrm{d}y.$$

The assertion now follows by using

$$[\alpha\partial_t - \Delta_x] U^\star(x-y,t) = 0$$

for $(x,t) \in \mathbb{R}^n \times (0,T)$ and $y \in \Omega$. □

Theorem 4.1. *The initial potential $\widetilde{M}_0 : L^2(\Omega) \to \mathcal{V}(Q) \subset H^{1,1/2}(Q)$ is linear and bounded. That is, there exists a constant $c > 0$, such that*

$$\|\widetilde{M}_0 u_0\|_{\mathcal{V}(Q)} \le c\|u_0\|_{L^2(\Omega)} \quad \textit{for all } u_0 \in L^2(\Omega).$$

Proof. Follows the lines of the proof of [29, Lemma 7.10] with a restriction to the space $\mathcal{V}(Q)$ at the end. □

Due to Lemma 4.3 and the norm equivalence in Lemma 3.3, we conclude that there exists a constant $c_2^M > 0$, such that

$$\|\widetilde{M}_0 u_0\|_{H^{1,1/2}(Q)} \le c_2^M \|u_0\|_{L^2(\Omega)} \quad \text{for all } u_0 \in L^2(\Omega).$$

An important property of the initial potential is the continuity in time. That is, due to (3.15) and Theorem 4.1, we have $\widetilde{M}_0 u_0 \in C([0,T]; L^2(\Omega))$. Together with Lemma 4.2, this immediately implies $(\widetilde{M}_0 u_0)(x,0) = u_0(x)$ almost everywhere in Ω. Hence, the initial potential satisfies the initial condition.

Due to the mapping properties of the Dirichlet and Neumann trace operators, we finally conclude that the integral operators

$$M_0 := \gamma_0^{\text{int}} \widetilde{M}_0 : L^2(\Omega) \to H^{1/2,1/4}(\Sigma),$$
$$M_1 := \gamma_1^{\text{int}} \widetilde{M}_0 : L^2(\Omega) \to H^{-1/2,-1/4}(\Sigma)$$

are linear and bounded.

4.3 Newton potential

The Newton potential for a given function f defined on the space-time domain Q and $(x,t) \in \mathbb{R}^n \times (0,T)$ is defined as

$$(\widetilde{N}_0 f)(x,t) := \frac{1}{\alpha} \int_0^t \int_\Omega U^\star(x-y,t-\tau) f(y,\tau)\,\mathrm{d}y\,\mathrm{d}\tau. \tag{4.11}$$

Lemma 4.4. *The function $u(x,t) = (\widetilde{N}_0 f)(x,t)$ for $(x,t) \in \mathbb{R}^n \times (0,T)$ for f regular enough is a solution of the heat equation*

$$[\alpha\partial_t - \Delta_x]u(x,t) = \begin{cases} f(x,t), & \text{for } (x,t) \in \Omega \times (0,T), \\ 0, & \text{for } (x,t) \in \Omega^c \times (0,T). \end{cases}$$

Proof. First, let $(x,t) \in \Omega \times (0,T)$. Then

$$\begin{aligned} &[\alpha\partial_t - \Delta_x](\widetilde{N}_0 f)(x,t) \\ &\quad = [\alpha\partial_t - \Delta_x]\left(\lim_{\varepsilon\to 0}\frac{1}{\alpha}\int_0^{t-\varepsilon}\int_\Omega U^\star(x-y,t-\tau)f(y,\tau)\,\mathrm{d}y\,\mathrm{d}\tau\right). \end{aligned}$$

Applying the Leibniz integral rule yields

$$\begin{aligned} &[\alpha\partial_t - \Delta_x]\frac{1}{\alpha}\int_0^{t-\varepsilon}\int_\Omega U^\star(x-y,t-\tau)f(y,\tau)\,\mathrm{d}y\,\mathrm{d}\tau \\ &\quad = \int_0^{t-\varepsilon}\int_\Omega \partial_t U^\star(x-y,t-\tau)f(y,\tau)\,\mathrm{d}y\,\mathrm{d}\tau + \int_\Omega U^\star(x-y,\varepsilon)f(y,t-\varepsilon)\,\mathrm{d}y \\ &\qquad - \frac{1}{\alpha}\int_0^{t-\varepsilon}\int_\Omega \Delta_x U^\star(x-y,t-\tau)f(y,\tau)\,\mathrm{d}y\,\mathrm{d}\tau \\ &\quad = \frac{1}{\alpha}\int_0^{t-\varepsilon}\int_\Omega [\alpha\partial_t - \Delta_x]U^\star(x-y,t-\tau)f(y,\tau)\,\mathrm{d}y\,\mathrm{d}\tau \\ &\qquad + \int_\Omega U^\star(x-y,\varepsilon)f(y,t-\varepsilon)\,\mathrm{d}y. \end{aligned}$$

Since $U^\star(\cdot - y, \cdot - \tau)$ is, for $(y,\tau) \in \Omega \times (0, t-\varepsilon)$, a solution of the homogeneous heat equation, the first integral on the right hand side vanishes. Additionally, Lemma 4.2 implies

$$\int_\Omega U^\star(x-y,\varepsilon)f(y,t-\varepsilon)\,\mathrm{d}y \xrightarrow{\varepsilon\to 0} f(x,t) \quad \text{for } x \in \Omega.$$

Note that $[\alpha\partial_t - \Delta_x]u(x,t) = 0$ for $x \in \Omega^c$ follows analogously by considering a ball $B_R \subset \mathbb{R}^n$ with radius $R > 0$, such that $\Omega \cup \{x\} \subset B_R$, and by choosing a zero extension of f in $B_R \setminus \Omega$. □

The following theorem is essential to derive the mapping properties of the Newton potential and subsequently of the single and double layer potentials in Subsections 4.4 and 4.6. The theorem provides the mapping properties of the convolution with the fundamental solution of the heat equation, see [5, Section 3] and [30, 31].

Theorem 4.2. *The convolution with the fundamental solution* $U^\star$

$$A : \widetilde{H}^{r,r/2}_{\text{comp}}(\mathbb{R}^n \times (0,T)) \to \widetilde{H}^{r+2,r/2+1}_{\text{loc}}(\mathbb{R}^n \times (0,T))$$
$$f \mapsto U^\star * f$$

is linear and continuous for any $r \in \mathbb{R}$.

Here, $\widetilde{H}^{r,r/2}_{\text{comp}}(\mathbb{R}^n \times (0,T))$ denotes the space of functions with compact support in space, whereas the subscript "loc" refers to the local behavior in the spatial variables [5]. Hence, we immediately get the continuity of the Newton potential

$$\widetilde{N}_0 : [H^{1,1/2}_{;,0}(Q)]' \to \widetilde{H}^{1,1/2}_{\text{loc}}(\mathbb{R}^n \times (0,T)),$$

and by restriction, we obtain the following mapping properties:

Theorem 4.3. *The Newton potential* $\widetilde{N}_0 : [H^{1,1/2}_{;,0}(Q)]' \to H^{1,1/2}_{;0,}(Q)$ *is linear and bounded. That is, there exists a constant* $c_2^N > 0$, *such that*

$$\|\widetilde{N}_0 f\|_{H^{1,1/2}_{;0,}(Q)} \le c_2^N \|f\|_{[H^{1,1/2}_{;,0}(Q)]'} \quad \textit{for all } f \in [H^{1,1/2}_{;,0}(Q)]'.$$

Proof. Follows by applying Theorem 4.2 with $r = -1$ and, by restriction, to the space-time domain Q. □

The application of the interior Dirichlet trace operator to the Newton potential defines a linear bounded operator

$$N_0 := \gamma_0^{\text{int}} \widetilde{N}_0 : [H^{1,1/2}_{;,0}(Q)]' \to H^{1/2,1/4}(\Sigma),$$

satisfying

$$\|N_0 f\|_{H^{1/2,1/4}(\Sigma)} \le c_2^{N_0} \|f\|_{[H^{1,1/2}_{;,0}(Q)]'} \quad \text{for all } f \in [H^{1,1/2}_{;,0}(Q)]'$$

with some constant $c_2^{N_0} > 0$. Moreover, the application of the Neumann trace operator yields the bounded operator

$$N_1 := \gamma_1^{\text{int}} \widetilde{N}_0 : L^2(Q) \to H^{-1/2,-1/4}(\Sigma).$$

That is, there exists $c_2^{N_1} > 0$, such that

$$\|N_1 f\|_{H^{-1/2,-1/4}(\Sigma)} \le c_2^{N_1} \|f\|_{L^2(Q)} \quad \text{for all } f \in L^2(Q).$$

Here, we have to restrict the domain to the space $L^2(Q)$ due to the definition of the Neumann trace operator γ_1^{int}.

4.4 Single layer potential

We introduce the single layer potential with density $w \in L^1(\Sigma)$ as

$$(\widetilde{V}w)(x,t) := \frac{1}{\alpha}\int_0^t\int_\Gamma U^\star(x-y,t-\tau)w(y,\tau)\,ds_y\,d\tau \quad \text{for } (x,t) \in \mathcal{D}_\Gamma, \tag{4.12}$$

where $\mathcal{D}_\Gamma := (\mathbb{R}^n \setminus \Gamma) \times (0,T)$. The fundamental solution $U^\star(x-\cdot,t-\cdot)$ is smooth on Σ for $(x,t) \in \mathcal{D}_\Gamma$. Hence, the single layer potential is well defined for $w \in L^1(\Sigma)$.

Theorem 4.4. *For $w \in L^1(\Sigma)$, the single layer potential $\widetilde{V}w$ satisfies the homogeneous heat equation. That is,*

$$[\alpha\partial_t - \Delta_x](\widetilde{V}w)(x,t) = 0 \quad \textit{for all } (x,t) \in \mathcal{D}_\Gamma.$$

Proof. For $(x,t) \in \mathcal{D}_\Gamma$, there exists a compact neighborhood O of (x,t), such that $O \subset \mathcal{D}_\Gamma$. Hence, $\operatorname{dist}(O,\Sigma) > 0$. Therefore, the restriction of $U^\star(x-y,t-\tau)$ to $(x,t) \in O$ and $(y,\tau) \in \Sigma$ is bounded and differentiable on O for $(y,\tau) \in \Sigma$. Moreover, $U^\star$ is integrable over Σ for $(x,t) \in O$. Hence, we can apply the Leibniz integral rule and get

$$\begin{aligned}[\alpha\partial_t - \Delta_x](\widetilde{V}w)(x,t) =& \frac{1}{\alpha}\int_0^t\int_\Gamma [\alpha\partial_t - \Delta_x]U^\star(x-y,t-\tau)w(y,\tau)\,ds_y\,d\tau \\ &+ \lim_{\varepsilon\to 0}\int_\Gamma U^\star(x-y,\varepsilon)w(y,t-\varepsilon)\,ds_y.\end{aligned}$$

We then use $[\alpha\partial_t - \Delta_x]U^\star(x-y,t-\tau) = 0$ for $(x,t) \in \mathcal{D}_\Gamma$ and $(y,\tau) \in \Sigma$, and the dominated convergence theorem to conclude

$$[\alpha\partial_t - \Delta_x](\widetilde{V}w)(x,t) = \lim_{\varepsilon\to 0}\int_\Gamma U^\star(x-y,\varepsilon)w(y,t-\varepsilon)\,ds_y = 0. \qquad \square$$

The explicit representation (4.12) of the operator $\widetilde{V}$ is only suited for $w \in L^1(\Sigma)$. However, the domain of the single layer potential can be extended by using the previously defined convolution operator A in Theorem 4.2. We define the linear and bounded operator $\gamma_0' : H^{-1/2,-1/4}(\Sigma) \to \widetilde{H}^{-1,-1/2}_{\text{comp}}(\mathbb{R}^n \times (0,T))$ by

$$\langle \gamma_0' w, v\rangle = \langle w, \gamma_0^{\text{int}} v\rangle_\Sigma \quad \text{for all } v \in [\widetilde{H}^{-1,-1/2}_{\text{comp}}(\mathbb{R}^n \times (0,T))]'.$$

The single layer potential is then given by

$$\widetilde{V} := A\gamma_0' : H^{-1/2,-1/4}(\Sigma) \to \widetilde{H}^{1,1/2}_{\text{loc}}(\mathbb{R}^n \times (0,T)). \tag{4.13}$$

Due to the boundedness of the operators A and γ_0', the operator

$$\widetilde{V} : H^{-1/2,-1/4}(\Sigma) \to H^{1,1/2}_{;0,}(Q)$$

is, by restriction, bounded as well. That is, there exists a positive constant $c_2^{\widetilde{V}} > 0$, such that

$$\|\widetilde{V}w\|_{H^{1,1/2}_{;0,}(Q)} \leq c_2^{\widetilde{V}} \|w\|_{H^{-1/2,-1/4}(\Sigma)} \quad \text{for all } w \in H^{-1/2,-1/4}(\Sigma). \tag{4.14}$$

Recall that due to construction, the single layer potential is a solution of the homogeneous heat equation on $\mathcal{D}_\Gamma$. That is, for $w \in H^{-1/2,-1/4}(\Sigma)$, we have

$$[\alpha\partial_t - \Delta_x]\widetilde{V}w = 0 \quad \text{on } \mathcal{D}_\Gamma.$$

Hence, $\widetilde{V}w \in H^{1,1/2}_{;0,}(Q, \mathcal{L})$ for $w \in H^{-1/2,-1/4}(\Sigma)$. Therefore, the Dirichlet trace and the Neumann trace of the single layer potential are well defined. To show the jump relations, we proceed as follows: Let $B_R \subset \mathbb{R}^n$ be a ball with radius $R > 0$, such that $\Omega \subset B_R$, and set $\Omega^c := B_R \setminus \overline{\Omega}$. Moreover, $Q^c := \Omega^c \times (0, T)$. As before, we obtain the continuity of the mapping

$$\widetilde{V} : H^{-1/2,-1/4}(\Sigma) \to H^{1,1/2}_{;0,}(Q^c, \mathcal{L}).$$

Thus, the Dirichlet and Neumann traces are defined from both sides of Σ. Let γ_0^{ext} and γ_1^{ext} denote the exterior Dirichlet trace operator and the exterior Neumann trace operator, respectively. Then the jumps on Σ are defined as

$$\begin{aligned} [\gamma_0 u] &:= \gamma_0^{\text{ext}} u - \gamma_0^{\text{int}} u, \\ [\gamma_1 u] &:= \gamma_1^{\text{ext}} u - \gamma_1^{\text{int}} u. \end{aligned} \tag{4.15}$$

Theorem 4.5. *The single layer potential $\widetilde{V}w$ satisfies the jump relations*

$$[\gamma_0 \widetilde{V}w] = 0, \quad [\gamma_1 \widetilde{V}w] = -w, \quad \textit{for all } w \in H^{-1/2,-1/4}(\Sigma).$$

Proof. For $w \in H^{-1/2,-1/4}(\Sigma)$, we have $u := \widetilde{V}w \in H^{1,1/2}_{;0,}(B_R \times (0, T))$; therefore, $\gamma_0^{\text{int}} u = \gamma_0^{\text{ext}} u$. Moreover, we have $[\alpha\partial_t - \Delta_x]u = 0$ on $Q \cup Q^c$. By using Green's second formula (3.6) with a test function $\varphi \in C_0^\infty(B_R \times [0, T))$, we obtain

$$\begin{aligned} -\langle u, [-\alpha\partial_t - \Delta_x]\varphi\rangle_{L^2(Q)} &= \langle \gamma_0^{\text{int}} u, \gamma_1^{\text{int}} \varphi\rangle_\Sigma - \langle \gamma_1^{\text{int}} u, \gamma_0^{\text{int}} \varphi\rangle_\Sigma, \\ -\langle u, [-\alpha\partial_t - \Delta_x]\varphi\rangle_{L^2(Q^c)} &= -\langle \gamma_0^{\text{ext}} u, \gamma_1^{\text{ext}} \varphi\rangle_\Sigma + \langle \gamma_1^{\text{ext}} u, \gamma_0^{\text{ext}} \varphi\rangle_\Sigma. \end{aligned}$$

Adding both equations and using $\gamma_0^{\text{int}} \varphi = \gamma_0^{\text{ext}} \varphi$ and $\gamma_1^{\text{int}} \varphi = \gamma_1^{\text{ext}} \varphi$ yields

$$-\langle u, [-\alpha\partial_t - \Delta_x]\varphi\rangle_{L^2(B_R \times (0,T))} = -\langle [\gamma_0 u], \gamma_1^{\text{int}} \varphi\rangle_\Sigma + \langle [\gamma_1 u], \gamma_0^{\text{int}} \varphi\rangle_\Sigma.$$

Since $[\gamma_0 u] = 0$, we conclude

$$-\langle u, [-\alpha\partial_t - \Delta_x]\varphi\rangle_{L^2(B_R \times (0,T))} = \langle [\gamma_1 u], \gamma_0^{\text{int}} \varphi\rangle_\Sigma. \tag{4.16}$$

From the representation (4.13) of the single layer potential $\widetilde{V}w$ follows that

$$[\alpha\partial_t - \Delta_x]\widetilde{V}w = [\alpha\partial_t - \Delta_x]A\gamma_0'w = \gamma_0'w$$

holds in $B_R \times (0,T)$ in the distributional sense. Hence, we obtain

$$\begin{aligned}\langle u, [-\alpha\partial_t - \Delta_x]\varphi\rangle_{L^2(B_R\times(0,T))} &= \langle [\alpha\partial_t - \Delta_x]u, \varphi\rangle_{B_R\times(0,T)}\\ &= \langle [\alpha\partial_t - \Delta_x]\widetilde{V}w, \varphi\rangle_{B_R\times(0,T)}\\ &= \langle \gamma_0'w, \varphi\rangle_{B_R\times(0,T)} = \langle w, \gamma_0^{\text{int}}\varphi\rangle_\Sigma.\end{aligned}$$

Combined with (4.16), we get

$$\langle [\gamma_1\widetilde{V}w], \gamma_0^{\text{int}}\varphi\rangle_\Sigma = -\langle w, \gamma_0^{\text{int}}\varphi\rangle_\Sigma.$$

The assertion follows since $\gamma_0^{\text{int}}C_0^\infty(B_R \times [0,T))$ is dense in $H^{1/2,1/4}(\Sigma)$. □

The continuity of $\widetilde{V}$ and γ_0^{int} imply that the single layer boundary integral operator

$$V := \gamma_0^{\text{int}}\widetilde{V} : H^{-1/2,-1/4}(\Sigma) \to H^{1/2,1/4}(\Sigma)$$

is linear and bounded. That is, there exists a positive constant $c_2^V > 0$, such that

$$\|Vw\|_{H^{1/2,1/4}(\Sigma)} \le c_2^V \|w\|_{H^{-1/2,-1/4}(\Sigma)} \quad \text{for all } w \in H^{-1/2,-1/4}(\Sigma). \tag{4.17}$$

4.5 Adjoint double layer potential

The adjoint double layer potential $K'w$ with density $w \in H^{-1/2,-1/4}(\Sigma)$ is defined as

$$K'w := \frac{1}{2}(\gamma_1^{\text{int}}\widetilde{V}w + \gamma_1^{\text{ext}}\widetilde{V}w).$$

Due to the boundedness of the single layer operator $\widetilde{V}$ and the Neumann trace operators, the operator $K' : H^{-1/2,-1/4}(\Sigma) \to H^{-1/2,-1/4}(\Sigma)$ is bounded as well. For w regular enough, we have the representation

$$(K'w)(x,t) = \frac{1}{\alpha}\int_0^t\int_\Gamma \partial_{n_x}U^\star(x-y,t-\tau)w(y,\tau)\,\mathrm{d}s_y\,\mathrm{d}\tau$$

for $(x,t) \in \Sigma$ and Γ smooth in $x \in \Gamma$.

4.6 Double layer potential

We introduce the double layer potential with density $v \in L^1(\Sigma)$ as

$$(Wv)(x,t) := \frac{1}{\alpha} \int_0^t \int_\Gamma \partial_{n_y} U^\star(x-y, t-\tau) v(y,\tau)\, ds_y\, d\tau \quad \text{for } (x,t) \in \mathcal{D}_\Gamma. \tag{4.18}$$

The fundamental solution $U^\star(x - \cdot, t - \cdot)$ is smooth on Σ for $(x,t) \in \mathcal{D}_\Gamma$. Hence, the double layer potential is well defined for $v \in L^1(\Sigma)$.

Theorem 4.6. *For $v \in L^1(\Sigma)$, the double layer potential Wv satisfies the homogeneous heat equation. That is,*

$$[\alpha\partial_t - \Delta_x](Wv)(x,t) = 0 \quad \textit{for all } (x,t) \in \mathcal{D}_\Gamma.$$

Proof. For $(x,t) \in \mathcal{D}_\Gamma$, there exists a compact neighborhood O of (x,t), such that $O \subset \mathcal{D}_\Gamma$. Hence, $\operatorname{dist}(O,\Sigma) > 0$. Therefore, the restriction of $\partial_{n_y} U^\star(x-y, t-s)$ to $(x,t) \in O$ and $(y,s) \in \Sigma$ is bounded and differentiable on O for $(y,s) \in \Sigma$. Moreover, $\partial_{n_y} U^\star$ is integrable over Σ for $(x,t) \in O$. Hence, we can apply the Leibniz integral rule and additionally interchange the operators $\alpha\partial_t - \Delta_x$ and ∂_{n_y} under the integral sign to get

$$\begin{aligned}[\alpha\partial_t - \Delta_x](Wv)(x,t) =& \frac{1}{\alpha} \int_0^t \int_\Gamma \partial_{n_y} [\alpha\partial_t - \Delta_x] U^\star(x-y, t-\tau) v(y,\tau)\, ds_y\, d\tau \\ &+ \lim_{\varepsilon\to 0} \int_\Gamma \partial_{n_y} U^\star(x-y, \varepsilon) v(y, t-\varepsilon)\, ds_y.\end{aligned}$$

We then use $[\alpha\partial_t - \Delta_x] U^\star(x-y, t-\tau) = 0$ for $(x,t) \in \mathcal{D}_\Gamma$ and $(y,\tau) \in \Sigma$, and the dominated convergence theorem to conclude

$$[\alpha\partial_t - \Delta](Wv)(x,t) = \lim_{\varepsilon\to 0} \int_\Gamma \partial_{n_y} U^\star(x-y,\varepsilon) v(y, t-\varepsilon)\, ds_y = 0. \qquad \square$$

As in the case of the single layer potential $\widetilde{V}$, the representation (4.18) is only valid for $v \in L^1(\Sigma)$, and again, we can extend the domain of the double layer operator W by using the properties of the convolution operator A. For $v \in H^{1/2,1/4}(\Sigma)$, we have the representation $Wv = A\gamma_1' v$. Here, $\gamma_1' v$ is the distribution defined by

$$\langle \gamma_1' v, \varphi \rangle = \langle v, \gamma_1^{\text{int}} \varphi \rangle_\Sigma \quad \text{for all } \varphi \in C_0^\infty(\mathbb{R}^n \times \mathbb{R}).$$

The proof of the continuity of the operator

$$W : H^{1/2,1/4}(\Sigma) \to H^{1,1/2}_{;0,}(Q)$$

follows the lines of [5, Proposition 3.3]. We conclude that there exists a positive constant $c_2^W > 0$, such that

$$\|Wv\|_{H^{1,1/2}_{;0,}(Q)} \leq c_2^W \|v\|_{H^{1/2,1/4}(\Sigma)} \quad \text{for all } v \in H^{1/2,1/4}(\Sigma).$$

The double layer potential Wv for $v \in H^{1/2,1/4}(\Sigma)$ is a solution of the homogeneous heat equation on $\mathcal{D}_\Gamma$. That is, we have

$$[\alpha\partial_t - \Delta_x]Wv = 0 \quad \text{on } \mathcal{D}_\Gamma.$$

Hence, $Wv \in H^{1,1/2}_{;0,}(Q, \mathcal{L})$ for $v \in H^{1/2,1/4}(\Sigma)$; therefore, the traces are well defined. Analogously, as in the case of the single layer potential (see Section 4.4), we can define the interior and exterior Dirichlet and Neumann traces of Wv and obtain the following jump relations:

Theorem 4.7. *The double layer potential Wv satisfies the jump relations*

$$[\gamma_0 Wv] = v, \quad [\gamma_1 Wv] = 0, \quad \textit{for all } v \in H^{1/2,1/4}(\Sigma).$$

Proof. For $v \in H^{1/2,1/4}(\Sigma)$, we define $u := Wv$. Hence, $[\alpha\partial_t - \Delta_x]u = 0$ on $Q \cup Q^c$. By using Green's second formula (3.6) with $\varphi \in C_0^\infty(B_R \times [0, T))$, we get

$$-\langle u, [-\alpha\partial_t - \Delta_x]\varphi\rangle_{L^2(B_R\times(0,T))} = -\langle [\gamma_0 u], \gamma_1^{\text{int}}\varphi\rangle_\Sigma + \langle [\gamma_1 u], \gamma_0^{\text{int}}\varphi\rangle_\Sigma.$$

From the definition of the double layer potential W follows that

$$[\alpha\partial_t - \Delta_x]Wv = \gamma_1' v$$

holds in $B_R \times (0, T)$ in the distributional sense. Hence, we obtain

$$\begin{aligned} \langle u, [-\alpha\partial_t - \Delta_x]\varphi\rangle_{L^2(B_R\times(0,T))} &= \langle [\alpha\partial_t - \Delta_x]u, \varphi\rangle_{B_R\times(0,T)} \\ &= \langle [\alpha\partial_t - \Delta_x]Wv, \varphi\rangle_{B_R\times(0,T)} \\ &= \langle \gamma_1' v, \varphi\rangle_{B_R\times(0,T)} = \langle v, \gamma_1^{\text{int}}\varphi\rangle_\Sigma, \end{aligned}$$

and conclude

$$\langle [\gamma_1 Wv], \gamma_0^{\text{int}}\varphi\rangle_\Sigma = \langle [\gamma_0 Wv] - v, \gamma_1^{\text{int}}\varphi\rangle_\Sigma. \tag{4.19}$$

Remark 3.1 then implies that each side in (4.19) has to be zero. That is, $[\gamma_1 Wv] = 0$, and $[\gamma_0 Wv] = v$. □

The double layer boundary integral operator K for $v \in H^{1/2,1/4}(\Sigma)$ is defined as

$$Kv := \frac{1}{2}(\gamma_0^{\text{int}} Wv + \gamma_0^{\text{ext}} Wv). \tag{4.20}$$

Due to the boundedness of the double layer potential W and the Dirichlet trace operators, the operator $K : H^{1/2,1/4}(\Sigma) \to H^{1/2,1/4}(\Sigma)$ is bounded as well. That is, there exists a positive constant $c_2^K > 0$, such that

$$\|Kv\|_{H^{1/2,1/4}(\Sigma)} \leq c_2^K \|v\|_{H^{1/2,1/4}(\Sigma)} \quad \text{for all } v \in H^{1/2,1/4}(\Sigma).$$

4.7 Hypersingular operator

The hypersingular operator D defined as

$$D := -\gamma_1^{\text{int}} W : H^{1/2,1/4}(\Sigma) \to H^{-1/2,-1/4}(\Sigma)$$

is linear and bounded, satisfying

$$\|Dv\|_{H^{-1/2,-1/4}(\Sigma)} \leq c_2^D \|v\|_{H^{1/2,1/4}(\Sigma)} \quad \text{for all } v \in H^{1/2,1/4}(\Sigma) \tag{4.21}$$

with some positive constant $c_2^D > 0$. If the density v is smooth enough, we have the representation

$$(Dv)(x,t) = -\frac{1}{\alpha}\gamma_{1,x}^{\text{int}} \int_0^t \int_\Gamma \gamma_{1,y}^{\text{int}} U^\star(x-y,t-\tau)v(y,\tau)\, ds_y\, d\tau$$

for $(x,t) \in \Sigma$. When assuming that the boundary Γ, for $n = 2,3$, is piecewise smooth, we can derive an alternative representation of the bilinear form, which is induced by the hypersingular boundary integral operator D. That is,

$$\langle Du, v\rangle_\Sigma = -\frac{1}{\alpha} \int_\Sigma v(x,t)\gamma_{1,x}^{\text{int}} \int_\Sigma \gamma_{1,y}^{\text{int}} U^\star(x-y,t-\tau)u(y,\tau)\, ds_y\, d\tau\, ds_x\, dt.$$

In this case, the bilinear form can be written by means of the single layer boundary integral operator V. That means, we have weakly singular representations. For $n = 2$ (see, for example, [5, Theorem 6.1]), we obtain

$$\begin{aligned}\langle Du, v\rangle_\Sigma =& \frac{1}{\alpha} \int_\Sigma \operatorname{curl}_\Gamma v(x,t) \int_\Sigma U^\star(x-y,t-\tau)\operatorname{curl}_\Gamma u(y,\tau)\, ds_y\, d\tau\, ds_x\, dt \\ &- \int_\Sigma n_x^\top v(x,t) \int_\Sigma \partial_\tau U^\star(x-y,t-\tau) n_y u(y,\tau)\, ds_y\, d\tau\, ds_x\, dt,\end{aligned}$$

where

$$\operatorname{curl}_\Gamma v(x,t) := n_{x,1}\frac{\partial}{\partial x_2} v(x,t) - n_{x,2}\frac{\partial}{\partial x_1} v(x,t) \quad \text{for } (x,t) \in \Sigma.$$

Whereas for $n = 3$, we have the representation [27, Theorem 2.1]

$$\begin{aligned}\langle Du, v\rangle_\Sigma =& \frac{1}{\alpha} \int_\Sigma \underline{\operatorname{curl}}_\Gamma^\top v(x,t) \int_\Sigma U^\star(x-y,t-\tau)\underline{\operatorname{curl}}_\Gamma u(y,\tau)\, ds_y\, d\tau\, ds_x\, dt \\ &- \int_\Sigma n_x^\top v(x,t) \int_\Sigma \partial_\tau U^\star(x-y,t-\tau) n_y u(y,\tau)\, ds_y\, d\tau\, ds_x\, dt,\end{aligned}$$

with $\underline{\operatorname{curl}}_\Gamma v(x,t) := n_x \times \nabla_x v(x,t)$ for $(x,t) \in \Sigma$.

5 Boundary integral equations

In the following section, we introduce the Calderón projection operator and deduce related properties of the boundary integral operators, including the definition of the Steklov–Poincaré operator in Subsection 5.1. In Subsection 5.2, we discuss the unique solvability of the model problem (1.1) by means of analyzing related boundary integral equations.

The solution $u \in H^{1,1/2}(Q)$ of problem (1.1) with initial datum $u_0 \in L^2(\Omega)$ and source term $f \in L^2(Q)$ is given by the representation formula

$$u = (\widetilde{V}\gamma_1^{\mathrm{int}}u) - (W\gamma_0^{\mathrm{int}}u) + (\widetilde{M}_0 u_0) + (\widetilde{N}_0 f) \quad \text{in } Q. \tag{5.1}$$

By applying the Dirichlet trace operator to (5.1) and recalling the jump relations of the heat potentials, we obtain the first boundary integral equation:

$$\gamma_0^{\mathrm{int}}u = (V\gamma_1^{\mathrm{int}}u) + \frac{1}{2}\gamma_0^{\mathrm{int}}u - (K\gamma_0^{\mathrm{int}}u) + (M_0 u_0) + (N_0 f) \quad \text{on } \Sigma. \tag{5.2}$$

The application of the Neumann trace operator to (5.1) yields the second boundary integral equation:

$$\gamma_1^{\mathrm{int}}u = \frac{1}{2}\gamma_1^{\mathrm{int}}u + (K'\gamma_1^{\mathrm{int}}u) + (D\gamma_0^{\mathrm{int}}u) + (M_1 u_0) + (N_1 f) \quad \text{on } \Sigma. \tag{5.3}$$

Together these equations lead to the so-called Calderón system of boundary integral equations. We have

$$\begin{pmatrix} \gamma_0^{\mathrm{int}}u \\ \gamma_1^{\mathrm{int}}u \end{pmatrix} = \underbrace{\begin{pmatrix} \frac{1}{2}I - K & V \\ D & \frac{1}{2}I + K' \end{pmatrix}}_{=:\mathcal{C}} \begin{pmatrix} \gamma_0^{\mathrm{int}}u \\ \gamma_1^{\mathrm{int}}u \end{pmatrix} + \begin{pmatrix} M_0 u_0 \\ M_1 u_0 \end{pmatrix} + \begin{pmatrix} N_0 f \\ N_1 f \end{pmatrix}. \tag{5.4}$$

The operator $\mathcal{C}$ is called the Calderón projection operator.

Lemma 5.1. *$\mathcal{C}$ is a projection; that is, $\mathcal{C} = \mathcal{C}^2$.*

Proof. Let $(\psi, \varphi) \in H^{-1/2,-1,4}(\Sigma) \times H^{1/2,1/4}(\Sigma)$. Then the function

$$u := \widetilde{V}\psi - W\varphi$$

is a solution of the homogeneous heat equation. By applying the trace operators, we get the boundary integral equations

$$\begin{aligned} \gamma_0^{\mathrm{int}}u &= V\psi + \left(\frac{1}{2}I - K\right)\varphi, \\ \gamma_1^{\mathrm{int}}u &= \left(\frac{1}{2}I + K'\right)\psi + D\varphi. \end{aligned} \tag{5.5}$$

Additionally, u is a solution of the homogeneous heat equation with Cauchy data $\gamma_0^{\text{int}}u, \gamma_1^{\text{int}}u$ and initial condition $u_0 = 0$. Hence we have

$$\begin{pmatrix} \gamma_0^{\text{int}}u \\ \gamma_1^{\text{int}}u \end{pmatrix} = \begin{pmatrix} \frac{1}{2}I - K & V \\ D & \frac{1}{2}I + K' \end{pmatrix} \begin{pmatrix} \gamma_0^{\text{int}}u \\ \gamma_1^{\text{int}}u \end{pmatrix}.$$

Inserting (5.5) yields

$$\begin{pmatrix} \frac{1}{2}I - K & V \\ D & \frac{1}{2}I + K' \end{pmatrix} \begin{pmatrix} \psi \\ \varphi \end{pmatrix} = \begin{pmatrix} \frac{1}{2}I - K & V \\ D & \frac{1}{2}I + K' \end{pmatrix}^2 \begin{pmatrix} \psi \\ \varphi \end{pmatrix}.$$

Since the functions ψ, φ were arbitrarily chosen, we conclude $\mathcal{C} = \mathcal{C}^2$. □

As a consequence of the projection property of the Calderón operator $\mathcal{C}$, we obtain the following relations:

Corollary 5.1. *The boundary integral operators satisfy*

$$\begin{aligned} VD &= \left(\frac{1}{2}I + K\right)\left(\frac{1}{2}I - K\right), \\ DV &= \left(\frac{1}{2}I + K'\right)\left(\frac{1}{2}I - K'\right), \\ VK' &= KV, \\ K'D &= DK. \end{aligned}$$

Proof. Follows from $\mathcal{C} = \mathcal{C}^2$. □

Now we state the main theorem of this section.

Theorem 5.1. *The operator*

$$\mathcal{A} : H^{1/2,1/4}(\Sigma) \times H^{-1/2,-1/4}(\Sigma) \to H^{1/2,1/4}(\Sigma) \times H^{-1/2,-1/4}(\Sigma),$$

defined as

$$\mathcal{A} := \begin{pmatrix} -K & V \\ D & K' \end{pmatrix}$$

is an isomorphism, and there exists a constant $c_1 > 0$, such that

$$\left\langle \begin{pmatrix} \psi \\ \varphi \end{pmatrix}, \begin{pmatrix} V & -K \\ K' & D \end{pmatrix} \begin{pmatrix} \psi \\ \varphi \end{pmatrix} \right\rangle_{\Sigma \times \Sigma} \geq c_1 \left(\|\psi\|^2_{H^{-1/2,-1/4}(\Sigma)} + \|\varphi\|^2_{H^{1/2,1/4}(\Sigma)} \right)$$

for all $(\psi, \varphi) \in H^{-1/2,-1/4}(\Sigma) \times H^{1/2,1/4}(\Sigma)$.

Proof. Follows the lines of [5, Corollary 3.10, Theorem 3.11]. □

The ellipticity of the operator in Theorem 5.1 then immediately implies the ellipticity of the single layer boundary integral operator V and the hypersingular operator D.

Lemma 5.2. *The single layer boundary integral operator V defines an isomorphism, and there exists a positive constant $c_1^V > 0$, such that*

$$\langle Vw, w\rangle_\Sigma \geq c_1^V \|w\|^2_{H^{-1/2,-1/4}(\Sigma)} \quad \textit{for all } w \in H^{-1/2,-1/4}(\Sigma).$$

Proof. Follows from Theorem 5.1 with $\varphi = 0$. □

Lemma 5.3. *The hypersingular operator D defines an isomorphism, and there exists a positive constant $c_1^D > 0$, such that*

$$\langle Dv, v\rangle_\Sigma \geq c_1^D \|v\|^2_{H^{1/2,1/4}(\Sigma)} \quad \textit{for all } v \in H^{1/2,1/4}(\Sigma).$$

Proof. Follows from Theorem 5.1 with $\psi = 0$. □

5.1 Steklov–Poincaré operator

We consider the system of boundary integral equations with source term $f = 0$ and with homogeneous initial conditions; that is, $u_0 = 0$. Hence,

$$\begin{pmatrix} \gamma_0^{\text{int}} u \\ \gamma_1^{\text{int}} u \end{pmatrix} = \begin{pmatrix} \frac{1}{2}I - K & V \\ D & \frac{1}{2}I + K' \end{pmatrix} \begin{pmatrix} \gamma_0^{\text{int}} u \\ \gamma_1^{\text{int}} u \end{pmatrix}.$$

Using the first integral equation, we can define the Dirichlet to Neumann map

$$\gamma_1^{\text{int}} u = V^{-1}\left(\frac{1}{2}I + K\right)\gamma_0^{\text{int}} u. \tag{5.6}$$

The operator

$$S := V^{-1}\left(\frac{1}{2}I + K\right) : H^{1/2,1/4}(\Sigma) \to H^{-1/2,-1/4}(\Sigma) \tag{5.7}$$

is called Steklov–Poincaré operator for the heat equation. When inserting (5.6) into the second boundary integral equation, we obtain

$$\gamma_1^{\text{int}} u = \left[D + \left(\frac{1}{2}I + K'\right)V^{-1}\left(\frac{1}{2}I + K\right)\right]\gamma_0^{\text{int}} u.$$

Hence, we get a *symmetric* representation of the Steklov–Poincaré operator:

$$S = D + \left(\frac{1}{2}I + K'\right)V^{-1}\left(\frac{1}{2}I + K\right). \tag{5.8}$$

Due to the boundedness of the operators K, K', D, and V^{-1}, the operator S is bounded as well.

Lemma 5.4. *The Steklov–Poincaré operator S is elliptic; that is, there exists a positive constant $c_1^S > 0$, such that*

$$\langle Sv, v\rangle_\Sigma \geq c_1^S \|v\|^2_{H^{1/2,1/4}(\Sigma)} \quad \textit{for all } v \in H^{1/2,1/4}(\Sigma).$$

Proof. For $v \in H^{1/2,1/4}(\Sigma)$, we define $\psi := V^{-1}(\frac{1}{2}I + K)v \in H^{-1/2,-1/4}(\Sigma)$ and get

$$\begin{aligned}
&\left\langle \begin{pmatrix} \psi \\ v \end{pmatrix}, \begin{pmatrix} V & -K \\ K' & D \end{pmatrix} \begin{pmatrix} \psi \\ v \end{pmatrix} \right\rangle_{\Sigma\times\Sigma} \\
&= \frac{1}{2}\left\langle V^{-1}\left(\frac{1}{2}I + K\right)v, v\right\rangle_\Sigma + \left\langle v, K'V^{-1}\left(\frac{1}{2}I + K\right)v + Dv\right\rangle_\Sigma \\
&= \left\langle v, \left(\frac{1}{2}I + K'\right)V^{-1}\left(\frac{1}{2}I + K\right)v + Dv\right\rangle_\Sigma \\
&= \langle v, Sv\rangle_\Sigma.
\end{aligned}$$

The assertion now follows with Theorem 5.1. □

5.2 Initial Dirichlet boundary value problem

We consider the initial Dirichlet boundary value problem (1.1) with source term $f \in L^2(Q)$, boundary datum $g \in H^{1/2,1/4}(\Sigma)$, and initial datum $u_0 \in L^2(\Omega)$. The solution is given by the representation formula

$$u = (\widetilde{V}\gamma_1^{\text{int}}u) - (Wg) + (\widetilde{M}_0 u_0) + (\widetilde{N}_0 f) \quad \text{in } Q.$$

It remains to determine the unknown conormal derivative $\gamma_1^{\text{int}}u \in H^{-1/2,-1/4}(\Sigma)$. This can be done, for example, by using the first boundary integral equation in (5.4). We have to find $\gamma_1^{\text{int}}u \in H^{-1/2,-1/4}(\Sigma)$, such that

$$V\gamma_1^{\text{int}}u = \left(\frac{1}{2}I + K\right)g - M_0 u_0 - N_0 f \quad \text{on } \Sigma.$$

The corresponding variational formulation is to find $\gamma_1^{\text{int}}u \in H^{-1/2,-1/4}(\Sigma)$, such that

$$\langle V\gamma_1^{\text{int}}u, \tau\rangle_\Sigma = \left\langle \left(\frac{1}{2}I + K\right)g - M_0 u_0 - N_0 f, \tau \right\rangle_\Sigma \tag{5.9}$$

for all $\tau \in H^{-1/2,-1/4}(\Sigma)$. Since the boundary integral operators K, M_0, N_0, and V are bounded and V is elliptic, there exists a unique solution $\gamma_1^{\text{int}}u \in H^{-1/2,-1/4}(\Sigma)$ according to the lemma of Lax–Milgram. The solution $\gamma_1^{\text{int}}u$ then satisfies

$$\begin{aligned}
\|\gamma_1^{\text{int}}u\|_{H^{-1/2,-1/4}(\Sigma)} &\leq \frac{1}{c_1^V}\left\|\left(\frac{1}{2}I + K\right)g - M_0 u_0 - N_0 f\right\|_{H^{1/2,1/4}(\Sigma)} \\
&\leq \frac{1}{c_1^V}\left(\tilde{c}_2^W \|g\|_{H^{1/2,1/4}(\Sigma)} + c_2^{M_0}\|u_0\|_{L^2(\Omega)} + c_2^{N_0}\|f\|_{[H^{1,1/2}_{;,0}(Q)]'}\right).
\end{aligned}$$

Another approach is using an indirect formulation with the single layer potential $\widetilde{V}$. A solution of the heat equation with source term f and initial condition u_0 is given by

$$u = (\widetilde{V}w) + (\widetilde{M}_0 u_0) + (\widetilde{N}_0 f) \quad \text{in } Q \tag{5.10}$$

with an unknown density $w \in H^{-1/2,-1/4}(\Sigma)$ to be determined. By applying the Dirichlet trace operator to (5.10), we obtain

$$g = (Vw) + (M_0 u_0) + (N_0 f) \quad \text{on } \Sigma.$$

Thus, we have to find $w \in H^{-1/2,-1/4}(\Sigma)$, such that

$$Vw = g - M_0 u_0 - N_0 f \quad \text{on } \Sigma.$$

The corresponding variational formulation is to find $w \in H^{-1/2,-1/4}(\Sigma)$, such that

$$\langle Vw, \tau\rangle_\Sigma = \langle g - M_0 u_0 - N_0 f, \tau\rangle_\Sigma \quad \text{for all } \tau \in H^{-1/2,-1/4}(\Sigma). \tag{5.11}$$

As in the case of the direct formulation with the first boundary integral equation, the unique solvability follows with the lemma of Lax–Milgram.

In this paper, we only consider the Dirichlet boundary value problem. The analysis of the Neumann boundary value problem will be addressed in future work (see, for example, [6]). In this case, one can, for example, use the second boundary integral equation in (5.4) to obtain the unknown Dirichlet trace $\gamma_0^{\text{int}} u$ of the solution u. Due to the ellipticity of the hypersingular operator D, the second boundary integral equation is uniquely solvable as well. Another approach would be an indirect formulation with the double layer potential W.

6 Space-time discretization

In this section, we discuss two different space-time discretization techniques in order to compute an approximation of the unknown Neumann datum $\partial_n u_{|\Sigma}$ and derive related approximation properties. The first one is the so-called tensor product approach, where we consider separate decompositions of the spatial boundary Γ and the time interval $(0, T)$, and use space-time tensor product spaces to compute an approximation of $\partial_n u_{|\Sigma}$. The second one is using boundary element spaces, which are defined with respect to a shape-regular triangulation of the whole space-time boundary $\Sigma = \Gamma \times (0, T)$ into boundary elements, allowing us to apply adaptive refinement in space and time simultaneously, while maintaining the regularity of the boundary element mesh.

We assume, for $n = 2, 3$, that the spatial Lipschitz boundary $\Gamma = \partial\Omega$ is piecewise smooth; thus, $\overline{\Gamma} = \bigcup_{j=1}^J \overline{\Gamma}_j$. With $\Sigma_j := \Gamma_j \times (0, T)$, $j = 1, \ldots, J$, we then obtain $\overline{\Sigma} = \bigcup_{j=1}^J \overline{\Sigma}_j$. For the Galerkin boundary element discretization of the variational formulations (5.9) or (5.11), we consider a family $\{\Sigma_N\}_{N\in\mathbb{N}}$ of decompositions $\Sigma_N := \{\sigma_\ell\}_{\ell=1}^N$ of the space-time boundary Σ into boundary elements σ_ℓ,

$$\overline{\Sigma} = \bigcup_{\ell=1}^{N} \overline{\sigma}_\ell \,. \tag{6.1}$$

6.1 One-dimensional problem

In the spatially one-dimensional case, we have $\Gamma = \{a, b\}$, assuming $\Omega = (a, b)$, inducing that $\Sigma = \Sigma_a \cup \Sigma_b$ with $\Sigma_a = \{a\} \times (0, T)$ and $\Sigma_b = \{b\} \times (0, T)$. Hence, the boundary elements σ_ℓ are line segments in temporal dimension with fixed spatial coordinate $x_\ell \in \{a, b\}$ as shown in Figure 1.1.

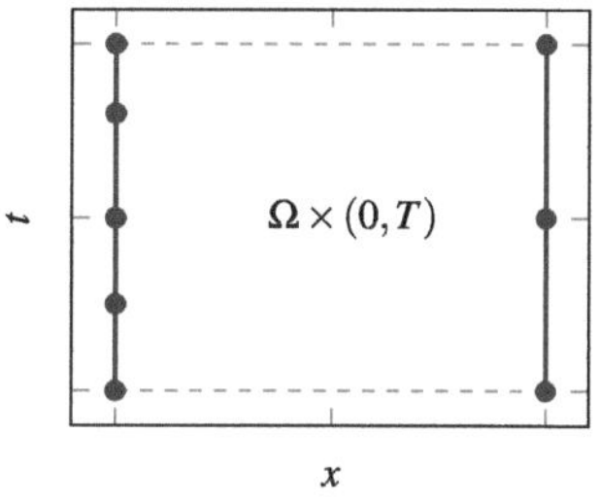

Figure 1.1: Sample BE mesh. We consider an arbitrary decomposition of the space-time boundary Σ. Note that there is no time-stepping scheme involved.

Remark 6.1. In the one-dimensional case, the spatial component of the space-time boundary Σ collapses to the points $\{a, b\}$, assuming $\Omega = (a, b)$; therefore, we can identify the anisotropic Sobolev spaces $H^{r,s}(\Sigma)$ with the isotropic version $H^s(\Sigma)$.

Let (x_ℓ, t_{ℓ_1}) and (x_ℓ, t_{ℓ_2}) be the nodes of the boundary element σ_ℓ. The local mesh size is then given as $h_\ell := |t_{\ell_2} - t_{\ell_1}|$, whereas $h := \max_{\ell=1,\dots,N} h_\ell$ is the global mesh size. The family $\{\Sigma_N\}_{N\in\mathbb{N}}$ is said to be globally quasi-uniform if there exists a constant $c_G \geq 1$ independent of Σ_N, such that

$$\frac{h_{\max}}{h_{\min}} \leq c_G.$$

For the approximation of the unknown Neumann datum $w = \gamma_1^{\text{int}} u \in H^{-1/4}(\Sigma)$, we consider the space $S_h^0(\Sigma) := \text{span}\{\varphi_\ell^0\}_{\ell=1}^N$ of piecewise constant basis functions φ_ℓ^0, which is defined with respect to the decomposition Σ_N. According to Remark 6.1, we can identify $H^{r,s}(\Sigma)$ with $H^s(\Sigma)$; hence, we have the same approximation properties as in the case of standard Sobolev spaces $H^s(\Sigma)$, see, for example, [33].

Approximation properties

The L^2 projection $Q_h^0 u \in S_h^0(\Sigma)$ of $u \in L^2(\Sigma)$ is defined as the unique solution of the variational problem

$$\langle Q_h^0 u, v_h \rangle_{L^2(\Sigma)} = \langle u, v_h \rangle_{L^2(\Sigma)} \quad \text{for all } v_h \in S_h^0(\Sigma).$$

The operator $Q_h^0 : L^2(\Sigma) \to L^2(\Sigma)$ satisfies the trivial stability estimate

$$\|Q_h^0 u\|_{L^2(\Sigma)} \le \|u\|_{L^2(\Sigma)} \quad \text{for all } u \in L^2(\Sigma).$$

Next, we summarize some error estimates for the L^2 projection $Q_h^0 u$.

Theorem 6.1 ([33, Theorem 10.2]). *Let $u \in H^s(\Sigma)$ for some $s \in [0,1]$ and $Q_h^0 u \in S_h^0(\Sigma)$ be the L^2 projection of u. Then there holds the error estimate*

$$\|u - Q_h^0 u\|_{L^2(\Sigma)} \le c h^s |u|_{H^s(\Sigma)}.$$

Lemma 6.1 ([33, Corollary 10.3]). *Let $u \in H^s(\Sigma)$ for some $s \in [0,1]$. For $\sigma \in [-1,0]$, there holds the error estimate*

$$\|u - Q_h^0 u\|_{\widetilde{H}^\sigma(\Sigma)} \le c h^{s-\sigma} |u|_{H^s(\Sigma)}.$$

Lemma 6.2 ([33, Lemma 10.10]). *Assume that the boundary decomposition Σ_N is globally quasi-uniform. For $\sigma \in [-1,0]$, there holds the inverse inequality*

$$\|\tau_h\|_{L^2(\Sigma)} \le c h^\sigma \|\tau_h\|_{\widetilde{H}^\sigma(\Sigma)} \quad \textit{for all } \tau_h \in S_h^0(\Sigma).$$

6.2 Two- and three-dimensional problems

We consider two different decomposition approaches. The first one is a separate decomposition of the spatial boundary $\Gamma = \partial\Omega$ and the time interval $(0,T)$, also discussed in, for example, [15, 25, 29]. In this case, we use space-time tensor product spaces to discretize the variational formulations (5.9) or (5.11). We derive error estimates simply by combining approximation properties of the spatial and temporal discretizations. The second approach is considering an arbitrary triangulation of the full space-time boundary $\Sigma = \Gamma \times (0,T)$ into boundary elements.

6.2.1 Space-time tensor product decompositions

Let $\{\Gamma_{N_x}\}_{N_x \in \mathbb{N}}$ be a family of admissible decompositions $\Gamma_{N_x} := \{\gamma_\ell\}_{\ell=1}^{N_x}$ of the boundary Γ into boundary elements γ_ℓ. That means, we have

$$\overline{\Gamma} = \bigcup_{\ell=1}^{N_x} \overline{\gamma}_\ell. \tag{6.2}$$

We assume that there are no curved elements and that there is no approximation of the boundary Γ. The boundary elements γ_ℓ are line segments for $n = 2$ and plane triangles for $n = 3$. For each boundary element γ_ℓ, there exists $j \in \{1, \dots, J\}$ such that $\gamma_\ell \subset \Gamma_j$. The boundary elements γ_ℓ can be described as $\gamma_\ell = \chi_\ell(\gamma)$, where γ is some reference element in $\mathbb{R}^{n-1}$. For each boundary element γ_ℓ, we define its volume

$$\Delta_\ell := \int_{\gamma_\ell} ds_x,$$

and its local mesh size

$$h_{\ell,x} := \Delta_\ell^{1/(n-1)}.$$

The global mesh size is then given by

$$h_x := \max_{\ell=1,\dots,N_x} h_{\ell,x}.$$

Moreover, we define the diameter of the element γ_ℓ as

$$d_{\ell,x} := \sup_{x,y\in\gamma_\ell} |x-y|.$$

The family $\{\Gamma_{N_x}\}_{N_x\in\mathbb{N}}$ of decompositions is said to be globally quasi-uniform if there exists a constant $c_{G,x} \geq 1$ independent of Γ_{N_x}, such that

$$\frac{h_{x,\max}}{h_{x,\min}} \leq c_{G,x}.$$

We assume that the boundary elements γ_ℓ are shape regular. That is, there exists a constant c_B independent of Γ_{N_x}, such that

$$d_{\ell,x} \leq c_B h_{\ell,x} \quad \text{for all } \ell = 1, \dots, N_x.$$

Moreover, we consider a family $\{I_{N_t}\}_{N_t\in\mathbb{N}}$ of decompositions $I_{N_t} := \{\tau_k\}_{k=1}^{N_t}$ of the time interval $I = (0, T)$ into line segments τ_k. That means, we have

$$\bar{I} = [0, T] = \bigcup_{k=1}^{N_t} \bar{\tau}_k. \tag{6.3}$$

The local mesh size of an element $\tau_k = (t_{k_1}, t_{k_2})$ is then given by $h_{k,t} := t_{k_2} - t_{k_1}$, whereas the global mesh size is defined as $h_t := \max_{k=1,\dots,N_t} h_{k,t}$. Again, the family $\{I_{N_t}\}_{N_t\in\mathbb{N}}$ of decompositions is said to be globally quasi-uniform, if there exists a constant $c_{G,t} \geq 1$ independent of I_{N_t}, such that

$$\frac{h_{t,\max}}{h_{t,\min}} \leq c_{G,t}.$$

The set $\mathcal{B}_N := \{\sigma_\ell\}_{l=1}^N$ of the boundary elements σ_ℓ in (6.1) is then given by

$$\mathcal{B}_N := \{\sigma = \gamma_i \times \tau_j, i \in \{1,\ldots,N_x\}, j \in \{1,\ldots,N_t\}\}.$$

These space-time boundary elements are rectangles for $n = 2$ and triangular prisms for $n = 3$. A sample decomposition of the space-time boundary of $Q = (0,1)^3$ is shown in Figure 1.2(a).

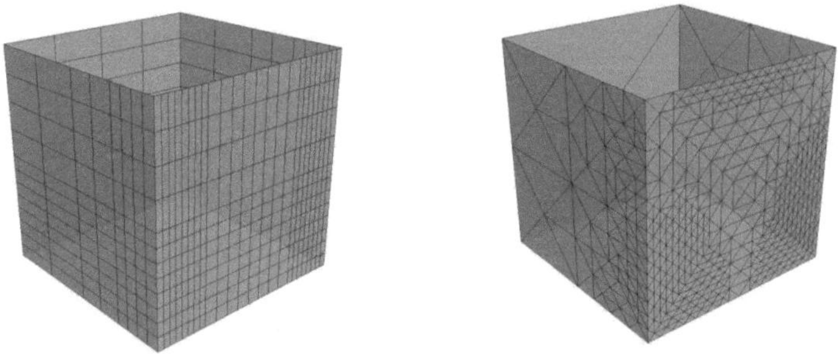

(a) Tensor product decomposition (b) Triangulation

Figure 1.2: Sample space-time boundary decompositions of $Q = (0,1)^3$.

Since the normal derivative of u on Σ could be discontinuous, depending on the spatial boundary Γ, it is reasonable to approximate the conormal derivative $w = \gamma_1^{\text{int}} u$ by discontinuous functions. Thus, we use the space of piecewise constant basis functions for the approximation of w. Let $S_h^0(I) := \text{span}\{\varphi_\ell^0\}_{\ell=1}^{N_t}$ be the space of piecewise constant basis functions on $(0,T)$ corresponding to the temporal decomposition I_{N_t}, and let $S_h^0(\Gamma) := \text{span}\{\psi_i^0\}_{i=1}^{N_x}$ be the space of piecewise constant basis functions on Γ, which is defined with respect to the spatial decomposition Γ_{N_x}. The boundary element space is then given as

$$S_{h_x,h_t}^{0,0}(\Sigma) := S_{h_x}^0(\Gamma) \otimes S_{h_t}^0(I). \tag{6.4}$$

Due to the structure of the decomposition, we can combine the approximation properties in spatial and temporal dimension to derive the approximation properties of the boundary element space $S_{h_x,h_t}^{0,0}(\Sigma)$.

Approximation properties

The L^2 projection $Q_{h_x} u \in S_{h_x}^0(\Gamma)$ of $u \in L^2(\Gamma)$ is defined as the unique solution of the variational problem

$$\langle Q_{h_x} u, v_h \rangle_{L^2(\Gamma)} = \langle u, v_h \rangle_{L^2(\Gamma)} \quad \text{for all } v_h \in S_{h_x}^0(\Gamma). \tag{6.5}$$

Analogously, the L^2 projection $Q_{h_t} u \in S^0_{h_t}(I)$ of $u \in L^2(0,T)$ is defined as the unique solution of the variational problem

$$\langle Q_{h_t} u, v_h \rangle_{L^2(I)} = \langle u, v_h \rangle_{L^2(I)} \quad \text{for all } v_h \in S^0_{h_t}(I). \tag{6.6}$$

The L^2 projection operator $Q_{h_x,h_t} : L^2(\Sigma) \to S^{0,0}_{h_x,h_t}(\Sigma)$, which is defined as the unique solution of the variational problem

$$\langle Q_{h_x,h_t} u, v_h \rangle_{L^2(\Sigma)} = \langle u, v_h \rangle_{L^2(\Sigma)} \quad \text{for all } v_h \in S^{0,0}_{h_x,h_t}(\Sigma), \tag{6.7}$$

has the representation $Q_{h_x,h_t} = Q^\Sigma_{h_x} Q^\Sigma_{h_t}$, where for $u \in L^2(\Sigma)$

$$(Q^\Sigma_{h_x} u)(x,t) := (Q_{h_x} u(\cdot,t))(x),$$
$$(Q^\Sigma_{h_t} u)(x,t) := (Q_{h_t} u(x,\cdot))(t).$$

Hence, we can use the well-known approximation properties of the operators Q_{h_x} and Q_{h_t} to derive estimates for the L^2 projection $Q_{h_x,h_t} u$ (see, for example, [5, 29]).

Theorem 6.2. *Let $u \in H^{r,s}(\Sigma)$ for some $r,s \in [0,1]$, and let $Q_{h_x,h_t} u \in S^{0,0}_{h_x,h_t}(\Sigma)$ be the L^2 projection of u. Then there hold the error estimates*

$$\|u - Q_{h_x,h_t} u\|_{L^2(\Sigma)} \le \|u\|_{L^2(\Sigma)},$$
$$\|u - Q_{h_x,h_t} u\|_{L^2(\Sigma)} \le c(h_x^r + h_t^s)\|u\|_{H^{r,s}(\Sigma)}.$$

Lemma 6.3. *Let $u \in H^{r,s}(\Sigma)$ for some $r,s \in [0,1]$. For $\sigma, \mu \in [0,1]$, there holds the error estimate*

$$\|u - Q_{h_x,h_t} u\|_{\widetilde{H}^{-\sigma,-\mu}(\Sigma)} \le c(h_x^\sigma + h_t^\mu)(h_x^r + h_t^s)\|u\|_{H^{r,s}(\Sigma)}.$$

Lemma 6.4 (Global inverse inequality). *Assume that the decompositions Γ_{N_x} and I_{N_t} are globally quasi-uniform. For $r \in [0,1)$, there holds the global inverse inequality*

$$\|\tau_h\|_{L^2(\Sigma)} \le c(h_x^{-r} + h_t^{-r/2})\|\tau_h\|_{\widetilde{H}^{-r,-r/2}(\Sigma)} \quad \textit{for all } \tau_h \in S^{0,0}_{h_x,h_t}(\Sigma).$$

Proof. Let $\tau_h \in S^{0,0}_{h_x,h_t}(\Sigma)$ and $0 \le r < \frac{1}{2}$. By applying the inverse inequality in spatial and temporal direction, we get

$$\begin{aligned}
\|\tau_h\|^2_{H^{r,r/2}(\Sigma)} &\le c \int_\Gamma \|\tau_h(x,\cdot)\|^2_{H^{r/2}((0,T))} \, ds_x + c \int_0^T \|\tau_h(\cdot,t)\|^2_{H^r(\Gamma)} \, dt \\
&\le c h_t^{-r} \int_\Gamma \|\tau_h(x,\cdot)\|^2_{L^2((0,T))} \, ds_x + c h_x^{-2r} \int_0^T \|\tau_h(\cdot,t)\|^2_{L^2(\Gamma)} \, dt \\
&\le c(h_x^{-2r} + h_t^{-r})\|\tau_h\|^2_{L^2(\Sigma)}.
\end{aligned}$$

Applying this estimate yields

$$\|\tau_h\|^2_{L^2(\Sigma)} = \langle \tau_h, \tau_h \rangle_{L^2(\Sigma)} \leq \|\tau_h\|_{H^{r,r/2}(\Sigma)} \|\tau_h\|_{\widetilde{H}^{-r,-r/2}(\Sigma)}$$
$$\leq c(h_x^{-r} + h_t^{-r/2}) \|\tau_h\|_{L^2(\Sigma)} \|\tau_h\|_{\widetilde{H}^{-r,-r/2}(\Sigma)},$$

and we conclude

$$\|\tau_h\|_{L^2(\Sigma)} \leq c(h_x^{-r} + h_t^{-r/2}) \|\tau_h\|_{\widetilde{H}^{-r,-r/2}(\Sigma)} \quad \text{for all } \tau_h \in S^{0,0}_{h_x,h_t}(\Sigma).$$

It remains to prove the estimate for $r \in [\frac{1}{2}, 1)$. For $\tau_h \in S^{0,0}_{h_x,h_t}(\Sigma)$, we have

$$\|\tau_h\|_{L^2(\Sigma)} \leq c(h_x^{-r/2} + h_t^{-r/4}) \|\tau_h\|_{\widetilde{H}^{-r/2,-r/4}(\Sigma)}. \tag{6.8}$$

By using interpolation results (see, for example, [21, 22]), we get

$$\|\tau_h\|^2_{\widetilde{H}^{-r/2,-r/4}(\Sigma)} \leq c \|\tau_h\|_{L^2(\Sigma)} \|\tau_h\|_{\widetilde{H}^{-r,-r/2}(\Sigma)}$$
$$\leq c(h_x^{-r/2} + h_t^{-r/4}) \|\tau_h\|_{\widetilde{H}^{-r/2,-r/4}(\Sigma)} \|\tau_h\|_{\widetilde{H}^{-r,-r/2}(\Sigma)},$$

and together with (6.8), we conclude

$$\|\tau_h\|_{L^2(\Sigma)} \leq c(h_x^{-r} + h_t^{-r/2}) \|\tau_h\|_{\widetilde{H}^{-r,-r/2}(\Sigma)} \quad \text{for all } \tau_h \in S^{0,0}_{h_x,h_t}(\Sigma).$$ □

For a shape regular boundary element mesh, that is, $h_{k,x} \sim h_{k,t}$, the global mesh size is given by $h := \max_{\ell=1,\dots,N} h_\ell$, where $h_\ell := |\sigma_\ell|^{1/n}$ is the local mesh size. We define $Q_h^{0,0} := Q_{h,h}^{0,0}$ and $S_h^{0,0}(\Sigma) := S_{h,h}^{0,0}(\Sigma)$, and we obtain the following estimates:

Corollary 6.1. *Assume that Σ_N is a shape regular boundary element mesh, and let $u \in H^{r,s}(\Sigma)$ for some $r, s \in [0,1]$. For $\sigma, \mu \in [0,1]$, there holds the error estimate*

$$\|u - Q_h^{0,0} u\|_{\widetilde{H}^{-\sigma,-\mu}(\Sigma)} \leq c h^{\min(r,s) + \min(\sigma,\mu)} \|u\|_{H^{r,s}(\Sigma)}.$$

Proof. Follows by applying Lemma 6.3 with $h_x \sim h_t$. □

Corollary 6.2. *Assume that Σ_N is a shape regular and globally quasi-uniform boundary element mesh. For $r \in [0,1)$, there holds the inverse inequality*

$$\|\tau_h\|_{L^2(\Sigma)} \leq c h^{-r} \|\tau_h\|_{\widetilde{H}^{-r,-r/2}(\Sigma)} \quad \textit{for all } \tau_h \in S_h^{0,0}(\Sigma).$$

Proof. Follows by applying Lemma 6.4 with $h_x \sim h_t$. □

6.2.2 Triangulation of Σ

Let $\{\Sigma_N\}_{N \in \mathbb{N}}$ be a family of admissible triangulations of Σ into boundary elements σ_ℓ given by (6.1). Again, we assume that there are no curved elements and that there is no approximation of the space-time boundary Σ. For each boundary element σ_ℓ, there exists exactly one $j \in \{1,\dots,J\}$, such that $\sigma_\ell \subset \Sigma_j$. The boundary elements σ_ℓ can be described as $\sigma_\ell = \chi_\ell(\sigma)$, where σ is some reference element in $\mathbb{R}^n$. The elements σ_ℓ are plane triangles for $n = 2$, and tetrahedra for $n = 3$. For each boundary element σ_ℓ, we define its volume

$$\Delta_\ell := \int_{\sigma_\ell} \mathrm{d}s_x\,\mathrm{d}t$$

and its local mesh size $h_\ell := \Delta_\ell^{1/n}$. The global mesh size is then given by $h := \max_{\ell=1,\dots,N} h_\ell$. The family $\{\Sigma_N\}_{N\in\mathbb{N}}$ of decompositions is said to be globally quasi-uniform if there exists a constant $c_G \geq 1$ independent of Σ_N, such that

$$\frac{h_{\max}}{h_{\min}} \leq c_G.$$

We consider shape regular boundary elements only. That means, there exists a constant c_B independent of the boundary decomposition Σ_N, such that

$$d_\ell \leq c_B h_\ell \quad \text{for } \ell = 1,\dots,N \tag{6.9}$$

with diameter d_ℓ given by

$$d_\ell := \sup_{(x,t),(y,s)\in\sigma_\ell} |(x,t) - (x,s)|.$$

A sample triangulation of the boundary Σ of the space-time domain $Q = (0,1)^3$ is shown in Figure 1.2(b).

For the approximation of the conormal derivative $w = \gamma_1^{\mathrm{int}} u$, we consider the space of piecewise constant basis functions $S_h^0(\Sigma) := \operatorname{span}\{\varphi_\ell^0\}_{\ell=1}^N$, which is defined with respect to the decomposition Σ_N.

Approximation properties

The L^2 projection $Q_h u \in S_h^0(\Sigma)$ of $u \in L^2(\Sigma)$ is defined as the unique solution of the variational problem:

$$\langle Q_h u, v_h\rangle_{L^2(\Sigma)} = \langle u, v_h\rangle_{L^2(\Sigma)} \quad \text{for all } v_h \in S_h^0(\Sigma). \tag{6.10}$$

By using Lemma 2.1 and the well-known approximation properties in standard Sobolev spaces, for example, [33], we immediately obtain the following results:

Theorem 6.3. *Let $u \in H^{r,s}(\Sigma)$ for some $r, s \in [0,1]$, and let $Q_h u \in S_h^0(\Sigma)$ be the L^2 projection of u. Then there hold the error estimates*

$$\|u - Q_h u\|_{L^2(\Sigma)} \leq \|u\|_{L^2(\Sigma)},$$
$$\|u - Q_h u\|_{L^2(\Sigma)} \leq c h^{\min(r,s)} \|u\|_{H^{r,s}(\Sigma)}.$$

Proof. First let $u \in L^2(\Sigma)$. By using

$$\langle u - Q_h u, v_h \rangle_{L^2(\Sigma)} = 0 \quad \text{for all } v_h \in S_h^0(\Sigma),$$

we obtain

$$\begin{aligned} \|u - Q_h u\|_{L^2(\Sigma)}^2 &= \langle u - Q_h u, u - Q_h u \rangle_{L^2(\Sigma)} = \langle u - Q_h u, u \rangle_{L^2(\Sigma)} \\ &\leq \|u - Q_h u\|_{L^2(\Sigma)} \|u\|_{L^2(\Sigma)}, \end{aligned}$$

and we conclude the first error estimate. For $u \in H^{r,s}(\Sigma)$ for some $r, s \in [0, 1]$ and $m := \min(r, s)$, we argue as follows: Analogously to [33, Theorem 10.2], we get

$$\|u - Q_h u\|_{L^2(\Sigma)} \leq c h^m \|u\|_{H^m(\Sigma)}.$$

According to Lemma 2.1, we have $H^{r,s}(\Sigma) \hookrightarrow H^m(\Sigma)$, and we therefore conclude:

$$\|u - Q_h u\|_{L^2(\Sigma)} \leq c h^m \|u\|_{H^{r,s}(\Sigma)}.$$ □

Lemma 6.5. *Let $u \in H^{r,s}(\Sigma)$ for some $r, s \in [0, 1]$ and $\sigma, \mu \in [0, 1]$. Then there holds the error estimate*

$$\|u - Q_h u\|_{\widetilde{H}^{-\sigma,-\mu}(\Sigma)} \leq c h^{\min(r,s) + \min(\sigma,\mu)} \|u\|_{H^{r,s}(\Sigma)}.$$

Proof. Let $u \in H^{r,s}(\Sigma)$. Using (6.10), this yields

$$\begin{aligned} \|u - Q_h u\|_{\widetilde{H}^{-\sigma,-\mu}(\Sigma)} &= \sup_{0 \neq v \in H^{\sigma,\mu}(\Sigma)} \frac{\langle u - Q_h u, v \rangle_\Sigma}{\|v\|_{H^{\sigma,\mu}(\Sigma)}} \\ &= \sup_{0 \neq v \in H^{\sigma,\mu}(\Sigma)} \frac{\langle u - Q_h u, v - Q_h v \rangle_\Sigma}{\|v\|_{H^{\sigma,\mu}(\Sigma)}}. \end{aligned}$$

By applying the Cauchy–Schwarz inequality and Theorem 6.3, we obtain

$$\begin{aligned} \|u - Q_h u\|_{\widetilde{H}^{-\sigma,-\mu}(\Sigma)} &\leq \|u - Q_h u\|_{L^2(\Sigma)} \sup_{0 \neq v \in H^{\sigma,\mu}(\Sigma)} \frac{\|v - Q_h v\|_{L^2(\Sigma)}}{\|v\|_{H^{\sigma,\mu}(\Sigma)}} \\ &\leq c h^{\min(r,s)} h^{\min(\sigma,\mu)} \|u\|_{H^{r,s}(\Sigma)}. \end{aligned}$$ □

Since we consider shape regular boundary elements, the following inverse inequality holds:

Lemma 6.6 (Global inverse inequality). *For a globally quasi-uniform boundary decomposition Σ_N and for $\sigma, \mu \in [0, 1]$, there holds*

$$\|\tau_h\|_{L^2(\Sigma)} \leq c h^{-\max(\sigma,\mu)} \|\tau_h\|_{\widetilde{H}^{-\sigma,-\mu}(\Sigma)} \quad \text{for all } \tau_h \in S_h^0(\Sigma). \tag{6.11}$$

Proof. Let $\tau_h \in S_h^0(\Sigma)$ and $\sigma, \mu \in [0,1]$. Application of the standard inverse inequality (see, for example, [33, Section 10.2]), yields

$$\|\tau_h\|_{L^2(\Sigma)} \leq ch^{-\max(\sigma,\mu)} \|\tau_h\|_{\widetilde{H}^{-\max(\sigma,\mu)}(\Sigma)}. \tag{6.12}$$

Since $H^{\max(\sigma,\mu)}(\Sigma) \hookrightarrow H^{\sigma,\mu}(\Sigma)$, see Lemma 2.1, we obtain

$$\begin{aligned} \|\tau_h\|_{\widetilde{H}^{-\max(\sigma,\mu)}(\Sigma)} &= \sup_{0 \neq v \in H^{\max(\sigma,\mu)}(\Sigma)} \frac{\langle \tau_h, v\rangle_\Sigma}{\|v\|_{H^{\max(\sigma,\mu)}(\Sigma)}} \\ &\leq c \sup_{0 \neq v \in H^{\sigma,\mu}(\Sigma)} \frac{\langle \tau_h, v\rangle_\Sigma}{\|v\|_{H^{\sigma,\mu}(\Sigma)}} = c\|\tau_h\|_{\widetilde{H}^{-\sigma,-\mu}(\Sigma)}, \end{aligned} \tag{6.13}$$

and the assertion follows from combining (6.12) and (6.13). □

Remark 6.2. For $r = 1/2$ and $s = 1/4$, we have $\widetilde{H}^{-1/2,-1/4}(\Sigma) = H^{-1/2,-1/4}(\Sigma)$; therefore, we obtain

$$\|\tau_h\|_{L^2(\Sigma)} \leq ch^{-1/2} \|\tau_h\|_{H^{-1/2,-1/4}(\Sigma)} \quad \text{for all } \tau_h \in S_h^0(\Sigma).$$

7 Boundary element methods

In this section we discretize the variational formulation (5.9) by using the previously introduced boundary element spaces, and we derive a priori error estimates for the Galerkin approximation of the Neumann datum $w = \gamma_1^{\text{int}} u$, see Subsection 7.1. The numerical analysis of the discretized indirect formulation (5.11) follows exactly the same path. In Subsection 7.2, we prove error estimates for the related approximation of the solution u in the space-time domain Q.

For the discretization of the variational formulation (5.9), we consider the space of piecewise constant basis functions $X_h \in \{S_h^{0,0}(\Sigma), S_h^0(\Sigma)\}$ defined with respect to a shape regular boundary element mesh Σ_N. The Galerkin–Bubnov variational formulation of (5.9) is to find $w_h \in X_h$, such that

$$\langle Vw_h, \tau_h\rangle_\Sigma = \left\langle \left(\frac{1}{2}I + K\right)g - M_0 u_0 - N_0 f, \tau_h \right\rangle_\Sigma \quad \text{for all } \tau_h \in X_h. \tag{7.1}$$

Due to the ellipticity of the single layer boundary integral operator V and the boundedness of the integral operators, problem (7.1) admits a unique solution.

Note that we only consider shape regular boundary element meshes Σ_N, both for an arbitrary triangulation of Σ and for a tensor product decomposition, since we want to compare the theoretical and practical results of the two discretization techniques. Hence, for the tensor product approach, we choose $h_t \sim h_x$. A priori error estimates and numerical experiments for a different refinement strategy, for example, $h_t \sim h_x^2$, can be found in [5, 29].

7.1 Error estimates

Since the operator V is elliptic and bounded, we can apply Cea's lemma to conclude quasi-optimality of the Galerkin approximation $w_h \in X_h$. That means, we have

$$\|w - w_h\|_{H^{-1/2,-1/4}(\Sigma)} \leq \frac{c_2^V}{c_1^V} \inf_{\tau_h \in X_h} \|w - \tau_h\|_{H^{-1/2,-1/4}(\Sigma)},$$

where $w \in H^{-1/2,-1/4}(\Sigma)$ is the unique solution of the variational problem (5.9). Hence, we can use the approximation properties of the boundary element space X_h to derive error estimates for the solution w_h of (7.1). Recall that Γ is assumed to be piecewise smooth. That is, we have the representation $\overline{\Sigma} = \bigcup_{j=1}^{J} \overline{\Sigma}_j$ with $\Sigma_j = \Gamma_j \times (0,T)$. Due to the local definition of the trial space X_h and by applying Lemma 2.3, we obtain

$$\|w - w_h\|_{H^{-1/2,-1/4}(\Sigma)} \leq \frac{c_2^V}{c_1^V} \sum_{j=1}^{J} \inf_{\tau_h^j \in X_{h|\Sigma_j}} \|w_{|\Sigma_j} - \tau_h^j\|_{\widetilde{H}^{-1/2,-1/4}(\Sigma_j)}. \tag{7.2}$$

Note that all the approximation properties shown in the previous section also hold for an open part $\Sigma_j \subset \Sigma$ of the space-time boundary Σ. Hence we can replace the space $H^{r,s}(\Sigma)$ with the larger space $H^{r,s}_{\mathrm{pw}}(\Sigma)$, and we still get the same error estimates in the appropriate norms.

7.1.1 One-dimensional problem

Recall that in the one-dimensional case we can identify the Sobolev spaces $H^{r,s}(\Sigma)$ with $H^s(\Sigma)$.

Theorem 7.1. *Let $w_h \in S_h^0(\Sigma)$ be the unique solution of the Galerkin variational problem (7.1). For $w \in H^s_{pw}(\Sigma)$ with $s \in [0,1]$, there holds the error estimate*

$$\|w - w_h\|_{H^{-1/4}(\Sigma)} \leq c h^{s+1/4} |w|_{H^s_{pw}(\Sigma)}.$$

Proof. Follows by applying Lemma 6.1 in (7.2). □

Moreover, we can derive an error estimate in the $L^2(\Sigma)$ norm, assuming that the family of boundary decompositions $\{\Sigma_N\}_{N\in\mathbb{N}}$ is globally quasi-uniform.

Theorem 7.2. *Let $w_h \in S_h^0(\Sigma)$ be the unique solution of the Galerkin variational problem (7.1). For $w \in H^s_{pw}(\Sigma)$ for some $s \in [0,1]$, there holds*

$$\|w - w_h\|_{L^2(\Sigma)} \leq c h^s |w|_{H^s_{pw}(\Sigma)}.$$

Proof. The assertion follows by using the triangle inequality

$$\|w - w_h\|_{L^2(\Sigma)} \leq \|w - Q_h^0 w\|_{L^2(\Sigma)} + \|Q_h^0 w - w_h\|_{L^2(\Sigma)},$$

and by applying Lemma 6.1, Lemma 6.2, and Theorem 7.1. □

7.1.2 Two- and three-dimensional problem

Theorem 7.3. *Let $w_h \in X_h$ be the unique solution of the Galerkin–Bubnov variational formulation* (7.1). *For $w \in H^{r,s}_{pw}(\Sigma)$ for some $r, s \in [0,1]$, there holds*

$$\|w - w_h\|_{H^{-1/2,-1/4}(\Sigma)} \le ch^{\min(r,s)+1/4}\|w\|_{H^{r,s}_{pw}(\Sigma)}.$$

Proof. The assertion follows by applying Corollary 6.1 if $X_h = S^{0,0}_h(\Sigma)$, and Lemma 6.5 if $X_h = S^0_h(\Sigma)$ in (7.2). □

Theorem 7.4. *Assume that the boundary decomposition Σ_N is globally quasi-uniform. Let $w_h \in X_h$ be the unique solution of the Galerkin–Bubnov variational problem* (7.1). *For $w \in H^{r,s}_{pw}(\Sigma)$ for some $r, s \in [1/4, 1]$, there holds*

$$\|w - w_h\|_{L^2(\Sigma)} \le ch^{\min(r,s)-1/4}\|w\|_{H^{r,s}_{pw}(\Sigma)}.$$

Proof. By using the triangle inequality, Theorem 6.2, and Corollary 6.2 for $X_h = S^{0,0}_h(\Sigma)$, and Theorem 6.3 and Remark 6.2 for $X_h = S^0_h(\Sigma)$, respectively. In both cases, we get

$$\begin{aligned}\|w - w_h\|_{L^2(\Sigma)} &\le \|w - Q_h w\|_{L^2(\Sigma)} + \|Q_h w - w_h\|_{L^2(\Sigma)}\\ &\le ch^{\min(r,s)}\|w\|_{H^{r,s}_{\mathrm{pw}}(\Sigma)} + ch^{-1/2}\|Q_h w - w_h\|_{H^{-1/2,-1/4}(\Sigma)}.\end{aligned}$$

Here, Q_h is either the L^2 projection onto $S^{0,0}_h(\Sigma)$, or onto $S^0_h(\Sigma)$. The assertion follows with

$$\|Q_h w - w_h\|_{H^{-1/2,-1/4}(\Sigma)} \le \|Q_h w - w\|_{H^{-1/2,-1/4}(\Sigma)} + \|w - w_h\|_{H^{-1/2,-1/4}(\Sigma)},$$

Theorem 7.3, Corollary 6.1 for $X_h = S^{0,0}_h(\Sigma)$, and Lemma 6.5 for $X_h = S^0_h(\Sigma)$, respectively. □

Hence, we can prove the same convergence rates for $X_h = S^{0,0}_h(\Sigma)$ and $X_h = S^0_h(\Sigma)$ of the Galerkin approximation w_h in the energy norm and in the $L^2(\Sigma)$ norm, assuming that the boundary element mesh Σ_N is shape regular. However, the numerical results in Section 8 show that the $L^2(\Sigma)$ error estimate is not optimal.

7.2 Domain error estimates

Let $w_h \in X_h$ be the unique solution of the Galerkin variational problem (7.1). We obtain an approximate solution of the initial Dirichlet boundary value problem (1.1) in Q by using the representation formula (5.1) with the approximation w_h. For $(x,t) \in Q$ we have

$$\tilde{u}(x,t) = (\widetilde{V}w_h)(x,t) - (Wg)(x,t) + (\widetilde{M}_0 u_0)(x,t) + (\widetilde{N}_0 f)(x,t). \qquad (7.3)$$

For the related error, we obtain for $(x,t) \in Q$

$$|u(x,t) - \tilde{u}(x,t)| = |(\widetilde{V}(w - w_h))(x,t)| \\ = \frac{1}{\alpha}\left|\int_\Sigma U^\star(x-y, t-\tau)(w - w_h)(y,\tau)\, ds_y\, d\tau\right|.$$

Since $(x,t) \in Q$ and $(y,\tau) \in \Sigma$, the fundamental solution $U^\star(x-y,t-\tau)$ is smooth and we therefore conclude $U^\star(x-\cdot,t-\cdot) \in H^{-\sigma,-\sigma/2}(\Sigma)$ for any $\sigma \in \mathbb{R}$. Hence,

$$|u(x,t) - \tilde{u}(x,t)| \le \frac{1}{\alpha}\|U^\star(x-\cdot,t-\cdot)\|_{H^{-\sigma,-\sigma/2}(\Sigma)}\|w - w_h\|_{\widetilde{H}^{\sigma,\sigma/2}(\Sigma)}. \tag{7.4}$$

Thus, in order to derive an error estimate for the pointwise error $|u(x,t) - \tilde{u}(x,t)|$, $(x,t) \in Q$, we need an error estimate for $\|w - w_h\|_{\widetilde{H}^{\sigma,\sigma/2}(\Sigma)}$ where $\sigma \in \mathbb{R}$ is minimal. In the following, $Q_h : L^2(\Sigma) \to X_h$ denotes the L^2 projection onto the space X_h.

Theorem 7.5 (Aubin–Nitsche Trick). *Let $w \in H^{r,s}_{pw}(\Sigma)$ for some $r \in [-1/2,1]$ and $s \in [-1/4,1]$ be the unique solution of (5.9), and let $w_h \in X_h$ be the unique solution of the Galerkin variational problem (7.1). Assume that the adjoint single layer operator*

$$V^* : H^{-1-\sigma,-1/2-\mu}(\Sigma) \to H^{-\sigma,-\mu}(\Sigma)$$

is continuous and bijective for some $-2 \le \sigma \le -1/2$ and $\mu = \sigma/2$. Then there holds the error estimate

$$\|w - w_h\|_{\widetilde{H}^{\sigma,\sigma/2}(\Sigma)} \le c h^{\min(r,s)+1/2+\min(-1-\sigma,-1/2-\mu)}\|w\|_{H^{r,s}_{pw}(\Sigma)}.$$

Proof. For $\sigma < -1/2$ and $\mu = \sigma/2$ we have

$$\|w - w_h\|_{\widetilde{H}^{\sigma,\mu}(\Sigma)} = \sup_{0 \ne v \in H^{-\sigma,-\mu}(\Sigma)} \frac{\langle w - w_h, v\rangle_\Sigma}{\|v\|_{H^{-\sigma,-\mu}(\Sigma)}}.$$

By assumption, the adjoint single layer operator

$$V^* : H^{-1-\sigma,-1/2-\mu}(\Sigma) \to H^{-\sigma,-\mu}(\Sigma)$$

is continuous and bijective. Hence, for $v \in H^{-\sigma,-\mu}(\Sigma)$ there exists a unique $z \in H^{-1-\sigma,-1/2-\mu}(\Sigma)$, such that $v = V^* z$. Therefore, and by applying the Galerkin orthogonality

$$\langle V(w - w_h), \tau_h\rangle_\Sigma = 0 \quad \text{for all } \tau_h \in X_h,$$

we obtain

$$\|w - w_h\|_{\widetilde{H}^{\sigma,\mu}(\Sigma)} = \sup_{0 \ne z \in H^{-1-\sigma,-1/2-\mu}(\Sigma)} \frac{\langle w - w_h, V^* z\rangle_\Sigma}{\|V^* z\|_{H^{-\sigma,-\mu}(\Sigma)}} \\ = \sup_{0 \ne z \in H^{-1-\sigma,-1/2-\mu}(\Sigma)} \frac{\langle V(w - w_h), z - Q_h z\rangle_\Sigma}{\|V^* z\|_{H^{-\sigma,-\mu}(\Sigma)}}.$$

Since V^* is bijective, there exists a constant $c > 0$, such that [11, Lemma A.40]

$$\|V^* z\|_{H^{-\sigma,-\mu}(\Sigma)} \geq c\|z\|_{H^{-1-\sigma,-1/2-\mu}(\Sigma)} \quad \text{for all } z \in H^{-1-\sigma,-1/2-\mu}(\Sigma).$$

Thus, by using the boundedness of $V : H^{-1/2,-1/4}(\Sigma) \to H^{1/2,1/4}(\Sigma)$, we conclude

$$\begin{aligned}
&\|w - w_h\|_{\widetilde{H}^{\sigma,\mu}(\Sigma)} \\
&\quad \leq \tilde{c}\|w - w_h\|_{H^{-1/2,-1/4}(\Sigma)} \sup_{0 \neq z \in H^{-1-\sigma,-1/2-\mu}(\Sigma)} \frac{\|z - Q_h z\|_{H^{-1/2,-1/4}(\Sigma)}}{\|z\|_{H^{-1-\sigma,-1/2-\mu}(\Sigma)}}.
\end{aligned}$$

When considering $-1-\sigma \leq 1$, i. e., $\sigma \geq -2$, we obtain from the approximation properties of the operator Q_h the error estimate

$$\|w - w_h\|_{\widetilde{H}^{\sigma,\mu}(\Sigma)} \leq \hat{c} h^{\min(-1-\sigma,-1/2-\mu)+1/4} \|w - w_h\|_{H^{-1/2,-1/4}(\Sigma)},$$

and the assertion follows by applying the error estimate for the Galerkin approximation w_h in the energy norm. □

Now, assume that the solution w of the variational formulation (5.9) is sufficiently smooth; that is, $w \in H^{1,1}_{\mathrm{pw}}(\Sigma)$. From estimate (7.4), and by choosing $\sigma = -2$ in Theorem (7.5), we get, for $(x,t) \in Q$, the pointwise error estimate

$$\begin{aligned}
|u(x,t) - \tilde{u}(x,t)| &\leq \tilde{c}\|U^\star(x-\cdot, t-\cdot)\|_{H^{2,1}(\Sigma)} \|w - w_h\|_{\widetilde{H}^{-2,-1}(\Sigma)} \\
&\leq c h^2 \|U^\star(x-\cdot, t-\cdot)\|_{H^{2,1}(\Sigma)} \|w\|_{H^{1,1}_{\mathrm{pw}}(\Sigma)}.
\end{aligned} \tag{7.5}$$

Let us now consider problem (3.7) with source term $f = 0$. To estimate the global error $\|u - \tilde{u}\|_{H^{1,1/2}(Q)}$, we proceed as follows: We first consider the Dirichlet trace of the discretized representation formula (7.3),

$$\hat{g} := V w_h + \frac{1}{2} g - K g.$$

Moreover, the first boundary integral equation in (5.4) gives

$$g = V w + \frac{1}{2} g - K g,$$

and we therefore conclude the relation

$$g - \hat{g} = V(w - w_h). \tag{7.6}$$

Theorem 7.6 (Domain error estimate). *Let $u \in H^{1,1/2}_{;0,}(Q)$ be the unique solution of the Dirichlet boundary value problem (3.7) with source term $f = 0$, and let $\tilde{u} \in H^{1,1/2}_{;0,}(Q)$ be the corresponding approximation given by (7.3), where $f = 0$ and $u_0 = 0$. Then there holds the error estimate*

$$\|u - \tilde{u}\|_{H^{1,1/2}_{;0,}(Q)} \leq c\|w - w_h\|_{H^{-1/2,-1/4}(\Sigma)}.$$

Proof. The solution $u = \bar{u} + \mathcal{E}_0 g \in H^{1,1/2}_{;0,}(Q)$ of problem (3.7) with source term $f = 0$ is given as the unique solution of the variational problem

$$a(\bar{u}, v) = -a(\mathcal{E}_0 g, v) \quad \text{for all } v \in H^{1,1/2}_{0;,0}(Q).$$

For the approximation $\tilde{u}$, we consider the decomposition $\tilde{u} = \hat{u} + \mathcal{E}_0 \hat{g} \in H^{1,1/2}_{;0,}(Q)$, which satisfies

$$a(\hat{u}, v) = -a(\mathcal{E}_0 \hat{g}, v) \quad \text{for all } v \in H^{1,1/2}_{0;,0}(Q).$$

By subtracting the last two equations, we obtain

$$a(\bar{u} - \hat{u}, v) = a(\mathcal{E}_0(\hat{g} - g), v) \quad \text{for all } v \in H^{1,1/2}_{0;,0}(Q).$$

Since $\bar{u} - \hat{u} \in H^{1,1/2}_{0;0,}(Q)$, we can apply the stability estimate (3.11) to get

$$\begin{aligned} \frac{1}{2}\|\bar{u} - \hat{u}\|_{H^{1,1/2}_{0;0,}(Q)} &\leq \sup_{0 \neq v \in H^{1,1/2}_{0;,0}(Q)} \frac{a(\bar{u} - \hat{u}, v)}{\|v\|_{H^{1,1/2}_{0;,0}(Q)}} \\ &= \sup_{0 \neq v \in H^{1,1/2}_{0;,0}(Q)} \frac{a(\mathcal{E}_0(\hat{g} - g), v)}{\|v\|_{H^{1,1/2}_{0;,0}(Q)}} \leq c\|\mathcal{E}_0(\hat{g} - g)\|_{H^{1,1/2}_{;0,}(Q)}. \end{aligned}$$

Hence, by using the triangle inequality, the Poincaré inequality, and the boundedness of the inverse trace operator $\mathcal{E}_0$, we obtain

$$\begin{aligned} \|u - \tilde{u}\|_{H^{1,1/2}_{;0,}(Q)} &\leq \|\bar{u} - \hat{u}\|_{H^{1,1/2}_{;0,}(Q)} + \|\mathcal{E}_0(\hat{g} - g)\|_{H^{1,1/2}_{;0,}(Q)} \\ &\leq \tilde{c}\|\bar{u} - \hat{u}\|_{H^{1,1/2}_{0;0,}(Q)} + \|\mathcal{E}_0(\hat{g} - g)\|_{H^{1,1/2}_{;0,}(Q)} \\ &\leq \hat{c}\|\mathcal{E}_0(\hat{g} - g)\|_{H^{1,1/2}_{;0,}(Q)} \leq \bar{c}\|\hat{g} - g\|_{H^{1/2,1/4}(\Sigma)}, \end{aligned}$$

and the assertion follows with the relation (7.6). □

Note that for $w \in H^{1,1}_{pw}(\Sigma)$, we finally conclude the error estimate

$$\|u - \tilde{u}\|_{H^{1,1/2}_{;0,}(Q)} \leq ch^{5/4}\|w\|_{H^{1,1}_{pw}(\Sigma)}.$$

8 Numerical results

We consider the model problem (1.1) with source term $f = 0$, final time $T = 1$, and with the heat capacity constant $\alpha = 10$. We present examples for the one- and two-dimensional case, and compare the tensor product decomposition with a triangulation of the space-time boundary Σ. All of the following examples refer to a shape regular boundary decomposition. The Galerkin boundary element discretization of the variational formulation (5.9) is done by using piecewise constant basis functions φ_ℓ. The resulting system of linear equations $V_h \underline{w} = \underline{f}$ with

$$V_h[\ell,k] := \langle V\varphi_k, \varphi_\ell\rangle_\Sigma, \quad f[\ell] = \left\langle \left(\frac{1}{2}I + K\right)g - M_0 u_0, \varphi_\ell \right\rangle_\Sigma$$

for $\ell, k = 1, \ldots, N$, is solved by using the GMRES method with a relative accuracy of 10^{-8} as stopping criteria.

8.1 One-dimensional problem

Let us start with the simple one-dimensional problem. We consider the spatial domain $\Omega = (0,1)$ and homogeneous Dirichlet conditions $g = 0$. Recall that the boundary elements are line segments in temporal dimension.

Uniform refinement

The first example corresponds to the initial datum

$$u_0(x) = \sin(2\pi x) \quad \text{for } x \in \Omega = (0,1)$$

and a globally uniform boundary element mesh of mesh size $h = 2^{-L}$. Table 1.1 shows the error $\|w - w_h\|_{L^2(\Sigma)}$ and the estimated order of convergence (eoc), which is linear as expected, according to Theorem 7.2. Moreover, the iteration numbers of the GMRES method are given.

Table 1.1: $L^2(\Sigma)$ error and convergence rate of the Galerkin approximation w_h, and iteration numbers of the GMRES method in the case of uniform refinement.

L	N	$\|w - w_h\|_{L^2(\Sigma)}$	eoc	Iter
5	64	7.950×10^{-2}	1.01	31
6	128	3.959×10^{-2}	1.01	41
7	256	1.976×10^{-2}	1.00	50
8	512	9.872×10^{-3}	1.00	59
9	1,024	4.929×10^{-3}	1.00	70
10	2,048	2.468×10^{-3}	1.00	82
11	4,096	1.233×10^{-3}	1.00	96

Adaptive refinement

For the second example, we consider the initial datum

$$u_0(x) = 5\exp(-10x)\sin(\pi x) \quad \text{for } x \in \Omega = (0,1),$$

which motivates the use of a locally quasi-uniform boundary element mesh resulting from some adaptive refinement strategy. The Galerkin approximation w_h is shown in

Figure 1.3. In Figure 1.4, the convergence history of the approximation for uniform and adaptive refinement is given.

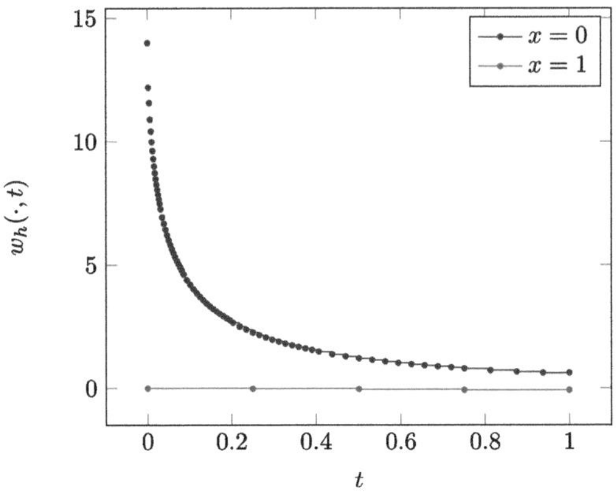

Figure 1.3: Galerkin approximation w_h in the case of adaptive refinement in 1D.

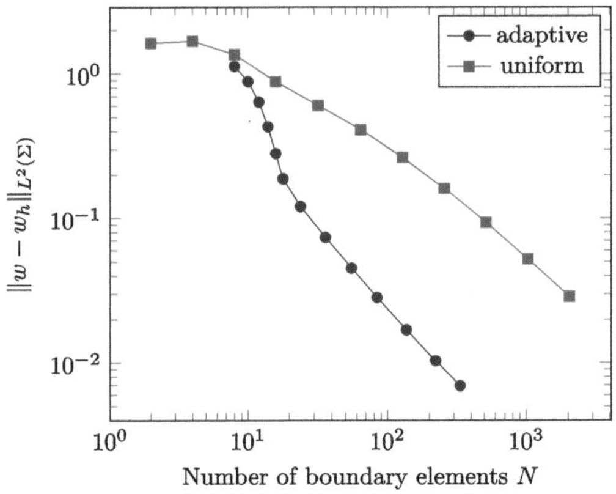

Figure 1.4: Convergence of the Galerkin approximation w_h for uniform and adaptive refinement in 1D.

8.2 Two-dimensional problem

For the following numerical examples we choose $\Omega = (0,1)^2$, i. e., $Q = (0,1)^3$.

Uniform refinement

We consider the exact solution

$$u(x,t) = \exp\left(-\frac{t}{\alpha}\right)\sin\left(x_1\cos\frac{\pi}{8} + x_2\sin\frac{\pi}{8}\right) \quad \text{for } (x,t) = (x_1,x_2,t) \in Q,$$

and determine the Dirichlet datum g, and the initial datum u_0 accordingly. We use a globally quasi-uniform boundary element mesh with mesh size $h = \mathcal{O}(2^{-L})$, both for the tensor product approach and for a triangulation of the space-time boundary Σ. Table 1.2 and Table 1.3 show the error $\|w - w_h\|_{L^2(\Sigma)}$ of the Galerkin approximation w_h and the pointwise error $|(u - \tilde{u})(x, t)|$ in $x = (0.5, 0.5)$, $t = 0.5$, and the corresponding convergence rates (eoc). Additionally, the iteration numbers of the GMRES method are listed. Whereas the convergence rate of the pointwise error is quadratic and therefore in line with the theoretical findings (7.5), we obtain linear convergence of the Galerkin approximation w_h in the $L^2(\Sigma)$ norm, which is, according to Theorem 7.4, better than expected.

Table 1.2: Error and convergence rates of the Galerkin approximation w_h and the approximated solution $\tilde{u}$ in the interior, and iteration numbers of the GMRES method in the case of uniform refinement for a tensor product decomposition of Σ.

L	N	$\|w - w_h\|_{L^2(\Sigma)}$	eoc	$\|(u - \tilde{u})(x, t)\|$	eoc	Iter
0	4	2.795×10^{-1}	–	2.597×10^{-2}	–	2
1	16	1.413×10^{-1}	0.98	5.544×10^{-3}	2.23	9
2	64	6.882×10^{-2}	1.04	9.146×10^{-4}	2.60	14
3	256	3.353×10^{-2}	1.04	2.485×10^{-4}	1.88	18
4	1,024	1.650×10^{-2}	1.02	6.315×10^{-5}	1.98	24
5	4,096	8.172×10^{-3}	1.01	1.563×10^{-5}	2.01	35
6	16,384	4.066×10^{-3}	1.01	3.748×10^{-6}	2.06	50
7	65,536	2.030×10^{-3}	1.00	8.468×10^{-7}	2.15	67

Table 1.3: Error and convergence rates of the Galerkin approximation w_h and of the approximated solution $\tilde{u}$ in the interior, and iteration numbers of the GMRES method in the case of uniform refinement for a triangulation of Σ.

L	N	$\|w - w_h\|_{L^2(\Sigma)}$	eoc	$\|(u - \tilde{u})(x, t)\|$	eoc	Iter
0	16	1.588×10^{-1}	–	2.046×10^{-2}	–	9
1	64	6.326×10^{-2}	1.33	5.395×10^{-3}	1.92	16
2	256	2.502×10^{-2}	1.34	1.337×10^{-3}	2.01	23
3	1,024	1.084×10^{-2}	1.21	3.336×10^{-4}	2.00	32
4	4,096	5.040×10^{-3}	1.11	8.348×10^{-5}	2.00	44
5	16,384	2.447×10^{-3}	1.04	2.093×10^{-5}	2.00	62
6	65,536	1.233×10^{-3}	0.99	5.265×10^{-6}	1.99	85

As already mentioned, we consider shape regular boundary elements only. In case of the tensor product approach we therefore choose $h_x \sim h_t$. Although the relation $h_t \sim h_x^2$ is recommended to obtain optimal convergence results of the Galerkin approximation

w_h in the energy norm [5, 29], we get linear convergence of the approximation in the $L^2(\Sigma)$ norm in our experiments. Note that numerical results in [5, Section 6] indicate that the relation $h_t \sim h_x^2$ is not necessary for an optimal convergence rate in the $L^2(\Sigma)$ norm.

Adaptive refinement

As a second example, we consider the initial datum

$$u_0(x_1, x_2) = 40 \exp(-10(x_1 + x_2)) \sin(\pi x_1) \sin(\pi x_2) \quad \text{for } (x_1, x_2) \in \Omega$$

(see Figure 1.5), and we use a globally quasi-uniform and a locally quasi-uniform triangulation of the space-time boundary resulting from some adaptive refinement strategy. In Figure 1.7, the convergence history of the approximation for uniform and adaptive refinement is given, whereas the resulting boundary element mesh is shown in Figure 1.6.

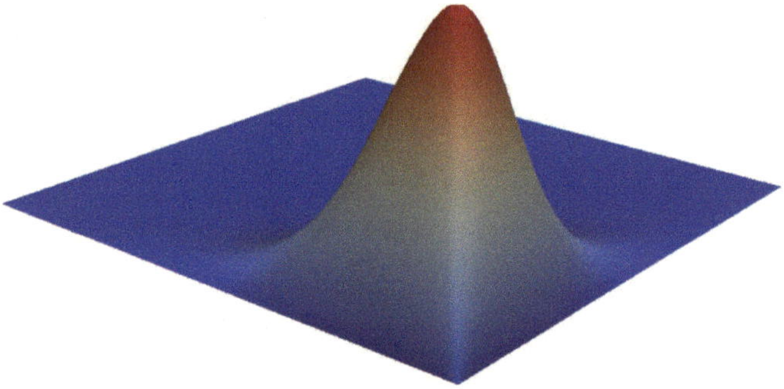

Figure 1.5: Initial datum u_0 for the 2D problem.

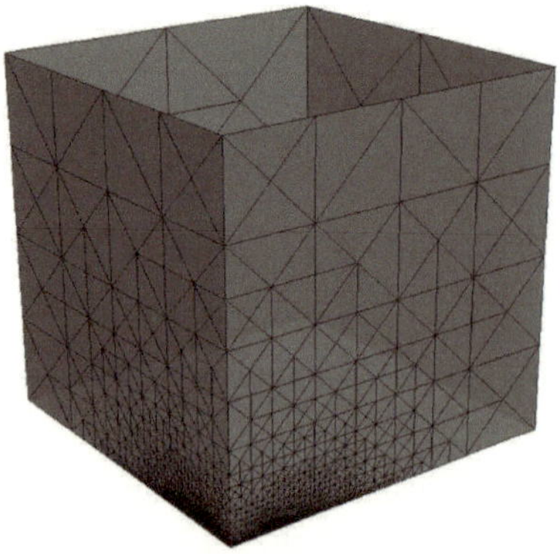

Figure 1.6: Triangular boundary element mesh in the case of adaptive refinement in 2D.

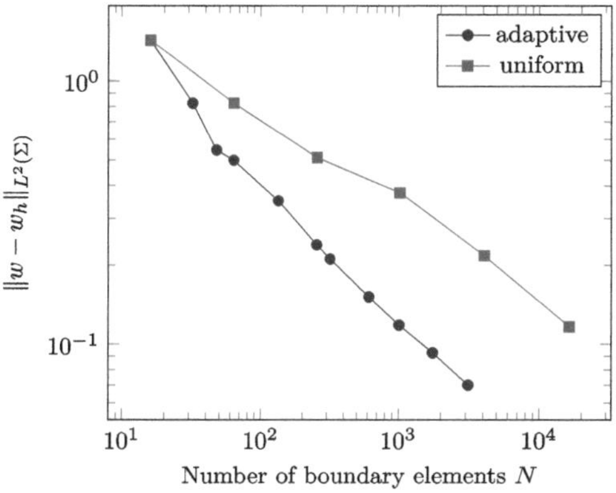

Figure 1.7: Convergence of the Galerkin approximation w_h for uniform and adaptive refinement in 2D.

9 Conclusion and outlook

In this work, we have described space-time boundary element discretizations for the initial Dirichlet boundary value problem for the heat equation. After the derivation of the representation formula for the solution of the model problem (1.1), we summarized the mapping properties of the heat potentials and of the resulting boundary integral operators and discussed the unique solvability of related boundary integral equations in the setting of anisotropic Sobolev spaces. The unknown Neumann datum $\partial_n u_{|\Sigma}$ can be determined by solving a weakly singular boundary integral equation. The ellipticity of the single layer operator ensures unique solvability of the problem. We compared two different space-time discretization techniques to compute an approximation of the Neumann datum $\partial_n u_{|\Sigma}$, namely, a tensor product decomposition, and an arbitrary triangulation of the space-time boundary Σ. Both methods allow us to parallelize the computation of the global solution of the whole space-time system, which leads to improved parallel scalability in distributed memory systems in contrast to, for example, time-stepping schemes. A parallel solver for space-time boundary element methods for the heat equation was introduced in [10]. One possible drawback of the tensor product approach is that we can only apply adaptive refinement in space and time separately. This can be resolved, for example, by allowing hanging nodes in the mesh, which is reasonable if the discretization of the integral equation is done by using piecewise constant basis functions, as we did. However, an arbitrary triangulation of the space-time boundary Σ allows for adaptive refinement in space and time simultaneously, while maintaining the admissibility of the mesh. We derived a priori error estimates for both discretization techniques and provided numerical experiments to confirm the theoretical findings.

In the numerical experiments, we used the exact solution to compute the errors of the Galerkin approximation and for the application of adaptive refinement. Of course, in general, we do not know the exact solution. Thus, we have to establish a posteriori error estimators for space-time boundary element methods to define suitable adaptive refinement strategies. One possible approach is the method described in [32], in case of the Laplace equation, which is based on an approximation of a second kind Fredholm integral equation by a Neumann series to compute the error. However, this method utilizes the contraction property of the double layer potential, which is, in the case of the heat equation, not yet proven for a general Lipschitz domain Ω. The development of a posteriori error estimators for space-time boundary element methods for the heat equation is left for future work.

As already mentioned, one advantage of space-time discretization methods is the ability to use parallel iterative solution strategies for time-dependent problems. But to get a competitive space-time solver, an efficient iterative solution technique for the global space-time system is necessary; that is, the solution requires an application of space-time preconditioners. A popular preconditioning strategy in boundary element methods is operator preconditioning [17, 35], which is based on boundary integral operators of opposite order, such as the single layer operator V and the hypersingular operator D, but which requires a related stability condition for the boundary element spaces used for the discretization to be satisfied. In [8], we analyzed this robust preconditioning strategy for space-time boundary element methods for the heat equation and discussed suitable choices of boundary element spaces. The parallel solver introduced in [10] is also applicable to the preconditioned space-time system, see [9].

The matrices related to the discretized space-time integral equations are dense; thus, fast methods are necessary to tackle large scale problems, especially for space-time systems. Fast methods for solving boundary integral equations for the heat equations were introduced in [37, 39]. The parabolic fast multipole method applied to a space-time Galerkin discretization is discussed in [26], where the discretization is done with respect to a tensor product decomposition of the space-time boundary. The extension of fast methods, for example, adaptive cross approximation and the parabolic fast multipole method to an arbitrary triangulation of Σ is still open and left for future work.

An advantage of boundary element methods is the natural handling of problems in exterior, unbounded domains. Thus, boundary element methods are a popular choice when solving transmission problems. The introduced domain variational formulation (3.10) in the setting of anisotropic Sobolev spaces allows us to establish symmetric and nonsymmetric FEM–BEM coupling methods in an appropriate functional framework.

Bibliography

[1] W. Arendt and K. Urban. *Partielle Differenzialgleichungen*. Springer Spektrum, Heidelberg, 2010.

[2] D. N. Arnold and P. J. Noon. Boundary integral equations of the first kind for the heat equation. In *Boundary elements IX, Vol. 3 (Stuttgart, 1987)*, pages 213–29. Comput. Mech., Southampton, 1987.

[3] D. N. Arnold and P. J. Noon. Coercivity of the single layer heat potential. *J. Comput. Math.*, 7 (2): 100–4, 1989.

[4] R. Chapko and R. Kress. Rothe's method for the heat equation and boundary integral equations. *J. Integral Equations Appl.*, 9 (1): 47–69, 1997.

[5] M. Costabel. Boundary integral operators for the heat equation. *Integral Equations Operator Theory*, 13 (4): 498–552, 1990.

[6] S. Dohr. Distributed and preconditioned space–time boundary element methods for the heat equation. PhD thesis, Graz University of Technology, 2019.

[7] S. Dohr and O. Steinbach. Preconditioned space–time boundary element methods for the one–dimensional heat equation. In *Domain Decomposition Methods in Science and Engineering XXIV*, volume 125 of *Lect. Notes Comput. Sci. Eng.*, pages 243–51. Springer, Cham, 2018.

[8] S. Dohr, K. Niino, and O. Steinbach. Preconditioned space–time boundary element methods for the heat equation, in preparation.

[9] S. Dohr, M. Merta, G. Of, O. Steinbach, and J. Zapletal. A parallel solver for a preconditioned space–time boundary element method for the heat equation, submitted, 2018.

[10] S. Dohr, J. Zapletal, G. Of, M. Merta, and M. Kravcenko. A parallel space–time boundary element method for the heat equation. *Comput. Math. Appl.*, published online, 2019. https://doi.org/10.1016/j.camwa.2018.12.031.

[11] A. Ern and J. Guermond. *Theory and practice of finite elements*, volume 125 of *Applied Mathematical Sciences*, Springer, New York, 2004.

[12] L. C. Evans. *Partial differential equations*. Volume 19 of Graduate Studies in Mathematics, AMS, Providence, 1997.

[13] M. Fontes. Initial–boundary value problems for parabolic equations. *Ann. Acad. Sci. Fenn. Math.*, 34 (2): 583–605, 2009.

[14] M. Gander and M. Neumüller. Analysis of a new space-time parallel multigrid algorithm for parabolic problems. *SIAM J. Sci. Comput.*, 38 (4): A2173–208, 2016.

[15] M. Hamina. Some boundary element methods for heat conduction problems. PhD thesis, University of Oulu, 2000.

[16] H. Harbrecht and J. Tausch. A fast sparse grid based space-time boundary element method for the nonstationary heat equation. *Numer. Math.*, 140 (1): 239–64, 2018.

[17] R. Hiptmair. Operator preconditioning. *Comput. Math. Appl.*, 52: 699–706, 2006.

[18] G. C. Hsiao and J. Saranen. Boundary integral solution of the two-dimensional heat equation. *Math. Methods Appl. Sci.*, 16 (2): 87–114, 1993.

[19] G. C. Hsiao and W. L. Wendland. *Boundary integral equations*, volume 164 of *Applied Mathematical Sciences*, Springer, Berlin, 2008.

[20] S. Larsson and C. Schwab. Compressive space-time Galerkin discretizations of parabolic partial differential equations. Research Report 2015–04, Seminar für Angewandte Mathematik, ETH Zürich, 2015.

[21] J. L. Lions and E. Magenes. *Non-homogeneous boundary value problems and applications. Volume I*. Springer, Berlin, Heidelberg, New York, 1972.

[22] J. L. Lions and E. Magenes. *Non-homogeneous boundary value problems and applications. Volume II*. Springer, Berlin, Heidelberg, New York, 1972.

[23] C. Lubich and R. Schneider. Time discretization of parabolic boundary integral equations. *Numer. Math.*, 63 (4): 455–81, 1992.

[24] W. McLean. *Strongly elliptic systems and boundary integral equations*. Cambridge University

Press, 2000.

[25] M. Messner. *A fast multipole Galerkin boundary element method for the transient heat equation*, volume 23 of *Monographic series TU Graz: Computation in engineering and science*, 2014.

[26] M. Messner, M. Schanz, and J. Tausch. A fast Galerkin method for parabolic space-time boundary integral equations. *J. Comput. Phys.*, 258: 15–30, 2014.

[27] M. Messner, M. Schanz, and J. Tausch. An efficient Galerkin boundary element method for the transient heat equation. *SIAM J. Sci. Comput.*, 37 (3): A1554–76, 2015.

[28] M. Neumüller. *Space–time methods: Fast solvers and applications*, volume 20 of *Monographic series TU Graz: Computation in engineering and science*, 2013.

[29] P. Noon. The single layer heat potential and Galerkin boundary element methods for the heat equation. PhD thesis, University of Maryland, 1988.

[30] A. Piriou. Une classe d'opérateurs pseudo–différentiels du type de Volterra. *Ann. Inst. Fourier*, 20 (1): 77–94, 1970.

[31] A. Piriou. Problèmes aux limites généraux pour des opérateurs différentiels paraboliques dans un domaine borné. *Ann. Inst. Fourier*, 21 (1): 59–78, 1971.

[32] H. Schulz and O. Steinbach. A new a posteriori error estimator in adaptive direct boundary element methods: the Dirichlet problem. *Calcolo*, 37: 79–96, 2000.

[33] O. Steinbach. *Numerical Approximation Methods for Elliptic Boundary Value Problems*. Springer, New York, 2008.

[34] O. Steinbach. Space–time finite element methods for parabolic problems. *Comput. Meth. Appl. Math.*, 15: 551–66, 2015.

[35] O. Steinbach and W. L. Wendland. The construction of some efficient preconditioners in the boundary element method. *Adv. Comput. Math.*, 9: 191–216, 1998.

[36] O. Steinbach and M. Zank. Coercive space–time finite element methods for initial boundary value problems, submitted, 2018.

[37] J. Tausch. A fast method for solving the heat equation by layer potentials. *J. Comput. Phys.*, 224: 956–69, 2007.

[38] J. Tausch. Nyström discretization of parabolic boundary integral equations. *Appl. Numer. Math.*, 59 (11): 2843–56, 2009.

[39] J. Tausch. Fast Nyström methods for parabolic boundary integral equations. In *Fast Boundary Element Methods in Engineering and Industrial Applications*, pages 185–219, Springer, Heidelberg, 2012.

[40] E. Zeidler. *Nonlinear Functional Analysis and its Applications II/A: Linear Monotone Operators*. Springer, New York, 1990.

Willy Dörfler, Stefan Findeisen, Christian Wieners, and
Daniel Ziegler

2 Parallel adaptive discontinuous Galerkin discretizations in space and time for linear elastic and acoustic waves

Abstract: We introduce a space-time discretization for elastic and acoustic waves using a discontinuous Galerkin approximation in space, and a Petrov–Galerkin scheme in time. For the DG method, the upwind flux is evaluated by explicitly solving a Riemann problem. Then we show well-posedness and convergence of the discrete system. Based on goal-oriented dual-weighted error estimation, an adaptive strategy is introduced. The full space-time linear system is solved with a parallel multilevel preconditioner. Numerical experiments for acoustic and elastic waves underline the efficiency of the overall adaptive solution process.

Keywords: Space-time methods, discontinuous Galerkin finite elements, linear hyperbolic systems, elastic and acoustic wave equation, dual weighted residual error estimator

MSC 2010: 65M60, 65M15, 65M55

1 Introduction

Modern discretizations of time-dependent PDEs consider the full problem in the space-time cylinder and aim to overcome limitations of classical approaches, such as the method of lines (first discretize in space and then solve the resulting ODE) and the Rothe method (first discretize in time and then solve the PDE). A main advantage of a holistic space-time method is the direct access to space-time adaptivity and to the backward problem (required for the goal-oriented error control or the dual problem in optimization, see [13] for more details). Moreover, this allows for parallel solution strategies simultaneously in time and space.

Several space-time concepts were proposed (different conforming and non-conforming space-time finite elements [10, 21, 23, 24, 28, 31, 33, 34, 36], the parareal method [17, 25], wavefront relaxation [18], et cetera) and this topic has become a rapidly growing field in numerical analysis and scientific computing.

Acknowledgement: This work was supported by the German Research Foundation (DFG) by CRC 1173.

Willy Dörfler, Stefan Findeisen, Christian Wieners, Daniel Ziegler, Institut für Angewandte und Numerische Mathematik, Karlsruher Institut für Technologie, 76128 Karlsruhe, Germany, e-mails: Willy.Doerfler@kit.edu, Christian.Wieners@kit.edu, Daniel.Ziegler@kit.edu

https://doi.org/10.1515/9783110548488-002

A further motivation for developing space-time methods is the design of modern computer facilities with an enormous number of processor cores, where the parallel realization of conventional methods becomes inefficient. Since these machines allow a fully implicit space-time approach, new parallel solution techniques are required to solve the huge linear systems, particularly for time-dependent applications in three spatial dimensions. Iterative solution techniques for full space-time discretizations were investigated, for example, in [1, 2, 5, 14, 15, 16, 29, 35, 37, 38].

Here, we use in space a discontinuous Galerkin (DG) method for time-dependent first-order systems, see, for example, [19], where this discretization is coupled with explicit time integration. This is applied to acoustic and elastic waves in [9] in combination with an p-adaptive space-time strategy. We then extend these spatial DG discretization by a Petrov–Galerkin method in time with continuous ansatz space and discontinuous test space (cf. [3] for the implicit midpoint rule). The second-order formulation in space for elastic waves with implicit discontinuous Galerkin time discretization is considered in [22].

The DG approach uses the same variational space-time setting as discontinuous Petrov–Galerkin (DPG) methods for general linear first-order systems in space and time, see [6] for an overview and [7, 11] for space-time applications. For acoustic and elastic waves, the hybridization in space (applied to the second-order formulation) is presented in [30], and a hybrid space-time discontinuous Galerkin method is proposed in [39]. Both methods are implicit in every time slab, and only Dirichlet traces are used for the hybrid coupling. Space-time (Trefftz) discontinuous Galerkin methods for wave problems are analyzed in [10, 23].

Error estimation for linear wave equations require a backtracking of the error source, as it is provided by a dual-primal error estimator. This achieves a reliable error control by solving the adjoint problem together with a goal-oriented technique [3].

Here, we transfer our results for the linear transport equation and for the Maxwell system in [8] to acoustic and elastic waves. We start with an introduction of the first-order system for the wave equation and a suitable variational setting, which provides stability of the space-time operator in a Hilbert space setting. Then we review the construction of discontinuous Galerkin methods for linear systems of conservation laws, and we compute the numerical flux for acoustic and elastic waves by solving the corresponding Riemann problem. In the next section, we derive an explicit error representation (involving the solution of the dual problem), where we extend our approach in [8] by a different variant to estimate the interpolation error of the dual problem, which can be estimated without additional regularity assumptions. We shortly summarize the construction of a suitable space-time multigrid preconditioner for the fully coupled implicit space-time discretization. Finally, the convergence of the method and the efficiency of the adaptive strategy is demonstrated for examples, comparing the propagation of acoustic and elastic waves.

2 Linear elastic and acoustic waves

The prototype equation describing linear waves in homogeneous media is the second-order evolution equation for a scalar potential ϕ:

$$\partial_t^2 \phi = \Delta \phi,$$

subject to initial and boundary conditions. Introducing the pressure $p = \partial_t \phi$ and the velocity $\boldsymbol{v} = \nabla \phi$, we obtain the first-order system

$$\partial_t p = \nabla \cdot \boldsymbol{v},$$
$$\partial_t \boldsymbol{v} = \nabla p,$$

describing, for example, acoustic waves. This system is now extended to describe linear elastic waves.

Waves in solids

Let $\Omega \subseteq \mathbb{R}^D$ be a bounded Lipschitz domain, and let $[0, T]$ be a finite time interval. In dynamic models in continuum mechanics, the motion of a material point $\boldsymbol{x}$ in the reference configuration Ω at time t is described by the deformation vector $\boldsymbol{\varphi}(t, \boldsymbol{x})$. The velocity is denoted by $\boldsymbol{v} = \partial_t \boldsymbol{\varphi}$. Elastic waves are determined by Newton's law for the balance of momentum

$$\rho \partial_t \boldsymbol{v} = \operatorname{div} \boldsymbol{\sigma} + \boldsymbol{b},$$

with the mass density ρ, acceleration $\partial_t \boldsymbol{v}$, and the vector of body forces $\boldsymbol{b}$, together with a constitutive relation for the stress $\boldsymbol{\sigma}$, depending on the deformation gradient $\boldsymbol{F} = \mathrm{D}\boldsymbol{\varphi}$. For elastic materials a response function $\hat{\Sigma}(\cdot)$ exists so that the stress is determined by the response $\boldsymbol{\sigma} = \hat{\Sigma}(\boldsymbol{F})$. Then the stress rate is given by

$$\partial_t \boldsymbol{\sigma} = \mathrm{D}\hat{\Sigma}(\mathrm{D}\boldsymbol{\varphi})(\mathrm{D}\boldsymbol{v}).$$

Assuming small strains and $\boldsymbol{\varphi} \approx \mathrm{id}$, this is approximated by its linearization

$$\partial_t \boldsymbol{\sigma} = \boldsymbol{C}\boldsymbol{\varepsilon}(\boldsymbol{v}), \quad \boldsymbol{\varepsilon}(\boldsymbol{v}) = \mathrm{sym}(\mathrm{D}\boldsymbol{v})$$

with the elasticity tensor $\boldsymbol{C} = \mathrm{D}\hat{\Sigma}(\mathbf{I})$. The balance of torsional moments yields that the stress is symmetric, and that the stress rate only depends on the symmetric strain rate. In isotropic media, the elasticity tensor

$$\boldsymbol{C}\boldsymbol{\varepsilon} = 2\mu\boldsymbol{\varepsilon} + \lambda \operatorname{trace}(\boldsymbol{\varepsilon})\mathbf{I}$$

is characterized by the Lamé parameters $\lambda \geq 0$, $\mu > 0$. Introducing the compression modulus $\kappa = \frac{2\mu + 3\lambda}{3}$ and the deviatoric stress $\operatorname{dev}(\boldsymbol{\sigma}) = \boldsymbol{\sigma} - \frac{1}{3}\operatorname{trace}(\boldsymbol{\sigma})\mathbf{I}$, we obtain

$$\boldsymbol{C}\boldsymbol{\varepsilon} = 2\mu \operatorname{dev}(\boldsymbol{\varepsilon}) + \kappa \operatorname{trace}(\boldsymbol{\varepsilon})\mathbf{I}, \quad \boldsymbol{C}^{-1}\boldsymbol{\sigma} = \frac{1}{2\mu}\operatorname{dev}(\boldsymbol{\sigma}) + \frac{1}{9\kappa}\operatorname{trace}(\boldsymbol{\sigma})\mathbf{I}.$$

Acoustic waves in solids and fluids

In fluids we assume that shear forces can be neglected, that is, we consider the limit $\mu \to 0$. Then, the stress $\boldsymbol{\sigma} = p\mathbf{I}$ is isotropic with hydrostatic pressure $p = \frac{1}{3}\operatorname{trace}\boldsymbol{\sigma}$, and compressional waves are described by the system

$$\partial_t p = \kappa \operatorname{div} \mathbf{v}, \quad \rho \partial_t \mathbf{v} = \nabla p + \mathbf{b}.$$

In particular, this applies to acoustic waves in air or in a gas at fixed temperature. Note that this is only a formal derivation of the acoustic wave equation using the setting of continuum mechanics of solids, see Table 2.1 for comparing the elastic and acoustic setting. The linearization of conservation laws for compressible fluids with a pressure-dependent constitutive relation for the density results in the same system for acoustic waves.

Table 2.1: First-order differential systems for elastic waves in $(0, T) \times \Omega$ with initial conditions at $t = 0$, and static and kinematic boundary conditions on $\partial\Omega = \Gamma_{\text{stat}} \cup \Gamma_{\text{kin}}$.

Elastic waves		Acoustic waves	
$\partial_t \boldsymbol{\sigma} = \mathbf{C}\boldsymbol{\varepsilon}(\mathbf{v})$	in $(0, T) \times \Omega$	$\partial_t p = \kappa \operatorname{div} \mathbf{v}$	in $(0, T) \times \Omega$
$\rho \partial_t \mathbf{v} = \operatorname{div} \boldsymbol{\sigma} + \mathbf{b}$	in $(0, T) \times \Omega$	$\rho \partial_t \mathbf{v} = \nabla p + \mathbf{b}$	in $(0, T) \times \Omega$
$\boldsymbol{\sigma}(0) = \boldsymbol{\sigma}_0$	at $t = 0$ in Ω	$p(0) = p_0$	at $t = 0$ in Ω
$\mathbf{v}(0) = \mathbf{v}_0$	at $t = 0$ in Ω	$\mathbf{v}(0) = \mathbf{v}_0$	at $t = 0$ in Ω
$\boldsymbol{\sigma}\mathbf{n} = \mathbf{t}_{\text{stat}}$	on $(0, T) \times \Gamma_{\text{stat}}$	$p = p_{\text{stat}}$	on $(0, T) \times \Gamma_{\text{stat}}$
$\mathbf{v} = \mathbf{g}_{\text{kin}}$	on $(0, T) \times \Gamma_{\text{kin}}$	$\mathbf{n} \cdot \mathbf{v} = g_{\text{kin}}$	on $(0, T) \times \Gamma_{\text{kin}}$

First-order differential systems

The previous examples are instances of a system of J equations in $\mathbb{R}^D$,

$$M\partial_t \mathbf{u} + A\mathbf{u} = \mathbf{f},$$

with a first order differential operator A and a weighting operator M, see Table 2.2.

Table 2.2: First-order differential systems $M\partial_t \mathbf{u} + A\mathbf{u} = \mathbf{f}$, and suitable domains $\mathcal{D}(A)$ for linear waves. Here, we choose kinematic boundary conditions (Dirichlet b. c. for elastic waves and Neumann b. c. for acoustic waves).

Elastic waves	Acoustic waves
$\mathbf{u} = (\boldsymbol{\sigma}, \mathbf{v})$	$\mathbf{u} = (p, \mathbf{v})$
$M(\boldsymbol{\sigma}, \mathbf{v}) = (\mathbf{C}^{-1}\boldsymbol{\sigma}, \rho\mathbf{v})$	$M(p, \mathbf{v}) = (\kappa^{-1} p, \rho\mathbf{v})$
$A(\boldsymbol{\sigma}, \mathbf{v}) = -(\boldsymbol{\varepsilon}(\mathbf{v}), \operatorname{div} \boldsymbol{\sigma})$	$A(p, \mathbf{v}) = -(\operatorname{div} \mathbf{v}, \nabla p)$
$\mathbf{f} = (0, \mathbf{b})$	$\mathbf{f} = (0, \mathbf{b})$
$\mathcal{D}(A) = \mathrm{H}(\operatorname{div}, \Omega; \mathbb{R}^{D\times D}_{\text{sym}}) \times H^1_0(\Omega; \mathbb{R}^D)$	$\mathcal{D}(A) = H^1(\Omega) \times H_0(\operatorname{div}, \Omega)$

We introduce the Hilbert space $H = L_2(\Omega; \mathbb{R}^J)$ with weighted inner product

$$(\boldsymbol{v}, \boldsymbol{w})_H = (M\boldsymbol{v}, \boldsymbol{w})_{0,\Omega},$$

where we assume that the operator $M \in L_\infty(\Omega, \mathbb{R}^{J\times J}_{\mathrm{sym}})$ is uniformly positive. The analysis of the wave problems will be considered with homogeneous boundary conditions on $\partial\Omega$, which are realized by the choice of a suitable domain $\mathcal{D}(A) \subset H$. We assume that the operator A is skew-adjoint in the domain; that is,

$$(A\boldsymbol{v}, \boldsymbol{w})_{0,\Omega} = -(\boldsymbol{v}, A\boldsymbol{w})_{0,\Omega}, \quad \boldsymbol{v}, \boldsymbol{w} \in \mathcal{D}(A). \tag{2.1}$$

For the corresponding evolution operator $L = M\partial_t + A$ on the space-time cylinder $Q = (0, T) \times \Omega$, we also observe

$$(L\boldsymbol{v}, \boldsymbol{w})_{0,Q} = -(\boldsymbol{v}, L\boldsymbol{w})_{0,Q}, \quad \boldsymbol{v}, \boldsymbol{w} \in \mathrm{C}^1_{\mathrm{c}}(Q; \mathbb{R}^J),$$

where $\mathrm{C}^1_{\mathrm{c}}$ denotes the set of compactly supported differentiable mappings.

Depending on L, we define the space

$$\begin{aligned}\mathrm{H}(L, Q) = \{\boldsymbol{v} \in L_2(Q; \mathbb{R}^J) : \boldsymbol{g} \in L_2(Q; \mathbb{R}^J) \text{ exists with}\\ (\boldsymbol{g}, \boldsymbol{w})_{0,Q} = -(\boldsymbol{v}, L\boldsymbol{w})_{0,Q} \text{ for all } \boldsymbol{w} \in \mathrm{C}^1_{\mathrm{c}}(Q; \mathbb{R}^J)\}.\end{aligned}$$

Then, L can be extended to this space, and $\mathrm{H}(L, \Omega)$ is a Hilbert space with respect to the weighted graph norm $\|\boldsymbol{v}\|_{L,Q} = \sqrt{(M\boldsymbol{v}, \boldsymbol{v})^2_{0,Q} + (M^{-1}L\boldsymbol{v}, L\boldsymbol{v})^2_{0,Q}}$.

Let $V \subset \mathrm{H}(L, Q)$ be the closure of $\{\boldsymbol{v} \in C^1([0, T]; \mathcal{D}(A)) : \boldsymbol{v}(0) = \boldsymbol{0}\}$ with respect to the graph norm. In particular, the space V includes homogeneous initial conditions. Then we define $W = \overline{L(V)} \subseteq L_2(Q; \mathbb{R}^J)$ with the weighted norm $\|\boldsymbol{w}\|^2_W = (M\boldsymbol{w}, \boldsymbol{w})_{0,Q}$. On V, we use the weighted graph norm

$$\|\boldsymbol{v}\|^2_V = \|\boldsymbol{v}\|^2_W + \|M^{-1}L\boldsymbol{v}\|^2_W.$$

Since A is skew-adjoint, we obtain the operator estimate in weighted norms [8, Lemma 1]

$$\|\boldsymbol{v}\|_W \leq 2T\|M^{-1}L\boldsymbol{v}\|_W, \quad \boldsymbol{v} \in V. \tag{2.2}$$

This implies that $L \in \mathcal{L}(V, W)$ is injective and the range is closed. Moreover, for $\boldsymbol{f} \in W$, a unique solution $\boldsymbol{u} \in V$ of the evolution equation

$$L\boldsymbol{u} = \boldsymbol{f} \tag{2.3}$$

exists [8, Lemma 2]. This extends to initial values $\boldsymbol{u}(0) = \boldsymbol{u}_0 \neq \boldsymbol{0}$ by replacing $\boldsymbol{f}(t)$ with $\boldsymbol{f}(t) - A\boldsymbol{u}_0$. Also inhomogeneous boundary conditions can be analyzed by modifying the right-hand side, when the existence of a sufficiently smooth extension of the boundary data into the domain can be assumed.

Remark 2.1. Since L mixes the derivatives in space and time, more regularity is difficult to show in this Hilbert space framework. Therefore, one can check the assumptions of the Lumer–Phillips theorem [32, Theorem 12.22] for the operator A in $\mathcal{D}(A)$, so that semigroup theory with more regularity can be applied, see, for example, [12]. The application to wave equations is discussed in [20, Section 2.2].

3 Discontinuous Galerkin methods for linear systems of conservation laws

All wave equations discussed so far can be more specifically considered as a system of linear conservation laws:

$$M\partial_t \boldsymbol{u}(t) + \operatorname{div} \boldsymbol{F}(\boldsymbol{u}(t)) = \boldsymbol{f}(t) \quad \text{for } t \in [0,T], \quad \boldsymbol{u}(0) = \boldsymbol{u}_0, \tag{3.1}$$

with a linear *flux function* $\boldsymbol{F}(\boldsymbol{v}) = [B_1\boldsymbol{v}, \dots, B_D\boldsymbol{v}]$ defined by symmetric matrices $B_d \in \mathbb{R}^{J\times J}_{\mathrm{sym}}$ such that

$$A\boldsymbol{v} = \operatorname{div} \boldsymbol{F}(\boldsymbol{v}) = \sum_{d=1}^{D} B_d \partial_d \boldsymbol{v}.$$

Traveling waves

In the case of constant coefficients in $\Omega = \mathbb{R}^D$, special solutions can be constructed as follows. For a given unit vector $\boldsymbol{n} = (n_1, \dots, n_D)^\top \in \mathbb{R}^D$, we have $\boldsymbol{n} \cdot \boldsymbol{F}(\boldsymbol{u}) = B_{\boldsymbol{n}}\boldsymbol{u}$ with the symmetric matrix $B_{\boldsymbol{n}} = \sum_{d=1}^{D} n_d B_d$. Then, for all eigenpairs $(\lambda, \boldsymbol{w}) \in \mathbb{R} \times \mathbb{R}^J$ of $B_{\boldsymbol{n}}\boldsymbol{w} = \lambda M\boldsymbol{w}$, and all sufficiently smooth functions $a\colon \mathbb{R} \to \mathbb{R}$, the traveling wave propagating with velocity $c = |\lambda|$,

$$\boldsymbol{u}(t, \boldsymbol{x}) = a(\boldsymbol{n}\cdot\boldsymbol{x} - \lambda t)\boldsymbol{w},$$

is a solution of (3.1) with initial value $\boldsymbol{u}_0(\boldsymbol{x}) = a(\boldsymbol{n}\cdot\boldsymbol{x})\boldsymbol{w}$, and right-hand side $\boldsymbol{f} = \boldsymbol{0}$. This also applies to traveling waves with discontinuous amplitude: the piecewise constant function

$$\boldsymbol{u}(t, \boldsymbol{x}) = \begin{cases} a_{\mathrm{L}}\boldsymbol{w} & \text{in } Q_{\mathrm{L}} = \{(t, \boldsymbol{x}) \in [0,T] \times \mathbb{R}^D : \boldsymbol{n}\cdot\boldsymbol{x} - \lambda t < 0\}, \\ a_{\mathrm{R}}\boldsymbol{w} & \text{in } Q_{\mathrm{R}} = \{(t, \boldsymbol{x}) \in [0,T] \times \mathbb{R}^D : \boldsymbol{n}\cdot\boldsymbol{x} - \lambda t > 0\}, \end{cases} \tag{3.2}$$

with $a_{\mathrm{L}}, a_{\mathrm{R}} \in \mathbb{R}$, is a weak solution. That is, we have

$$\int_{\mathbb{R}} \int_{\mathbb{R}^D} \boldsymbol{u} \cdot L\boldsymbol{v} \, \mathrm{d}\boldsymbol{x} \, \mathrm{d}t = 0 \quad \text{for all } \boldsymbol{v} \in \mathrm{C}^1_{\mathrm{c}}(\mathbb{R} \times \mathbb{R}^D; \mathbb{R}^J).$$

The Riemann problem for linear conservation laws

We now construct a weak solution of the *Riemann problem*; that is, a piecewise constant weak solution with right-hand side $\boldsymbol{f} = \mathbf{0}$ and the discontinuous initial function

$$\boldsymbol{u}_0(\boldsymbol{x}) = \begin{cases} \boldsymbol{u}_\mathrm{L} & \text{in } \Omega_\mathrm{L} = \{\boldsymbol{x} \in \mathbb{R}^D : \boldsymbol{n}\cdot\boldsymbol{x} < 0\}, \\ \boldsymbol{u}_\mathrm{R} & \text{in } \Omega_\mathrm{R} = \{\boldsymbol{x} \in \mathbb{R}^D : \boldsymbol{n}\cdot\boldsymbol{x} > 0\}, \end{cases} \tag{3.3}$$

with $\boldsymbol{u}_\mathrm{L}, \boldsymbol{u}_\mathrm{R} \in \mathbb{R}^J$. Let $\{(\lambda_j, \boldsymbol{w}_j)\}_{j=1,\dots,J}$ be the (necessarily M-orthogonal) set of eigenpairs; that is,

$$B_{\boldsymbol{n}}\boldsymbol{w}_j = \lambda_j M \boldsymbol{w}_j \quad \text{with } \boldsymbol{w}_k \cdot M\boldsymbol{w}_j = 0 \quad \text{for } j \neq k. \tag{3.4}$$

This defines a decomposition $B_{\boldsymbol{n}} = B_{\boldsymbol{n}}^- + B_{\boldsymbol{n}}^+$ with

$$B_{\boldsymbol{n}}^-\boldsymbol{v} = \sum_{\lambda_j<0} \lambda_j \frac{\boldsymbol{w}_j\cdot M\boldsymbol{v}}{\boldsymbol{w}_j\cdot M\boldsymbol{w}_j} M\boldsymbol{w}_j, \quad B_{\boldsymbol{n}}^+\boldsymbol{v} = \sum_{\lambda_j>0} \lambda_j \frac{\boldsymbol{w}_j\cdot M\boldsymbol{v}}{\boldsymbol{w}_j\cdot M\boldsymbol{w}_j} M\boldsymbol{w}_j.$$

By superposition of traveling waves, we obtain a weak solution of the Riemann problem

$$\boldsymbol{u}(t,\boldsymbol{x}) = \sum_{j=1}^J a_j(\boldsymbol{x}\cdot\boldsymbol{n} - \lambda_j t)\boldsymbol{w}_j, \quad a_j(s) = \begin{cases} \frac{\boldsymbol{w}_j\cdot M\boldsymbol{u}_\mathrm{L}}{\boldsymbol{w}_j\cdot M\boldsymbol{w}_j}, & s < 0, \\ \frac{\boldsymbol{w}_j\cdot M\boldsymbol{u}_\mathrm{R}}{\boldsymbol{w}_j\cdot M\boldsymbol{w}_j}, & s > 0. \end{cases}$$

The solution of the Riemann problem at $(t, \mathbf{0})$ for $t > 0$ defines the upwind flux on the interface $\partial\Omega_\mathrm{L} \cap \partial\Omega_\mathrm{R}$ by

$$\begin{aligned} \boldsymbol{n}\cdot\boldsymbol{F}^{\mathrm{num}}(\boldsymbol{u}_0) &= \sum_{\lambda_j>0} \frac{\boldsymbol{w}_j\cdot M\boldsymbol{u}_\mathrm{L}}{\boldsymbol{w}_j\cdot M\boldsymbol{w}_j} B_{\boldsymbol{n}}\boldsymbol{w}_j + \sum_{\lambda_j<0} \frac{\boldsymbol{w}_j\cdot M\boldsymbol{u}_\mathrm{R}}{\boldsymbol{w}_j\cdot M\boldsymbol{w}_j} B_{\boldsymbol{n}}\boldsymbol{w}_j \\ &= B_{\boldsymbol{n}}\boldsymbol{u}_\mathrm{L} + B_{\boldsymbol{n}}^-[\boldsymbol{u}], \end{aligned} \tag{3.5}$$

depending on the jump term $[\boldsymbol{u}] = \boldsymbol{u}_\mathrm{R} - \boldsymbol{u}_\mathrm{L}$. By construction, the upwind flux is consistent. That is, for $B_{\boldsymbol{n}}\boldsymbol{u}_\mathrm{L} = B_{\boldsymbol{n}}\boldsymbol{u}_\mathrm{R}$, we obtain $\boldsymbol{n}\cdot\boldsymbol{F}^{\mathrm{num}}(\boldsymbol{u}_0) = B_{\boldsymbol{n}}\boldsymbol{u}_\mathrm{L} = B_{\boldsymbol{n}}\boldsymbol{u}_\mathrm{R}$.

This transfers to the solution of the Riemann problem and the evaluation of the numerical flux in case of different material parameters in Ω_L and Ω_R, that is, two different matrices M_L and M_R; see [20, Section 3.2] for details.

Application to wave equations

For *elastic waves* with $\operatorname{div}\boldsymbol{F}(\boldsymbol{\sigma},\boldsymbol{v}) = -(\boldsymbol{\varepsilon}(\boldsymbol{v}), \operatorname{div}\boldsymbol{\sigma})$, we have the normal flux

$$\boldsymbol{n}\cdot\boldsymbol{F}(\boldsymbol{\sigma},\boldsymbol{v}) = -\begin{pmatrix} \frac{1}{2}(\boldsymbol{n}\otimes\boldsymbol{v} + \boldsymbol{v}\otimes\boldsymbol{n}) \\ \boldsymbol{\sigma}\boldsymbol{n} \end{pmatrix}.$$

Now we consider the isotropic case, where $c_{\mathrm{P}} = \sqrt{(2\mu+\lambda)/\rho}$ is the velocity of compressional waves, and $c_{\mathrm{S}} = \sqrt{\mu/\rho}$ is the velocity of shear waves. The eigenvalues in (3.4) are $\pm c_{\mathrm{P}}$, $\pm c_{\mathrm{S}}$, and the corresponding eigenvectors are of the form

$$\begin{pmatrix} 2\mu\boldsymbol{n}\otimes\boldsymbol{n}+\lambda\mathbf{I} \\ \pm c_{\mathrm{P}}\boldsymbol{n} \end{pmatrix} \quad \text{and} \quad \begin{pmatrix} \mu(\boldsymbol{\tau}\otimes\boldsymbol{n}+\boldsymbol{n}\otimes\boldsymbol{\tau}) \\ \pm c_{\mathrm{S}}\boldsymbol{\tau} \end{pmatrix},$$

where $\boldsymbol{\tau}$ is a unit tangent vector; that is, $\boldsymbol{\tau}\cdot\boldsymbol{n} = 0$. This yields

$$\begin{aligned} B_{\boldsymbol{n}}^{-}[\boldsymbol{u}] = &-\frac{\boldsymbol{n}\otimes\boldsymbol{n}\,[\boldsymbol{\sigma}] + \rho c_{\mathrm{P}}\boldsymbol{n}\cdot[\boldsymbol{v}]}{2\rho c_{\mathrm{P}}}\begin{pmatrix} \boldsymbol{n}\otimes\boldsymbol{n} \\ \rho c_{\mathrm{P}}\boldsymbol{n} \end{pmatrix} \\ &-\frac{\frac{1}{2}(\boldsymbol{\tau}\otimes\boldsymbol{n}+\boldsymbol{n}\otimes\boldsymbol{\tau})[\boldsymbol{\sigma}] + \rho c_{\mathrm{S}}\boldsymbol{\tau}\cdot[\boldsymbol{v}]}{2\rho c_{\mathrm{S}}}\begin{pmatrix} \frac{1}{2}(\boldsymbol{\tau}\otimes\boldsymbol{n}+\boldsymbol{n}\otimes\boldsymbol{\tau}) \\ \rho c_{\mathrm{S}}\boldsymbol{\tau} \end{pmatrix}, \end{aligned}$$

and inserting (3.5) results in the 2D upwind flux:

$$\begin{aligned} \boldsymbol{n}\cdot\boldsymbol{F}^{\mathrm{num}}(\boldsymbol{u}_0) = &-\begin{pmatrix} \frac{1}{2}(\boldsymbol{n}\otimes\boldsymbol{v}_{\mathrm{L}}+\boldsymbol{v}_{\mathrm{L}}\otimes\boldsymbol{n}) \\ \boldsymbol{\sigma}_{\mathrm{L}}\boldsymbol{n} \end{pmatrix} - \frac{\boldsymbol{n}\cdot[\boldsymbol{\sigma}]\boldsymbol{n} + \rho c_{\mathrm{P}}[\boldsymbol{v}]\cdot\boldsymbol{n}}{2\rho c_{\mathrm{P}}}\begin{pmatrix} \boldsymbol{n}\otimes\boldsymbol{n} \\ \rho c_{\mathrm{P}}\boldsymbol{n} \end{pmatrix} \\ &-\frac{\boldsymbol{\tau}\cdot[\boldsymbol{\sigma}]\boldsymbol{n} + \rho c_{\mathrm{S}}[\boldsymbol{v}]\cdot\boldsymbol{\tau}}{2\rho c_{\mathrm{S}}}\begin{pmatrix} \frac{1}{2}(\boldsymbol{\tau}\otimes\boldsymbol{n}+\boldsymbol{n}\otimes\boldsymbol{\tau}) \\ \rho c_{\mathrm{S}}\boldsymbol{\tau} \end{pmatrix}. \end{aligned}$$

For *acoustic waves* with $\operatorname{div}\boldsymbol{F}(p,\boldsymbol{v}) = -(\operatorname{div}\boldsymbol{v}, \nabla p)$, we obtain

$$\boldsymbol{n}\cdot\boldsymbol{F}(p,\boldsymbol{v}) = -\begin{pmatrix} \boldsymbol{n}\cdot\boldsymbol{v} \\ p\boldsymbol{n} \end{pmatrix},$$

the velocity of sound $c = \sqrt{\kappa/\rho}$, and the eigenvectors

$$\boldsymbol{w}_{\pm} = \begin{pmatrix} \kappa \\ \mp c\boldsymbol{n} \end{pmatrix}.$$

This yields

$$B_{\boldsymbol{n}}^{-} = -\frac{1}{2}\begin{pmatrix} \frac{1}{\rho c} & \boldsymbol{n} \\ \boldsymbol{n} & \rho c\boldsymbol{n}\otimes\boldsymbol{n} \end{pmatrix},$$

and thus the upwind flux

$$\boldsymbol{n}\cdot\boldsymbol{F}^{\mathrm{num}}(\boldsymbol{u}_0) = -\begin{pmatrix} \boldsymbol{n}\cdot\boldsymbol{v}_{\mathrm{L}} \\ p_{\mathrm{L}}\boldsymbol{n} \end{pmatrix} - \frac{[p] + \rho c\boldsymbol{n}\cdot[\boldsymbol{v}]}{2\rho c}\begin{pmatrix} 1 \\ \rho c\boldsymbol{n} \end{pmatrix}.$$

This extends to boundary conditions as follows. For given values $(p_{\mathrm{L}}, \boldsymbol{v}_{\mathrm{L}})$, the Riemann solution in Ω_L takes the form

$$\boldsymbol{u}(t,\boldsymbol{x}) = \begin{cases} \begin{pmatrix} p_{\mathrm{L}} \\ \boldsymbol{v}_{\mathrm{L}} \end{pmatrix}, & \boldsymbol{x}\cdot\boldsymbol{n} < tc, \\ \begin{pmatrix} p_{\mathrm{L}} \\ \boldsymbol{v}_{\mathrm{L}} \end{pmatrix} + a\begin{pmatrix} \kappa \\ c\boldsymbol{n} \end{pmatrix}, & \boldsymbol{x}\cdot\boldsymbol{n} > tc, \end{cases}$$

and the parameter a is determined by boundary conditions. For $p = p_{\text{stat}}$ on $\partial\Omega_L$, we obtain

$$a = \frac{p_{\text{stat}} - p_{\text{L}}}{\kappa},$$

and for $\boldsymbol{n} \cdot \boldsymbol{v} = g_{\text{kin}}$ on $\partial\Omega_L$, we obtain

$$a = \frac{g_{\text{kin}} - \boldsymbol{n} \cdot \boldsymbol{v}_{\text{L}}}{c}.$$

This yields on the boundary

$$\begin{aligned}\boldsymbol{n} \cdot \boldsymbol{F}^{\text{num}}(\boldsymbol{u}) &= -\begin{pmatrix} \boldsymbol{n} \cdot \boldsymbol{v}_{\text{L}} \\ p_{\text{L}}\boldsymbol{n} \end{pmatrix} - c\frac{p_{\text{stat}} - p_{\text{L}}}{\kappa}\begin{pmatrix} 1 \\ \rho c\boldsymbol{n} \end{pmatrix} \\ &= -\begin{pmatrix} \boldsymbol{n} \cdot \boldsymbol{v}_{\text{L}} \\ p_{\text{L}}\boldsymbol{n} \end{pmatrix} - \frac{[p] + \rho c\boldsymbol{n} \cdot [\boldsymbol{v}]}{2\rho c}\begin{pmatrix} 1 \\ \rho c\boldsymbol{n} \end{pmatrix} - \frac{p_{\text{stat}}}{\rho c}\begin{pmatrix} 1 \\ \rho c\boldsymbol{n} \end{pmatrix}\end{aligned}$$

with $[p] = -2p_{\text{L}}$ and $\boldsymbol{n} \cdot [\boldsymbol{v}] = 0$ for the static case (Dirichlet b. c.), and

$$\begin{aligned}\boldsymbol{n} \cdot \boldsymbol{F}^{\text{num}}(\boldsymbol{u}) &= -\begin{pmatrix} \boldsymbol{n} \cdot \boldsymbol{v}_{\text{L}} \\ p_{\text{L}}\boldsymbol{n} \end{pmatrix} - c\frac{g_{\text{kin}} - \boldsymbol{n} \cdot \boldsymbol{v}_{\text{L}}}{c}\begin{pmatrix} 1 \\ \rho c\boldsymbol{n} \end{pmatrix} \\ &= -\begin{pmatrix} \boldsymbol{n} \cdot \boldsymbol{v}_{\text{L}} \\ p_{\text{L}}\boldsymbol{n} \end{pmatrix} - \frac{[p] + \rho c\boldsymbol{n} \cdot [\boldsymbol{v}]}{2\rho c}\begin{pmatrix} 1 \\ \rho c\boldsymbol{n} \end{pmatrix} + g_{\text{kin}}\begin{pmatrix} 1 \\ \rho c\boldsymbol{n} \end{pmatrix},\end{aligned}$$

with $[p] = 0$ and $\boldsymbol{n} \cdot [\boldsymbol{v}] = -2\boldsymbol{n} \cdot \boldsymbol{v}_{\text{L}}$ for the kinematic case (Neumann b. c.).

The discontinuous Galerkin discretization in space

We assume that Ω is a bounded polyhedral Lipschitz domain decomposed into a finite number of open elements $K \subset \Omega$, such that $\bar{\Omega} = \bigcup_{K \in \mathcal{K}} \bar{K}$, where $\mathcal{K}$ is the set of elements in space. Let $\mathcal{F}_K$ be the set of faces of $K \in \mathcal{K}$. For inner faces $f \in \mathcal{F}_K$, let K_f be the neighboring cell such that $f = \partial K \cap \partial K_f$, and let $\boldsymbol{n}_K$ be the outer unit normal vector on ∂K. The outer unit normal vector field on $\partial\Omega$ is denoted by $\boldsymbol{n}$.

We select polynomial degrees p_K, and define the local spaces $H_{h,K} = \mathbb{P}_{p_K}(K; \mathbb{R}^J)$ and the global discontinuous Galerkin space

$$H_h = \{\boldsymbol{v}_h \in L_2(\Omega)^J : \boldsymbol{v}_h|_K \in H_{h,K} \text{ for all } K \in \mathcal{K}\}.$$

For $\boldsymbol{v}_h \in H_h$, we define $\boldsymbol{v}_{h,K} = \boldsymbol{v}_{h|K} \in H_{h,K}$ for the restriction to K. In the semi-discrete problem

$$M_h\partial_t\boldsymbol{u}_h(t) + A_h\boldsymbol{u}_h(t) = \boldsymbol{f}_h(t), \quad t \in (0, T), \tag{3.6}$$

the discrete mass operator $M_h \in \mathcal{L}(H_h, H_h)$, and the right-hand side $\boldsymbol{f}_h \in H_h$, are the Galerkin approximations of M and $\boldsymbol{f}$ defined by

$$\begin{aligned}(M_h\boldsymbol{v}_h, \boldsymbol{w}_h)_{0,\Omega} &= (M\boldsymbol{v}_h, \boldsymbol{w}_h)_{0,\Omega}, \quad \boldsymbol{v}_h, \boldsymbol{w}_h \in H_h, \\ (\boldsymbol{f}_h, \boldsymbol{w}_h)_{0,\Omega} &= (\boldsymbol{f}, \boldsymbol{w}_h)_{0,\Omega}, \quad \boldsymbol{w}_h \in H_h.\end{aligned} \tag{3.7}$$

Note that M_h is represented by a block diagonal positive definite matrix.

The discrete operator $A_h \in \mathcal{L}(H_h, H_h)$ is constructed as follows: Integration by parts yields for smooth ansatz functions $\boldsymbol{v}$ and smooth test functions $\boldsymbol{\phi}_K$:

$$\begin{aligned}(A\boldsymbol{v}, \boldsymbol{\phi}_K)_{0,K} &= (\operatorname{div} \boldsymbol{F}(\boldsymbol{v}), \boldsymbol{\phi}_K)_{0,K} \\ &= -(\boldsymbol{F}(\boldsymbol{v}), \nabla\boldsymbol{\phi}_K)_{0,K} + \sum_{f\in\mathcal{F}_K} (\boldsymbol{n}_K \cdot \boldsymbol{F}(\boldsymbol{v}), \boldsymbol{\phi}_K)_{0,f}.\end{aligned}$$

We then define for $\boldsymbol{v}_h \in H_h$ and $\boldsymbol{\phi}_{h,K} \in H_{h,K}$:

$$(A_h\boldsymbol{v}_h, \boldsymbol{\phi}_{h,K})_{0,K} = -(\boldsymbol{F}(\boldsymbol{v}_{h,K}), \nabla\boldsymbol{\phi}_{h,K})_{0,K} + \sum_{f\in\mathcal{F}_K} (\boldsymbol{n}_K \cdot \boldsymbol{F}_K^{\mathrm{num}}(\boldsymbol{v}_h), \boldsymbol{\phi}_{h,K})_{0,f},$$

where $\boldsymbol{n}_K \cdot \boldsymbol{F}_K^{\mathrm{num}}(\boldsymbol{v}_h)$ is the upwind flux obtained from local solutions of Riemann problems. Again, using integration by parts, we obtain

$$\begin{aligned}(A_h\boldsymbol{v}_h, \boldsymbol{\phi}_{h,K})_{0,K} &= (\operatorname{div} \boldsymbol{F}(\boldsymbol{v}_{h,K}), \boldsymbol{\phi}_{h,K})_{0,K} \\ &\quad + \sum_{f\in\mathcal{F}_K} (\boldsymbol{n}_K \cdot (\boldsymbol{F}_K^{\mathrm{num}}(\boldsymbol{v}_h) - \boldsymbol{F}(\boldsymbol{v}_{h,K})), \boldsymbol{\phi}_{h,K})_{0,f}.\end{aligned} \tag{3.8}$$

On inner faces $f = \partial K \cap \partial K_f$, the difference $\boldsymbol{n}_K \cdot (\boldsymbol{F}_K^{\mathrm{num}}(\boldsymbol{v}_h) - \boldsymbol{F}(\boldsymbol{v}_{h,K}))$ only depends on the jump term $[\boldsymbol{v}_h]_{K,f} = \boldsymbol{v}_{h,K_f} - \boldsymbol{v}_{h,K}$, so that

$$\boldsymbol{n}_K \cdot (\boldsymbol{F}_K^{\mathrm{num}}(\boldsymbol{v}) - \boldsymbol{F}(\boldsymbol{v})) = 0$$

on all faces $f \in \mathcal{F}_K$ for $\boldsymbol{v} \in \mathcal{D}(A)$. On boundary faces, we define the jump term $[\boldsymbol{v}_h]_{K,f}$ depending on the boundary conditions, as in the last paragraph. On H_h, we define the operator A_h by

$$(A_h\boldsymbol{v}_h, \boldsymbol{\phi}_h)_{0,\Omega} = \sum_{K\in\mathcal{K}} (A_h\boldsymbol{v}_h, \boldsymbol{\phi}_{h,K})_{0,K}, \quad \boldsymbol{v}_h, \boldsymbol{\phi}_h \in H_h.$$

By construction, the operator A_h satisfies the consistency condition

$$(A\boldsymbol{v}, \boldsymbol{\phi}_h)_{0,\Omega} = (A_h\boldsymbol{v}, \boldsymbol{\phi}_h)_{0,\Omega}, \quad \boldsymbol{v} \in \mathcal{D}(A), \boldsymbol{\phi}_h \in H_h, \tag{3.9}$$

since the numerical flux $\boldsymbol{F}^{\mathrm{num}}$ satisfies

$$\sum_{K\in\mathcal{K}} (\boldsymbol{n}_K \cdot \boldsymbol{F}_K^{\mathrm{num}}(\boldsymbol{v}_{h,K}), \boldsymbol{v})_{0,\partial K} = 0, \quad \boldsymbol{v} \in \mathcal{D}(A) \cap H^1(\Omega; \mathbb{R}^J) \tag{3.10}$$

for $\boldsymbol{v}_h \in H_h$. For our applications, we can show that the upwind flux together with the described choice of the boundary flux guarantees that the discrete operator is nonnegative and controls the nonconformity. That is, a constant $C_{\mathrm{A}} > 0$ exists such that

$$(A_h\boldsymbol{v}_h, \boldsymbol{v}_h)_{0,\Omega} \geq C_{\mathrm{A}} \sum_{K\in\mathcal{K}} \sum_{f\in\mathcal{F}_K} \|\boldsymbol{n}_K \cdot (\boldsymbol{F}_K^{\mathrm{num}}(\boldsymbol{v}_h) - \boldsymbol{F}(\boldsymbol{v}_{h,K}))\|_{0,f}^2 \geq 0 \tag{3.11}$$

for all $\boldsymbol{v}_h \in H_h$.

For *elastic waves*, we obtain for $(\boldsymbol{\sigma}_h, \mathbf{v}_h) \in H_h$ and $(\boldsymbol{\varphi}_{K,h}, \boldsymbol{\psi}_{K,h}) \in H_{K,h}$

$$\begin{aligned}
&(A_h(\boldsymbol{\sigma}_h, \mathbf{v}_h), (\boldsymbol{\varphi}_{K,h}, \boldsymbol{\psi}_{K,h}))_{0,K} \\
&= -(\boldsymbol{\varepsilon}(\mathbf{v}_{K,h}), \boldsymbol{\varphi}_{K,h})_{0,K} - (\operatorname{div} \boldsymbol{\sigma}_{K,h}, \boldsymbol{\psi}_{K,h})_{0,K} \\
&\quad - \frac{1}{2\rho c_{\mathrm{S}}} \sum_{f \in \mathcal{F}_K} (\mathbf{n}_K \times ([\boldsymbol{\sigma}_h]_{K,f} \mathbf{n}_K + \rho c_{\mathrm{S}} [\mathbf{v}_h]_{K,f}), \mathbf{n}_K \times (\boldsymbol{\varphi}_{K,h} \mathbf{n}_K + \rho c_{\mathrm{S}} \boldsymbol{\psi}_{K,h}))_{0,f} \\
&\quad - \frac{1}{2\rho c_{\mathrm{P}}} \sum_{f \in \mathcal{F}_K} (\mathbf{n}_K \cdot ([\boldsymbol{\sigma}_h]_{K,f} \mathbf{n}_K + \rho c_{\mathrm{P}} [\mathbf{v}_h]_{K,f}), \mathbf{n}_K \cdot (\boldsymbol{\varphi}_{K,h} \mathbf{n}_K + \rho c_{\mathrm{P}} \boldsymbol{\psi}_{K,h}))_{0,f}.
\end{aligned}$$

On boundary faces $f = \partial K \cap \partial\Omega$, we set $[\mathbf{v}_h]_{K,f} = -2\mathbf{v}_{K,h}$ and $[\boldsymbol{\sigma}_h]_{K,f} = \mathbf{0}$ for Dirichlet boundary conditions. This yields

$$\begin{aligned}
&(A_h(\boldsymbol{\sigma}_h, \mathbf{v}_h), (\boldsymbol{\sigma}_{K,h}, \mathbf{v}_{K,h}))_{0,\Omega} \\
&= \sum_{K \in \mathcal{K}} \Big(- \sum_{f \in \mathcal{F}_K} (\mathbf{v}_{K,h}, \boldsymbol{\sigma}_{K,h} \mathbf{n}_K)_{0,f} \\
&\quad - \frac{1}{2\rho c_{\mathrm{S}}} \sum_{f \in \mathcal{F}_K} (\mathbf{n}_K \times ([\boldsymbol{\sigma}]_{K,f} \mathbf{n}_K + \rho c_{\mathrm{S}} [\mathbf{v}]_{K,f}), \mathbf{n}_K \times (\boldsymbol{\sigma}_{K,h} \mathbf{n}_K + \rho c_{\mathrm{S}} \mathbf{v}_{K,h}))_{0,f} \\
&\quad - \frac{1}{2\rho c_{\mathrm{P}}} \sum_{f \in \mathcal{F}_K} (\mathbf{n}_K \cdot ([\boldsymbol{\sigma}]_{K,f} \mathbf{n}_K + \rho c_{\mathrm{P}} [\mathbf{v}]_{K,f}), \mathbf{n}_K \cdot (\boldsymbol{\sigma}_{K,h} \mathbf{n}_K + \rho c_{\mathrm{P}} \mathbf{v}_{K,h}))_{0,f} \Big) \\
&= \frac{1}{2} \sum_{K \in \mathcal{K}} \sum_{f \in \mathcal{F}_K} \Big(\frac{1}{\rho c_{\mathrm{S}}} \|\mathbf{n}_K \times [\boldsymbol{\sigma}]_{K,f} \mathbf{n}_K\|_{0,f}^2 + \rho c_{\mathrm{S}} \|\mathbf{n}_K \times [\mathbf{v}_h]_{K,f}\|_{0,f}^2 \\
&\quad + \frac{1}{\rho c_{\mathrm{P}}} \|\mathbf{n}_K \cdot [\boldsymbol{\sigma}]_{K,f} \mathbf{n}_K\|_{0,f}^2 + \rho c_{\mathrm{P}} \|\mathbf{n}_K \cdot [\mathbf{v}_h]_{K,f}\|_{0,f}^2 \Big).
\end{aligned}$$

For *acoustic waves*, we obtain for $(p_h, \mathbf{v}_h) \in H_h$ and $(\varphi_{K,h}, \boldsymbol{\psi}_{K,h}) \in H_{K,h}$

$$\begin{aligned}
&(A_h(p_h, \mathbf{v}_h), (\varphi_{K,h}, \boldsymbol{\psi}_{K,h}))_{0,K} \\
&= -(\operatorname{div} \mathbf{v}_{K,h}, \varphi_{K,h})_{0,K} - (\nabla p_{K,h}, \boldsymbol{\psi}_{K,h})_{0,K} \\
&\quad - \frac{1}{2\rho c} \sum_{f \in \mathcal{F}_K} ([p_h]_{K,f} + \rho c \mathbf{n}_K \cdot [\mathbf{v}_h]_{K,f}, \varphi_{K,h} + \rho c \boldsymbol{\psi}_{K,h} \cdot \mathbf{n}_K)_{0,f}.
\end{aligned}$$

On boundary faces $f = \partial K \cap \partial\Omega$, we set $[p_h]_{K,f} = -2p_h$ and $[\mathbf{v}_h]_{K,f} \cdot \mathbf{n}_K = 0$ for Dirichlet boundary conditions, and $[p_h]_{K,f} = 0$ and $[\mathbf{v}_h]_{K,f} \cdot \mathbf{n}_K = -2\mathbf{v}_{K,h} \cdot \mathbf{n}_K$ for Neumann boundary conditions. This yields

$$\begin{aligned}
&(A_h(p_h, \mathbf{v}_h), (p_h, \mathbf{v}_h))_{0,\Omega} \\
&= \frac{1}{2} \sum_{K \in \mathcal{K}} \sum_{f \in \mathcal{F}_K} \Big(\frac{1}{\rho c} \|[p_h]_{K,f}\|_{0,f}^2 + \rho c \|\mathbf{n}_K \cdot [\mathbf{v}_h]_{K,f}\|_{0,f}^2 \Big).
\end{aligned}$$

Together with inhomogeneous boundary conditions $p = p_{\mathrm{stat}}$ on Γ_{stat} and $\mathbf{n} \cdot \mathbf{v} = g_{\mathrm{kin}}$ on Γ_{kin}, we obtain the semi-discrete equation

$$(M_h(\partial_t p_h, \partial_t \boldsymbol{v}_h) + A_h(p_h, \boldsymbol{v}_h), (\varphi_h, \boldsymbol{\psi}_h))_{0,\Omega} = (\boldsymbol{b}, \boldsymbol{\psi}_h)_{0,\Omega}$$
$$+ \frac{1}{\rho c} \sum_{K \in \mathcal{K}} \Big(\sum_{f \in \mathcal{F}_K \cap \Gamma_{\mathrm{stat}}} (p_{\mathrm{stat}}, \varphi_{K,h} + \rho c \boldsymbol{\psi}_{K,h} \cdot \boldsymbol{n}_K)_{0,f} + \sum_{f \in \mathcal{F}_K \cap \Gamma_{\mathrm{kin}}} (g_{\mathrm{kin}}, \varphi_{K,h} + \rho c \boldsymbol{\psi}_{K,h} \cdot \boldsymbol{n}_K)_{0,f} \Big).$$

4 A Petrov–Galerkin space-time discretization

Let $\overline{Q} = \bigcup_{R \in \mathcal{R}} \overline{R}$ be a decomposition of the space-time cylinder into space-time cells $R = I \times K$ with $K \in \mathcal{K}$ and $I \subset (0, T)$ an interval; $\mathcal{R}$ denotes the set of space-time cells. For every $R \in \mathcal{R}$, we choose local test spaces $W_{h,R} \subset L_2(R; \mathbb{R}^J)$, and we define the global test space

$$W_h = \{\boldsymbol{w}_h \in L_2((0, T); H) : \boldsymbol{w}_{h,R} = \boldsymbol{w}_h|_R \in W_{h,R}\}.$$

The functions in W_h are discontinuous in space and time. Now we construct $V_h \subset H^1((0, T); H)$ with $\dim V_h = \dim W_h$. Then, functions in V_h are continuous in time. That is, $\boldsymbol{v}_h(\cdot, \boldsymbol{x})$ is continuous on $[0, T]$ for almost all $\boldsymbol{x} \in \Omega$.

In the most simple case, this can be achieved for a tensor product space-time discretization with a fixed mesh $\mathcal{K}$ in space and a time series

$$0 = t_0 < t_1 < \cdots < t_N = T.$$

That is, $\mathcal{R} = \{I_n \times K : I_n := (t_{n-1}, t_n),\ n = 1, \ldots, N,\ K \in \mathcal{K}\}$. Then, we can select a discrete space H_h with $H_{h,K} = \mathbb{P}_p(K; \mathbb{R}^J)$ independently of t, and in every time slice, we define $W_{h,R} = H_{h,K}$ constant in time on $R = I_n \times K$. In the case of V_h, we use piecewise linear approximations in time:

$$V_h = \Big\{ \boldsymbol{v}_h \in H^1((0, T); H) :$$
$$\boldsymbol{v}_h(0, \boldsymbol{x}) = \boldsymbol{0}, \boldsymbol{v}_h(t_n, \boldsymbol{x}) \in H_h \text{ for a. a. } \boldsymbol{x} \in \Omega \text{ and } n = 1, \ldots, N, \text{ and}$$
$$\boldsymbol{v}_h(t, \boldsymbol{x}) = \frac{t_n - t}{t_n - t_{n-1}} \boldsymbol{v}_h(t_{n-1}, \boldsymbol{x}) + \frac{t - t_{n-1}}{t_n - t_{n-1}} \boldsymbol{v}_h(t_n, \boldsymbol{x}) \text{ for } t \in I_n \Big\}.$$

In the more general case, we consider a tensor product space-time mesh with a local selection of polynomial degrees in space and time p_R and q_R in every cell R, and we set for the local test space $W_{h,R} = \mathbb{P}_{q_R - 1}(I_n; \mathbb{R}^J) \otimes \mathbb{P}_{p_R}(K; \mathbb{R}^J)$. Then, the local ansatz spaces $V_{h,R} = V_{h|R}$ take the form

$$V_{h,R} = \Big\{ \boldsymbol{v}_{h,R} \in L_2(R; \mathbb{R}^J) :$$
$$\boldsymbol{v}_{h,R}(t, \boldsymbol{x}) = \frac{t_n - t}{t_n - t_{n-1}} \boldsymbol{v}_h(t_{n-1}, \boldsymbol{x}) + \frac{t - t_{n-1}}{t_n - t_{n-1}} \boldsymbol{w}_{h,R}(t, \boldsymbol{x}),$$
$$\boldsymbol{v}_h \in V_{h|[0,t_{n-1}]}, \boldsymbol{w}_{h,R} \in W_{h,R}, (t, \boldsymbol{x}) \in R = I_n \times K \Big\}.$$

The discontinuous Galerkin operator in space is extended to the space-time operator $A_h \mathbf{v}_h \in W_h$ by defining for $\mathbf{v}_h \in V_h$ and $\mathbf{w}_h \in W_h$:

$$(A_h \mathbf{v}_h, \mathbf{w}_h)_{0,Q} = \sum_{R=I\times K\in\mathcal{R}} \Big((\operatorname{div} \boldsymbol{F}(\mathbf{v}_{h,R}), \mathbf{w}_{h,R})_{0,R} + \sum_{f\in\mathcal{F}_K} (\boldsymbol{n}_K \cdot (\boldsymbol{F}_K^{\mathrm{num}}(\mathbf{v}_h) - \boldsymbol{F}(\mathbf{v}_{h,R})), \mathbf{w}_{h,R})_{0,I\times f} \Big). \tag{4.1}$$

The discrete space-time operator $L_h \in \mathcal{L}(V_h, W_h)$ and the corresponding discrete bilinear form $b_h(\cdot,\cdot) = (L_h \cdot, \cdot)_{0,Q}$ are defined by

$$(L_h \mathbf{v}_h, \mathbf{w}_h)_{0,Q} = (M_h \partial_t \mathbf{v}_h + A_h \mathbf{v}_h, \mathbf{w}_h)_{0,Q}.$$

To show that a solution to our Petrov–Galerkin scheme exists, we check the inf-sup stability of the discrete bilinear form $b_h(\cdot,\cdot)$ with respect to the discrete norm

$$\|\mathbf{v}_h\|_{V_h}^2 = \|\mathbf{v}_h\|_W^2 + \|M_h^{-1} L_h \mathbf{v}_h\|_W^2. \tag{4.2}$$

By construction, $b_h(\cdot,\cdot)$ is bounded in $V_h \times W_h$; that is,

$$\begin{aligned} b_h(\mathbf{v}_h, \mathbf{w}_h) &= (L_h \mathbf{v}_h, \mathbf{w}_h)_{0,Q} \\ &\le \|M_h^{-1} L_h \mathbf{v}_h\|_W \|\mathbf{w}_h\|_W \le \|\mathbf{v}_h\|_{V_h} \|\mathbf{w}_h\|_W, \quad \mathbf{v}_h \in V_h, \mathbf{w}_h \in W_h. \end{aligned}$$

For the verification of the inf-sup stability, we introduce the L_2 projection

$$\Pi_h \colon W \to W_h, \quad (\Pi_h \mathbf{v}, \mathbf{w}_h)_{0,Q} = (\mathbf{v}, \mathbf{w}_h)_{0,Q}, \quad \mathbf{w}_h \in W_h.$$

Then, by construction, $\Pi_h A_h = A_h$, and $\Pi_h L_h = L_h$. Moreover, we define the nonnegative weight function in time $d_T(t) = T - t$, and we observe

$$\int_0^T \int_0^t \phi(s)\,\mathrm{ds}\,\mathrm{dt} = \int_0^T d_T(t)\phi(t)\,\mathrm{dt}, \quad \phi \in L_1(0,T). \tag{4.3}$$

Lemma 4.1 ([8, Lemma 3]). *Assume that*

$$(M_h \partial_t \mathbf{v}_h, d_T \mathbf{v}_h)_{0,Q} \le (L_h \mathbf{v}_h, d_T \Pi_h \mathbf{v}_h)_{0,Q}, \quad \mathbf{v}_h \in V_h. \tag{4.4}$$

Then, the bilinear form $b_h(\cdot,\cdot)$ is inf-sup stable in $V_h \times W_h$ with $\beta = 1/\sqrt{1+4T^2}$. That is,

$$\sup_{\mathbf{w}_h \in W_h \setminus \{\mathbf{0}\}} \frac{b_h(\mathbf{v}_h, \mathbf{w}_h)}{\|\mathbf{w}_h\|_W} \ge \beta \|\mathbf{v}_h\|_{V_h}, \quad \mathbf{v}_h \in V_h.$$

Referring to [8, Theorem 4.2], we find that, for given $\boldsymbol{f} \in L_2(Q;\mathbb{R}^J)$, a unique solution $\boldsymbol{u}_h \in V_h$ exists, solving

$$(L_h \boldsymbol{u}_h, \boldsymbol{w}_h)_{0,Q} = (\boldsymbol{f}, \boldsymbol{w}_h)_{0,Q}, \quad \boldsymbol{w}_h \in W_h, \tag{4.5}$$

and satisfying the a priori bound $\|\boldsymbol{u}_h\|_{V_h} \le \sqrt{4T^2+1}\|M_h^{-1}\Pi_h \boldsymbol{f}\|_W$.

In the following example, we check assumption (4.4) in case of a tensor product discretization with homogeneous polynomial degrees in space and polynomial degree one in time ($q_R \equiv 1$). Note that for this case, the Petrov–Galerkin method in time is equivalent to the implicit midpoint rule. A general proof for tensor product discretizations with arbitrary polynomial degrees is given in [8, Lemma 4.4].

Example 4.1. Let $\mathcal{R}$ be a tensor product discretization and $p_R \equiv p$ and $q_R \equiv 1$ for all $R \in \mathcal{R}$. For $\boldsymbol{v}_h \in V_h$, we set $\boldsymbol{v}_h^n = \boldsymbol{v}_h(t_n, \cdot)$. This yields for $t \in I_n = (t_{n-1}, t_n)$

$$\begin{aligned} \boldsymbol{v}_h(t,\boldsymbol{x}) &= \frac{t_n - t}{t_n - t_{n-1}} \boldsymbol{v}_h^{n-1}(\boldsymbol{x}) + \frac{t - t_{n-1}}{t_n - t_{n-1}} \boldsymbol{v}_h^n(\boldsymbol{x}), \\ \partial_t \boldsymbol{v}_h(t,\boldsymbol{x}) &= \frac{1}{t_n - t_{n-1}} (\boldsymbol{v}_h^n(\boldsymbol{x}) - \boldsymbol{v}_h^{n-1}(\boldsymbol{x})), \end{aligned}$$

and thus $\partial_t \boldsymbol{v}_h = \Pi_h \partial_t \boldsymbol{v}_h \in W_h$, and $\Pi_h \boldsymbol{v}_h(\boldsymbol{x}, t) = \frac{1}{2}(\boldsymbol{v}_h^{n-1} + \boldsymbol{v}_h^n)(\boldsymbol{x})$. Due to

$$\Pi_h \boldsymbol{v}_h - \boldsymbol{v}_h = \frac{t_n + t_{n-1} - 2t}{2(t_n - t_{n-1})} (\boldsymbol{v}_h^n - \boldsymbol{v}_h^{n-1}),$$

we conclude

$$\begin{aligned} &(M_h \partial_t \boldsymbol{v}_h, d_T(\Pi_h \boldsymbol{v}_h - \boldsymbol{v}_h))_{0,Q} \\ &\quad = \sum_{n=1}^{N} (M_h(\boldsymbol{v}_h^n - \boldsymbol{v}_h^{n-1}), \boldsymbol{v}_h^n - \boldsymbol{v}_h^{n-1})_{0,\Omega} \int_{t_{n-1}}^{t_n} d_T(t) \frac{t_n + t_{n-1} - 2t}{2(t_n - t_{n-1})^2}\, dt \\ &\quad = \sum_{n=1}^{N} \frac{t_n - t_{n-1}}{12} (M_h(\boldsymbol{v}_h^n - \boldsymbol{v}_h^{n-1}), \boldsymbol{v}_h^n - \boldsymbol{v}_h^{n-1})_{0,\Omega} \ge 0 \end{aligned}$$

for all $n = 0, \dots, N$, since $(M_h \boldsymbol{w}_h, \boldsymbol{w}_h)_{0,\Omega} \ge 0$ for all $\boldsymbol{w}_h \in W_h$.

Furthermore, $A_h = \Pi_h A_h$ yields

$$\begin{aligned} &(A_h \boldsymbol{v}_h, d_T \Pi_h \boldsymbol{v}_h)_{0,Q} \\ &\quad = (\Pi_h A_h \boldsymbol{v}_h, d_T \Pi_h \boldsymbol{v}_h)_{0,Q} \\ &\quad = \sum_{n=1}^{N} \Big(T - \frac{t_{n-1} + t_n}{2}\Big) \frac{t_n - t_{n-1}}{4} (A_h(\boldsymbol{v}_h^{n-1} + \boldsymbol{v}_h^n), \boldsymbol{v}_h^{n-1} + \boldsymbol{v}_h^n)_{0,\Omega} \ge 0, \end{aligned}$$

since $T - \frac{1}{2}(t_{n-1} + t_n) \ge 0$ and $(A_h \boldsymbol{v}_h, \boldsymbol{v}_h)_{0,\Omega} \ge 0$ for all $\boldsymbol{v}_h \in V_h$ by (3.11). Combining both inequalities finally proves assumption (4.4).

Lemma 4.1 directly implies an a priori error estimate in the discrete graph norm (4.2). Let $h = \max_{R\in\mathcal{R}} \operatorname{diam}(R)$ be the mesh size:

$$\operatorname{diam}(R)^2 = |I|^2 + \operatorname{diam}(K)^2 \quad \text{for } R = I \times K.$$

For $1 \le m \le \min_R\{p_R + 1, q_R + 1\}$, we have

$$\inf_{\boldsymbol{v}_h\in V_h\cap H^1(Q;\mathbb{R}^J)} \|\boldsymbol{v} - \boldsymbol{v}_h\|_{1,Q} \le Ch^{m-1}\|\boldsymbol{v}\|_{m,Q}, \quad \boldsymbol{v} \in H^m(Q;\mathbb{R}^J) \tag{4.6}$$

with $C > 0$, depending on the mesh quality.

Theorem 4.1 ([8, Theorem 5]). *Let $\boldsymbol{u} \in V$ be the solution of* (2.3) *and $\boldsymbol{u}_h \in V_h$ its approximation, solving* (4.5). *If the solution satisfies $\boldsymbol{u} \in H^m(Q;\mathbb{R}^J)$ with $1 \le m \le \min_R\{p_R + 1, q_R + 1\}$, the error can be bounded by*

$$\|\boldsymbol{u} - \boldsymbol{u}_h\|_{V_h} \le Ch^{m-1}\|\boldsymbol{u}\|_{m,Q}.$$

Proof. Since M_h is the Galerkin projection of M in W, we have

$$b_h(\boldsymbol{u}, \boldsymbol{w}_h) = b(\boldsymbol{u}, \boldsymbol{w}_h) = (\boldsymbol{f}, \boldsymbol{w}_h)_{0,Q} = b_h(\boldsymbol{u}_h, \boldsymbol{w}_h), \quad \boldsymbol{v}_h \in V_h,$$

which yields

$$b_h(\boldsymbol{v}_h - \boldsymbol{u}_h, \boldsymbol{w}_h) = b_h(\boldsymbol{v}_h - \boldsymbol{u}, \boldsymbol{w}_h) \le \|\boldsymbol{v}_h - \boldsymbol{u}\|_{V_h}\|\boldsymbol{w}_h\|_W, \quad \boldsymbol{v}_h \in V_h,$$

and thus

$$\begin{aligned}\|\boldsymbol{u} - \boldsymbol{u}_h\|_{V_h} &\le \|\boldsymbol{u} - \boldsymbol{v}_h\|_{V_h} + \|\boldsymbol{v}_h - \boldsymbol{u}_h\|_{V_h}\\ &\le \|\boldsymbol{u} - \boldsymbol{v}_h\|_{V_h} + \beta^{-1} \sup_{\boldsymbol{w}_h\in W_h\setminus\{\boldsymbol{0}\}} \frac{b_h(\boldsymbol{v}_h - \boldsymbol{u}_h, \boldsymbol{w}_h)}{\|\boldsymbol{w}_h\|_W}\\ &\le (1 + \beta^{-1})\|\boldsymbol{u} - \boldsymbol{v}_h\|_{V_h}.\end{aligned}$$

Now the assertion follows from $\|\boldsymbol{v}\|_{V_h} \le C\|\boldsymbol{v}\|_{1,Q}$ for $\boldsymbol{v} \in H^1(Q;\mathbb{R}^J)$ and (4.6). □

5 Duality based goal-oriented error estimation

To develop an adaptive strategy for the selection of the local polynomial degrees p_R, q_R, we derive an error indicator with respect to a given linear goal functional $E \in W'$. Following the framework in [4], we define the adjoint problem and solve the dual problem. Then, the error is estimated in terms of the local residual and the dual weight.

The adjoint operator L^* in space and time is defined on the adjoint Hilbert space:

$$\begin{aligned}V^* = \{&\boldsymbol{w} \in W\colon \text{there exists } \boldsymbol{g} \in W \text{ such that}\\ &(L\boldsymbol{v}, \boldsymbol{w})_{0,Q} = (\boldsymbol{v}, \boldsymbol{g})_{0,Q} \text{ for all } \boldsymbol{v} \in V\},\end{aligned}$$

and is characterized by

$$(\boldsymbol{v}, L^*\boldsymbol{w})_{0,Q} = (L\boldsymbol{v}, \boldsymbol{w})_{0,Q}, \quad \boldsymbol{v} \in V, \boldsymbol{w} \in V^*.$$

We observe $\{\boldsymbol{v}^* \in C^1([0,T];\mathcal{D}(A^*)) \colon \boldsymbol{v}^*(T) = \boldsymbol{0}\} \subset V^*$ and $L^* = -L$ on $V \cap V^*$.

For the evaluation of the error functional E, we introduce the dual solution $\boldsymbol{u}^* \in V^*$ defined by

$$(\boldsymbol{w}, L^*\boldsymbol{u}^*)_{0,Q} = \langle E, \boldsymbol{w}\rangle, \quad \boldsymbol{w} \in W.$$

Let $\boldsymbol{u} \in V$ be the solution of (2.3) and $\boldsymbol{u}_h \in V_h$ its approximation, solving (4.5). Now we derive an exact error representation for the error functional in the case where the dual solution is sufficiently smooth, such that $\boldsymbol{u}^*(t,\cdot)|_f \in L_2(f;\mathbb{R}^J)$ for all faces $f \in \mathcal{F}_h$ and almost all $t \in (0,T)$. Inserting the consistency of the numerical flux (3.9) yields for all $\boldsymbol{w}_h \in W_h \cap V^*$

$$\begin{aligned}
\langle E, \boldsymbol{u} - \boldsymbol{u}_h\rangle &= (\boldsymbol{u} - \boldsymbol{u}_h, -M\partial_t\boldsymbol{u}^* - \operatorname{div}\boldsymbol{F}(\boldsymbol{u}^*))_{0,Q} \\
&= (\boldsymbol{u}, -M\partial_t\boldsymbol{u}^* - \operatorname{div}\boldsymbol{F}(\boldsymbol{u}^*))_{0,Q} - (\boldsymbol{u}_h, -M\partial_t\boldsymbol{u}^* - \operatorname{div}\boldsymbol{F}(\boldsymbol{u}^*))_{0,Q} \\
&= (M\partial_t\boldsymbol{u} + \operatorname{div}\boldsymbol{F}(\boldsymbol{u}), \boldsymbol{u}^*)_{0,Q} - (\boldsymbol{u}, \boldsymbol{n}\cdot\boldsymbol{F}(\boldsymbol{u}^*))_{0,\partial Q} \\
&\quad - \sum_{R\in\mathcal{R}} ((M\partial_t\boldsymbol{u}_h + \operatorname{div}\boldsymbol{F}(\boldsymbol{u}_h), \boldsymbol{u}^*)_{0,R} - (\boldsymbol{u}_h, \boldsymbol{n}_R\cdot\boldsymbol{F}(\boldsymbol{u}^*))_{0,\partial R}) \\
&= (\boldsymbol{f}, \boldsymbol{u}^*)_{0,Q} - \sum_{R=I\times K\in\mathcal{R}} ((M\partial_t\boldsymbol{u}_{h,R} + \operatorname{div}\boldsymbol{F}(\boldsymbol{u}_{h,R}), \boldsymbol{u}^*)_{0,R} \\
&\quad - (\boldsymbol{u}_h, \boldsymbol{n}_K\cdot\boldsymbol{F}(\boldsymbol{u}^*))_{0,I\times\partial K}) \\
&= \sum_{R=I\times K\in\mathcal{R}} ((\boldsymbol{f} - M\partial_t\boldsymbol{u}_{h,R} - \operatorname{div}\boldsymbol{F}(\boldsymbol{u}_{h,R}), \boldsymbol{u}^*)_{0,R} \\
&\quad + (\boldsymbol{n}_K\cdot\boldsymbol{F}(\boldsymbol{u}_{h,R}), \boldsymbol{u}^*)_{0,I\times\partial K}) \\
&= \sum_{R=I\times K\in\mathcal{R}} ((\boldsymbol{f} - M\partial_t\boldsymbol{u}_{h,R} - \operatorname{div}\boldsymbol{F}(\boldsymbol{u}_{h,R}), \boldsymbol{u}^* - \boldsymbol{w}_h)_{0,R} \\
&\quad + (\boldsymbol{n}_K\cdot\boldsymbol{F}(\boldsymbol{u}_{h,R}), \boldsymbol{u}^* - \boldsymbol{w}_h)_{0,I\times\partial K}).
\end{aligned}$$

However, this identity cannot be evaluated numerically, since it depends on the unknown function $\boldsymbol{u}^*$. In applications, the following heuristic error bound is used instead. Let $\boldsymbol{u}_h^* \in W_h$ be a numerical approximation of the dual solution, given by

$$b_h(\boldsymbol{v}_h, \boldsymbol{u}_h^*) = \langle E, \boldsymbol{v}_h\rangle, \quad \boldsymbol{v}_h \in V_h.$$

Inserting some interpolation $\boldsymbol{w}_h = I_h\boldsymbol{u}^*$, the interpolation error $\boldsymbol{u}^* - I_h\boldsymbol{u}^*$ has to be estimated in terms of $\boldsymbol{u}_h^*$. For this purpose, we use also the face jumps $[\boldsymbol{u}_h^*]_{K,f}$, which are also meaningful in case of piecewise constant approximations in W_h. In case of higher-order approximations in W_h, we use $[Q_h\boldsymbol{u}_h^*]_{K,f}$, where Q_h denotes the piecewise L_2 projection in space to $\mathbb{P}_0(K;\mathbb{R}^J)$.

Finally, $|\langle E, \boldsymbol{u}-\boldsymbol{u}_h\rangle|$ is estimated by $\sum_{R\in\mathcal{R}} \eta_R$ with local contributions η_R, depending on residual terms and jump terms of the discrete solution and on jump terms of the dual approximation.

For elastic waves, we obtain for $(\boldsymbol{\varphi}_h, \boldsymbol{\psi}_h) \in W_h \cap V^*$ the error representation:

$$\begin{aligned}
\langle E, (\boldsymbol{\sigma}-\boldsymbol{\sigma}_h, \boldsymbol{v}-\boldsymbol{v}_h)\rangle = & \sum_{R=I\times K\in\mathcal{R}} \big((-\boldsymbol{C}^{-1}\partial_t\boldsymbol{\sigma}_{h,R} + \boldsymbol{\varepsilon}(\boldsymbol{v}_{h,R}), \boldsymbol{\sigma}^* - \boldsymbol{\varphi}_h)_{0,R} \\
& + (\boldsymbol{b} - \rho\partial_t\boldsymbol{v}_{h,R} + \operatorname{div}\boldsymbol{\sigma}_{h,R}, \boldsymbol{v}^* - \boldsymbol{\psi}_h)_{0,R} \\
& + (\boldsymbol{v}_{h,R}, (\boldsymbol{\sigma}^* - \boldsymbol{\varphi}_h)\boldsymbol{n}_K)_{0,I\times\partial K} \\
& + (\boldsymbol{\sigma}_{R,h}\boldsymbol{n}_K, \boldsymbol{v}^* - \boldsymbol{\psi}_h)_{0,I\times\partial K}\big) \\
= & \sum_{R=I\times K\in\mathcal{R}} \Big((-\boldsymbol{C}^{-1}\partial_t\boldsymbol{\sigma}_{h,R} + \boldsymbol{\varepsilon}(\boldsymbol{v}_{h,R}), \boldsymbol{\sigma}^* - \boldsymbol{\varphi}_h)_{0,R} \\
& + (\boldsymbol{b} - \rho\partial_t\boldsymbol{v}_{h,R} + \operatorname{div}\boldsymbol{\sigma}_{h,R}, \boldsymbol{v}^* - \boldsymbol{\psi}_h)_{0,R} \\
& + \frac{1}{2}\sum_{f\in\mathcal{F}_K} \big(([\boldsymbol{v}_h]_{K,f}, (\boldsymbol{\sigma}^* - \boldsymbol{\varphi}_h)\boldsymbol{n}_K)_{0,I\times f} \\
& + ([\boldsymbol{\sigma}_h]_{K,f}\boldsymbol{n}_K, \boldsymbol{v}^* - \boldsymbol{\psi}_h)_{0,I\times f}\big)\Big).
\end{aligned}$$

This motivates the local error estimate:

$$\begin{aligned}
\eta_R = & \|-\boldsymbol{C}^{-1}\partial_t\boldsymbol{\sigma}_{h,R} + \boldsymbol{\varepsilon}(\boldsymbol{v}_{h,R})\|_{0,R} h_K^{1/2} \|[Q_h\boldsymbol{\sigma}_h^*]_K\boldsymbol{n}_K\|_{0,I\times\partial K} \\
& + \|\boldsymbol{b} - \rho\partial_t\boldsymbol{v}_{h,R} + \operatorname{div}\boldsymbol{\sigma}_{h,R}\|_{0,R} h_K^{1/2} \|[Q_h\boldsymbol{v}_h^*]_K\|_{0,I\times\partial K} \\
& + \frac{1}{2}\sum_{f\in\mathcal{F}_K} \big(\|[\boldsymbol{v}_h]_{K,f}\|_{0,I\times f} \|[Q_h\boldsymbol{\sigma}_h^*]_{K,f}\boldsymbol{n}_K\|_{0,I\times f} \\
& + \|[\boldsymbol{\sigma}_h]_{K,f}\boldsymbol{n}_K\|_{0,I\times f} \|[Q_h\boldsymbol{v}_h^*]_{K,f}\|_{0,I\times f}\big),
\end{aligned}$$

where the jump terms $[Q_h\boldsymbol{\sigma}_h^*]_{K,f}$ and $[Q_h\boldsymbol{v}_h^*]_{K,f}$ are used to estimate the best approximation error of $(\boldsymbol{\sigma}^* - \boldsymbol{\varphi}_h)\boldsymbol{n}_K$ and $\boldsymbol{v}^* - \boldsymbol{\psi}_h$.

In the same way, we obtain for acoustic waves the error representation:

$$\begin{aligned}
\langle E, (p-p_h, \boldsymbol{v}-\boldsymbol{v}_h)\rangle = & \sum_{R=I\times K\in\mathcal{R}} \Big[(-\kappa^{-1}\partial_t p_{h,R} + \operatorname{div}\boldsymbol{v}_{h,R}, p^* - \varphi_h)_{0,R} \\
& + (\boldsymbol{b} - \rho\partial_t\boldsymbol{v}_{h,R} + \nabla p_{h,R}, \boldsymbol{v}^* - \boldsymbol{\psi}_h)_{0,R} \\
& + \frac{1}{2}\sum_{f\in\mathcal{F}_K} \big((\boldsymbol{n}_K\cdot[\boldsymbol{v}_h]_{K,f}, p^* - \varphi_h)_{0,I\times f} \\
& + ([p_h]_{K,f}, \boldsymbol{n}_K\cdot(\boldsymbol{v}^* - \boldsymbol{\psi}_h))_{0,I\times f}\big)\Big]
\end{aligned}$$

and the local error estimate:

$$\begin{aligned}\eta_R = {} & \|-\kappa^{-1}\partial_t p_{h,R} + \operatorname{div} \mathbf{v}_{h,R}\|_{0,R} h_K^{1/2} \|[Q_h p_h^*]_K\|_{0,I\times\partial K} \\ & + \|\mathbf{b} - \rho\partial_t \mathbf{v}_{h,R} + \nabla p_{h,R}\|_{0,R} h_K^{1/2} \|\mathbf{n}_K \cdot [Q_h \mathbf{v}_h^*]_K\|_{0,I\times\partial K} \\ & + \frac{1}{2} \sum_{f\in\mathcal{F}_K} \big(\|[\mathbf{v}_h]_{K,f}\|_{0,I\times f} \|[Q_h p_h^*]_{K,f}\|_{0,I\times f} \\ & + \|[p_h]_{K,f} \mathbf{n}_K\|_{0,I\times f} \|\mathbf{n}_K \cdot [Q_h \mathbf{v}_h^*]_{K,f}\|_{0,I\times f}\big).\end{aligned}$$

In our examples, we use the adaptive strategy for p-refinement described in Algorithm 1. It depends on a parameter $\vartheta < 1$ for the adaptive selection criterion.

Algorithm 1: Adaptive algorithm.

```
choose low-order polynomial degrees on the initial mesh
while max_R(p_R) < p_max and max_R(q_R) < q_max do
    compute u_h
    compute u_h^* and the projection Q_h u_h^*
    compute η_R on every cell R
    if the error is small enough STOP
    mark space-time cell R if η_R > ϑ max_R' η_R'
    increase polynomial degrees on marked cells by one
    redistribute cells on processes for better load balancing
end while
```

6 Space-time multilevel preconditioner

In this section, we address the numerical aspects and, in particular, solution methods for the discrete hyperbolic space-time problem. First, we describe the realization of our discretization using nodal basis functions in space and time, and then a multilevel preconditioner is introduced.

Nodal discretization

Now we consider the structure of the linear system for the special case of a tensor product space-time mesh $\mathcal{R} = \bigcup_{n=1}^{N} \mathcal{R}^n$ with time slices

$$\mathcal{R}^n = \{I_n \times K : K \in \mathcal{K}\}$$

and variable polynomial degrees p_R, q_R in every space-time cell R, see Section 4. Let $\{\boldsymbol{\psi}_{R,j}^n\}_{j=1,\dots,\dim W_{h,R}}$ be a basis of $W_{h,R}$, and define

$$W_h^n = \operatorname{span}\Bigg\{\bigcup_{R\in\mathcal{R}^n} \bigcup_{j=1}^{\dim W_{h,R}} \boldsymbol{\psi}_{R,j}^n\Bigg\}.$$

Then, the solution $\boldsymbol{u}_h \in V_h$ is represented by finite element functions $\boldsymbol{u}_h^n \in W_h^n$, $n = 1,\dots,N$. Together with $\boldsymbol{u}_h^0 = \mathbf{0}$, we obtain

$$\boldsymbol{u}_h(t,\boldsymbol{x}) = \frac{t_n - t}{t_n - t_{n-1}}\boldsymbol{u}_h^{n-1}(t_{n-1},\boldsymbol{x}) + \frac{t - t_{n-1}}{t_n - t_{n-1}}\boldsymbol{u}_h^n(t,\boldsymbol{x}), \quad (t,\boldsymbol{x}) \in I_n \times K.$$

The corresponding coefficient vector of the solution is denoted by $\underline{u} = (\underline{u}^1,\dots,\underline{u}^N)^\top$, where $\underline{u}^n \in \mathbb{R}^{\dim W_h^n}$ is the coefficient vector of

$$\boldsymbol{u}_h^n = \sum_{R\in\mathcal{R}^n} \sum_{j=1}^{\dim W_{h,R}} \underline{u}_{R,j}^n \boldsymbol{\psi}_{R,j}^n.$$

With respect to this basis, the discrete space-time system (4.5) has the matrix representation $\underline{L}\,\underline{u} = \underline{f}$ with the block matrix

$$\underline{L} = \begin{pmatrix} \underline{D}^1 & & & \\ \underline{C}^1 & \underline{D}^2 & & \\ & \ddots & \ddots & \\ & & \underline{C}^{N-1} & \underline{D}^N \end{pmatrix},$$

and matrix entries

$$\underline{D}_{R',k,R,j}^n = \int_{t_{n-1}}^{t_n}\int_\Omega L_h\Big(\frac{\cdot - t_{n-1}}{t_n - t_{n-1}}\boldsymbol{\psi}_{R,j}^n\Big)(t,\boldsymbol{x})\boldsymbol{\psi}_{R',k}^n(t,\boldsymbol{x})\,\mathrm{d}\boldsymbol{x}\,\mathrm{d}t, \quad R,R' \in \mathcal{R}^n,$$

$$\underline{C}_{R',k,R,j}^n = \int_{t_{n-1}}^{t_n}\int_\Omega L_h\Big(\frac{t_n - \cdot}{t_n - t_{n-1}}\boldsymbol{\psi}_{R,j}^{n-1}(t_{n-1},\cdot)\Big)(t,\boldsymbol{x})\boldsymbol{\psi}_{R',k}^n(t,\boldsymbol{x})\,\mathrm{d}\boldsymbol{x}\,\mathrm{d}t, \quad R \in \mathcal{R}^{n-1}, R' \in \mathcal{R}^n,$$

and the right-hand side $\underline{f} = (\underline{f}^{1,\top},\dots,\underline{f}^{N,\top})^\top$ with $\underline{f}_{j,R}^n = (\boldsymbol{f},\boldsymbol{\psi}_{R,j}^n)_{0,R}$. Sequentially, this system can be solved by a block-Gauss–Seidel method (corresponding to implicit time integration),

$$\underline{D}^1\underline{u}^1 = \underline{f}^1, \quad \underline{D}^2\underline{u}^2 = \underline{f}^2 - \underline{C}^1\underline{u}^1,\dots,\underline{D}^N\underline{u}^N = \underline{f}^N - \underline{C}^{N-1}\underline{u}^{N-1},$$

provided that $\underline{D}^n$ can be inverted efficiently.

Multilevel methods

For space-time multilevel preconditioners, we consider hierarchies in space and time. Therefore, let $\mathcal{R}_{0,0}$ be the coarse space-time mesh, and let $\mathcal{R}_{l,k}$ be the discretization obtained by $l = 1,\dots,l_{\max}$ uniform refinements in space, and $k = 1,\dots,k_{\max}$ refinements in time. Let $V_{l,k}$ be the approximation spaces on $\mathcal{R}_{l,k}$ with fixed polynomial degrees

$p_R \equiv p$ and $q_R \equiv q$. Let $\underline{L}_{l,k}$ be the corresponding matrix representation of the discrete operator L_h in $V_{l,k}$.

The multilevel preconditioner combines smoothing operations on different levels and requires transfer matrices between the levels. Since the spaces are nested, we can define prolongation matrices $\underline{P}^{l,k}_{l-1,k}$ and $\underline{P}^{l,k}_{l,k-1}$ representing the natural injections $V_{l-1,k} \subset V_{l,k}$ in space, and $V_{l,k-1} \subset V_{l,k}$ in time. Correspondingly, the restriction matrices $\underline{R}^{l,k}_{l-1,k}$ and $\underline{R}^{l,k}_{l,k-1}$ represent the L_2 projections in space and in time of the test spaces $W_{l,k} \supset W_{l-1,k}$ and $W_{l,k} \supset W_{l,k-1}$.

For the smoothing operations on level (l, k), we consider the block-Jacobi preconditioner or the block-Gauss–Seidel preconditioner (where all components corresponding to a space-time cell R build a block):

$$\begin{aligned} \underline{B}^{\mathrm{J}}_{l,k} &= \theta_{l,k} \text{block\_diag}(\underline{L}_{l,k})^{-1}, \\ \underline{B}^{\mathrm{GS}}_{l,k} &= \theta_{l,k} \big(\text{block\_lower}(\underline{L}_{l,k}) + \text{block\_diag}(\underline{L}_{l,k})\big)^{-1} \end{aligned}$$

with damping parameter $\theta_{l,k} \in (0, 1]$. The corresponding iteration matrices are given by $\underline{S}^{\mathrm{J}}_{l,k} = \underline{\mathrm{Id}}_{l,k} - \underline{B}^{\mathrm{J}}_{l,k} \underline{L}_{l,k}$ and $\underline{S}^{\mathrm{GS}}_{l,k} = \underline{\mathrm{Id}}_{l,k} - \underline{B}^{\mathrm{GS}}_{l,k} \underline{L}_{l,k}$, and the number of pre- and post-smoothing steps are denoted by $\nu^{\mathrm{pre}}_{l,k}$ and $\nu^{\mathrm{post}}_{l,k}$.

Now, the multilevel preconditioner $\underline{B}^{\mathrm{ML}}_{l,k}$ is defined recursively. On the coarse level, we use a parallel direct linear solver $\underline{B}^{\mathrm{ML}}_{0,0} = (\underline{L}_{0,0})^{-1}$, see [26, 27]. Then, we have two options: restricting in time defines $\underline{B}^{\mathrm{ML}}_{l,k}$ by

$$\begin{aligned} \underline{\mathrm{Id}}_{l,k} - \underline{B}^{\mathrm{ML}}_{l,k} \underline{L}_{l,k} &= \big(\underline{\mathrm{Id}}_{l,k} - \underline{B}^{\mathrm{J}}_{l,k} \underline{L}_{l,k}\big)^{\nu^{\mathrm{pre}}_{l,k}} \\ &\quad \cdot \big(\underline{\mathrm{Id}}_{l,k} - \underline{P}^{l,k}_{l,k-1} \underline{B}^{\mathrm{ML}}_{l,k-1} \underline{R}^{l,k}_{l,k-1} \underline{L}_{l,k}\big) \big(\underline{\mathrm{Id}}_{l,k} - \underline{B}^{\mathrm{J}}_{l,k} \underline{L}_{l,k}\big)^{\nu^{\mathrm{post}}_{l,k}} \end{aligned}$$

with Jacobi smoothing, and restricting in space yields

$$\begin{aligned} \underline{\mathrm{Id}}_{l,k} - \underline{B}^{\mathrm{ML}}_{l,k} \underline{L}_{l,k} &= \big(\underline{\mathrm{Id}}_{l,k} - \underline{B}^{\mathrm{GS}}_{l,k} \underline{L}_{l,k}\big)^{\nu^{\mathrm{pre}}_{l,k}} \\ &\quad \cdot \big(\underline{\mathrm{Id}}_{l,k} - \underline{P}^{l,k}_{l-1,k} \underline{B}^{\mathrm{ML}}_{l-1,k} \underline{R}^{l,k}_{l-1,k} \underline{L}_{l,k}\big) \big(\underline{\mathrm{Id}}_{l,k} - \underline{B}^{\mathrm{GS}}_{l,k} \underline{L}_{l,k}\big)^{\nu^{\mathrm{post}}_{l,k}} \end{aligned}$$

with Gauss–Seidel smoothing. Our tests in [8] indicate that it is advantageous to start with refinement in time and then refinement in space. That is, we use the sequence of meshes $\mathcal{R}_{0,0}, \mathcal{R}_{0,1}, \dots, \mathcal{R}_{0,k_{\max}}, \mathcal{R}_{1,k_{\max}}, \dots, \mathcal{R}_{l_{\max},k_{\max}}$ (see Algorithm 2 for the recursive realization of the multilevel preconditioner).

7 Numerical experiments

We illustrate the numerical performance of the space-time method with two examples. The first test is a simple plane wave solution for the acoustic problem, where the

Algorithm 2: Multilevel preconditioner $\underline{c}_{l,k} = \underline{B}^{\mathrm{ML}}_{l,k}\underline{r}_{l,k}$ with Gauss–Seidel smoother $\underline{B}^{\mathrm{SM}}_{l,k} = \underline{B}^{\mathrm{GS}}_{l,k}$ in space for $l > 0$, or Jacobi smoother $\underline{B}^{\mathrm{SM}}_{0,k} = \underline{B}^{\mathrm{J}}_{0,k}$ in time.

$\underline{c}_{l,k} = \underline{0}$
for $\nu = 1, \dots, \nu^{\mathrm{pre}}_{lk}$ **do**
  $\underline{w}_{l,k} = \underline{B}^{\mathrm{SM}}_{l,k}\underline{r}_{l,k}$
  $\underline{c}_{l,k} := \underline{c}_{l,k} + \underline{w}_{l,k}$ and $\underline{r}_{l,k} := \underline{r}_{l,k} - \underline{L}_{l,k}\underline{w}_{l,k}$
end for
$\underline{r}_{l-1,k} = \underline{R}^{l,k}_{l-1,k}\underline{r}_{l,k}$ for $l > 0$ or $\underline{r}_{0,k-1} = \underline{R}^{l,k}_{0,k-1}\underline{r}_{0,k}$
$\underline{c}_{l-1,k} = \underline{B}^{\mathrm{ML}}_{l-1,k}\underline{r}_{l-1,k}$ for $l > 0$ or $\underline{c}_{0,k-1} = \underline{B}^{\mathrm{ML}}_{0,k-1}\underline{r}_{0,k-1}$
$\underline{w}_{l,k} = \underline{P}^{l,k}_{l-1,k}\underline{c}_{l-1,k}$ for $l > 0$ or $\underline{w}_{0,k} = \underline{P}^{l,k}_{0,k-1}\underline{c}_{0,k-1}$
$\underline{c}_{l,k} := \underline{c}_{l,k} + \underline{w}_{l,k}$ and $\underline{r}_{l,k} := \underline{r}_{l,k} - \underline{L}_{l,k}\underline{w}_{l,k}$
for $\nu = 1, \dots, \nu^{\mathrm{post}}_{lk}$ **do**
  $\underline{w}_{l,k} = \underline{B}^{\mathrm{SM}}_{l,k}\underline{r}_{l,k}$
  $\underline{c}_{l,k} := \underline{c}_{l,k} + \underline{w}_{l,k}$ and $\underline{r}_{l,k} := \underline{r}_{l,k} - \underline{L}_{l,k}\underline{w}_{l,k}$
end for

solution is known, so that we can test the convergence properties for uniform h- and p-refinement. The second example is application-oriented and shows the behavior of the p-adaptive algorithm for a configuration motivated from tunnel exploration.

In all cases, the linear systems are solved approximately with a GMRES iteration and the space-time multigrid preconditioner. As general multigrid parameters, we use for coarsening in time a damped block-Jacobi preconditioner ($\theta = 0.5$) with 2 pre- and post-smoothing steps, and for coarsening in space a block-Gauss–Seidel preconditioner with 5 pre- and post-smoothing steps. A V-cycle with coarsening the mesh first in space, and then in time, is applied. All computations use multigrid over three levels in space and in time.

The adaptive refinement starts with a finite volume discretization in space ($p = 0$), and linear ansatz and constant test functions in time on each space-time cell ($q = 1$). The algorithm increases in the first step adaptively; the polynomial degrees in space and later the polynomial degrees in space and time simultaneously. The approximation spaces $V_{l,k}$ are chosen such that the polynomial degrees on each cell is the maximum over all corresponding cells of the fine mesh. For the underlying 2D mesh in space, we use quadrilaterals.

7.1 A benchmark experiment

The first example is specially designed for a convergence test. We use the time interval $(0, T) = (0, 4)$ and the spatial domain $\Omega = (-2, 4) \times (0, 2) \subset \mathbb{R}^2$ with piecewise constant parameters

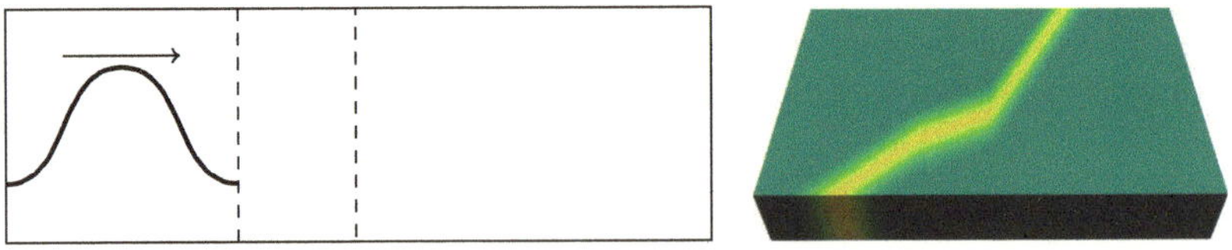

Figure 2.1: Benchmark experiment: The initial wave will travel from the left to the right. Sketch of the impulse (left) and pressure component of the space-time solution (right).

$$\rho(x_1, x_2) = \begin{cases} 1, & x_1 < 0, \\ 2, & 0 < x_1 < 1, \\ 1/2, & 1 < x_1. \end{cases} \quad \text{and} \quad \kappa(\boldsymbol{x}) = 1/\rho(\boldsymbol{x}).$$

Starting with

$$\boldsymbol{u}_0(\boldsymbol{x}) = A(x_1) \begin{pmatrix} 1 \\ 1 \\ 0 \end{pmatrix} \quad \text{for } A(x_1) = \begin{cases} \cos((x_1 - 1)\pi/2)^6, & -2 < x_1 < 0, \\ 0, & \text{else} \end{cases}$$

results in the plane-wave solution, see Figure 2.1, with

$$\boldsymbol{u}(t, x_1, x_2) = \begin{cases} \boldsymbol{u}_0(x_1 - t, x_2), & x_1 \leq 0, \\ \boldsymbol{u}_0(2x_1 - t, x_2), & 0 < x_1 \leq 1, \\ \boldsymbol{u}_0(2 + 0.5(x_1 - 1) - t, x_2), & 1 \leq x_1. \end{cases}$$

The computed experimental orders of convergence are shown in Table 2.3. We observe the expected order of convergence, as predicted in Theorem 4.1, for sufficiently smooth solutions.

7.2 A tunnel experiment

The second example illustrates seismic tunnel exploration: An artificially generated surface wave in the tunnel propagates into the solid, and the reflected waves are measured in a certain region. Here, we compare the results of acoustic and elastic waves.

We choose a rectangular domain $\Omega \subset (-2, 2) \times (-1.5, 2.5) \subset \mathbb{R}^2$, and we use density $\rho = 1$, Lamé parameters $\lambda = 0.5$, and $\mu = 0.25$ for the elastic wave equation. This results in compressional waves with velocity $c_P = \sqrt{(2\mu + \lambda)/\rho} = 1$ and shear waves with velocity $c_S = \sqrt{\mu/\rho} = 0.5$. In the acoustic case, we use the parameters $\rho = \kappa = 1$, so that the velocity of sound $c = \sqrt{\kappa/\rho} = 1$ is equal to the wave propagation speed of the elastic compressional waves.

At $t = 0$, we start with a smooth pulse located at $\boldsymbol{x}_{\text{mid}} = (0.5, 1) \in \partial\Omega$, defining the initial velocity

$$\boldsymbol{v}_0 = \begin{pmatrix} x_1 - 0.5 \\ x_2 - 1.0 \end{pmatrix} \phi$$

Table 2.3: Benchmark experiment: Convergence of the error $\boldsymbol{e}_h = \boldsymbol{u} - \boldsymbol{u}_h$ with respect to the norm $\|\cdot\|_{V_h}$ and extrapolated orders of convergence (EOC) for uniformly refined space-time meshes and different polynomial degrees.

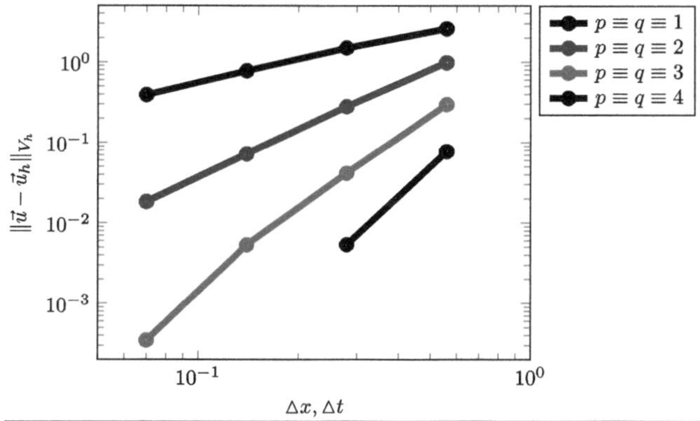

cells	$p = q = 1$		$p = q = 2$		$p = q = 3$		$p = q = 4$	
	DoFs	$\|e_h\|_{V_h}$	**DoFs**	$\|e_h\|_{V_h}$	**DoFs**	$\|e_h\|_{V_h}$	**DoFs**	$\|e_h\|_{V_h}$
1,536	18,432	2.5916	82,944	0.9841	221,184	0.2974	460,800	0.0775
12,288	147,456	1.5041	663,552	0.2796	1,769,472	0.0416	3,686,400	0.0054
98,304	1,179,648	0.7772	5,308,416	0.0722	14,155,776	0.0053	29,491,200	0.00035
786,432	9,437,184	0.3900	42,467,328	0.0182	out of memory		out of memory	
eoc		0.995		1.989		2.961		3.960

with

$$\phi(\boldsymbol{x}) = \begin{cases} \cos^6(2\pi|\boldsymbol{x}_{\mathrm{mid}} - \boldsymbol{x}|^2), & |\boldsymbol{x}_{\mathrm{mid}} - \boldsymbol{x}|^2 < 0.25, \\ 0, & \text{else.} \end{cases}$$

In the acoustic case, we set $p_0 \equiv 0$ and in the elastic case $\boldsymbol{\sigma}_0 \equiv \mathbf{0}$.

In applications, the velocity is measured at certain points within a region of interest RoI; here, we use RoI $= (0.5, 1) \times (0.5, 1)$, see Figure 2.2. Since we are interested in the velocity at the final time $T = 3$, we consider the linear goal functional:

$$E(\boldsymbol{v}) = \frac{1}{|\mathrm{RoI}|} \int_{\mathrm{RoI}\times\{T\}} v_1 \, \mathrm{d}\boldsymbol{x}.$$

The smooth pulse starts at $\boldsymbol{x}_{\mathrm{mid}}$ and expands through the domain. After being reflected at the right boundary, the wave reaches back to the region of interest. The visualization is obtained by slicing through the space-time mesh, see Figure 2.3.

The results for the uniform and adaptive refinement in the acoustic case are given in Table 2.4. We observe that the adaptive algorithm saves over 70 % of the degrees of freedom, whereas achieving the same accuracy compared with uniform refinement.

Comparing the acoustic wave in Figure 2.3 with the results in Figure 2.5 for the elastic wave, we can see the additional shear wave, which propagates with half of

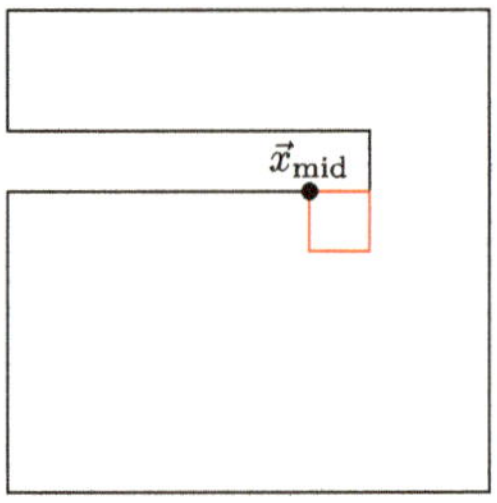

Figure 2.2: Tunnel experiment: Sketch of the computational domain Ω with marked region of interest RoI.

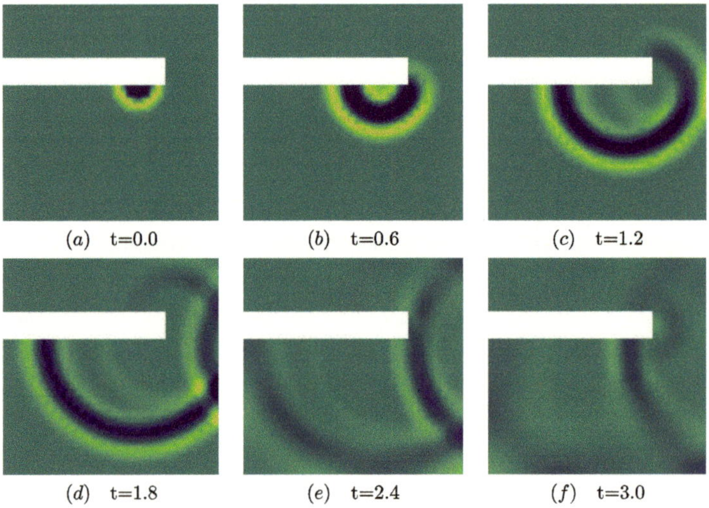

Figure 2.3: Acoustic wave: Slices through the space-time mesh of the pressure component.

Table 2.4: Acoustic wave: Uniform vs. adaptive refinement on 44,544 = 928 × 48 space-time cells distributed on 64 processor cores. The error $\triangle E_{\text{ex}}(\mathbf{u}_h) = |E(\mathbf{u}_h) - E_{\text{ex}}|$ of the goal functional is approximately estimated with respect to a linear extrapolation of the uniform results $E_{\text{ex}} = 4.9052 \times 10^{-3}$.

ref-step	(p,q)	DoFs (effort)	GMRES Iter with MG-PC	$E(\mathbf{u}_h)$	$\triangle E_{\text{ex}}(\mathbf{u}_h)$
uniform refinement					
$r=1$	(1, 1)	534,528	7	4.9961×10^{-3}	9.09×10^{-5}
$r=2$	(2, 2)	2,405,376	13	4.8946×10^{-3}	1.06×10^{-5}
$r=3$	(3, 3)	6,414,336	19	4.8810×10^{-3}	2.42×10^{-5}
$r=4$	(4, 4)	13,363,200	27	4.8931×10^{-3}	1.21×10^{-5}
adaptive refinement					
$r=0$	(0, 1)	133,632	5	4.4104×10^{-4}	4.46×10^{-3}
$r=1$		291,411 (55 %)	7	4.9677×10^{-3}	6.25×10^{-5}
$r=2$		819,279 (34 %)	13	4.8767×10^{-3}	2.85×10^{-5}
$r=3$		1,875,753 (29 %)	20	4.8779×10^{-3}	2.73×10^{-5}
$r=4$		3,594,969 (27 %)	28	4.8866×10^{-3}	1.86×10^{-5}

Table 2.5: Elastic wave: Uniform vs. adaptive refinement on 44,544 = 928 × 48 space-time cells distributed to 64 processor cores (for uniform computations $p = q \leq 3$ due to memory restrictions). The error of the goal functional is approximately estimated with respect to $E_{ex} = 1.9057 \times 10^{-3}$.

ref–step	(p, q)	DoFs (effort)	GMRES Iter with MG–PC	$E(u_h)$	$\triangle E_{ex}$
uniform refinement					
$r = 1$	$(1, 1)$	890,880	5	1.3625×10^{-3}	5.43×10^{-4}
$r = 2$	$(2, 2)$	4,008,960	7	1.7686×10^{-3}	1.37×10^{-4}
$r = 3$	$(3, 3)$	10,690,560	8	1.8371×10^{-3}	6.86×10^{-5}
adaptive refinement					
$r = 0$	$(0, 1)$	222,720	5	4.7218×10^{-4}	1.43×10^{-3}
$r = 1$		551,370 (62 %)	6	1.1362×10^{-3}	5.44×10^{-4}
$r = 2$		1,477,655 (37 %)	7	1.7687×10^{-3}	1.37×10^{-4}
$r = 3$		3,379,390 (32 %)	9	1.8371×10^{-3}	6.86×10^{-5}
$r = 4$		6,510,765 (29 %)	11	1.8723×10^{-3}	3.34×10^{-5}

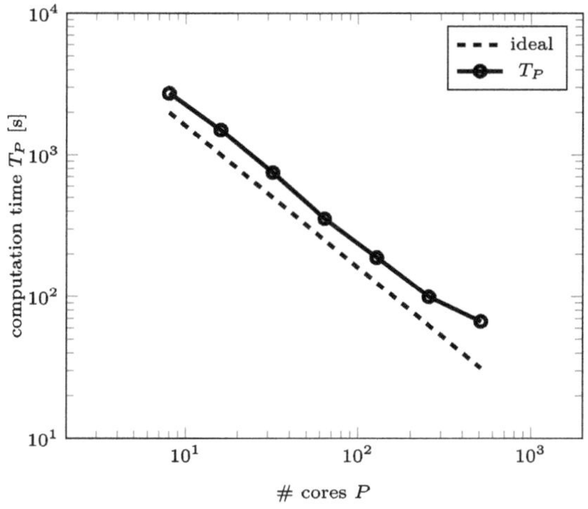

Figure 2.4: Tunnel experiment: Strong scaling for ~ 34 Mio. DoFs (acoustic wave).

the velocity behind the compressional wave. The acoustic wave equation in 2D has three components, and the elastic wave equation has five components. This results in more DoF and thus in larger matrices. To save random access memory, in this case, we use as approximation spaces $V_{l,k}$ on the coarser meshes a lowest-order finite volume discretization. The results for uniform and adaptive refinement in the elastic case are shown in Table 2.5 and illustrated in Figure 2.5, which demonstrates the excellent efficiency of the adaptive scheme.

The parallel scaling behavior of the parallel multilevel preconditioner is tested for different numbers of processes. On mesh level 4, we have 2,850,816 space-time cells, a linear discretization in space and time results in 34,209,792 DoFs for the acoustic case.

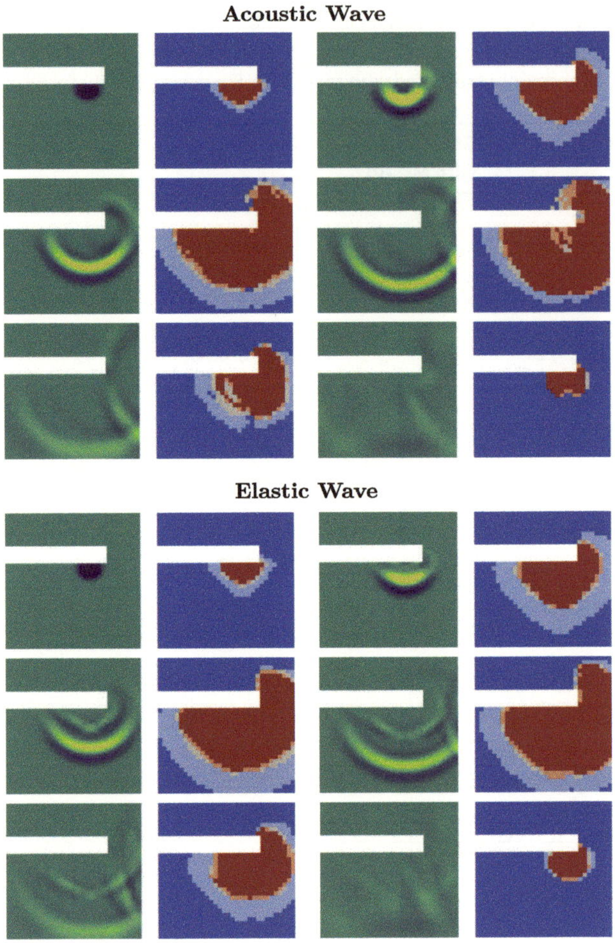

Figure 2.5: Acoustic and elastic waves: Velocity component v_2 and adaptive distribution of polynomial degrees.

The computing time for solving this huge linear system with the parallel multigrid method scales nearly optimal,[1] see Figure 2.4.

Bibliography

[1] R. Andreev and C. Tobler. Multilevel preconditioning and low-rank tensor iteration for space-time simultaneous discretizations of parabolic PDEs. *Numer. Linear Algebra Appl.*, 22 (2): 317–337, 2015.

1 http://www.scc.kit.edu/dienste/bwUniCluster.php (last access June 18, 2018).

[2] U. M. Ascher, S. J. Ruuth, and B. T. R. Wetton. Implicit-explicit methods for time-dependent partial differential equations. *SIAM J. Numer. Anal.*, 32 (3): 797–823, 1995.

[3] W. Bangerth and R. Rannacher. Finite element approximation of the acoustic wave equation: Error control and mesh adaptation. *East-West J. Numer. Math.*, 7 (4): 263–282, 1999.

[4] W. Bangerth and R. Rannacher. *Adaptive finite element methods for differential equations.* Birkhäuser, Basel, 2003.

[5] C. Cong, X.-C. Cai, and K. Gustafson. Implicit space-time domain decomposition methods for stochastic parabolic partial differential equations. *SIAM J. Sci. Comput.*, 36 (1): C1–C24, 2014.

[6] L. F. Demkowicz and J. Gopalakrishnan. An overview of the discontinuous Petrov–Galerkin method. In *Recent developments in discontinuous Galerkin finite element methods for partial differential equations*, volume 157 of IMA Vol. Math. Appl., pages 149–180. Springer, Cham, 2014.

[7] L. F. Demkowicz, J. Gopalakrishnan, S. Nagaraj, and P. Sepulveda. A spacetime DPG method for the Schrödinger equation. *SIAM J. Numer. Anal.*, 55 (4): 1740–1759, 2017.

[8] W. Dörfler, S. Findeisen, and C. Wieners. Space-time discontinuous Galerkin discretizations for linear first-order hyperbolic evolution systems. *Comput. Meth. Appl. Math.*, 16 (3): 409–428, 2016.

[9] M. Dumbser, M. Käser, and E. F. Toro. An arbitrary high-order discontinuous Galerkin method for elastic waves on unstructured meshes – V. Local time stepping and p-adaptivity. *Geophys. J. Int.*, 171: 695–717, 2007.

[10] H. Egger, F. Kretzschmar, S. M. Schnepp, and T. Weiland. A space-time discontinuous Galerkin–Trefftz method for time dependent Maxwell's equations. *SIAM J. Sci. Comp.*, 37 (5): B689–B711, 2015.

[11] T. E. Ellis, L. F. Demkowicz, J. L. Chan, and R. D. Moser. Space-time DPG: Designing a method for massively parallel CFD. ICES Report 14–32, The University of Texas at Austin, 2014.

[12] K.-J. Engel and R. Nagel. *A short course on operator semigroups*, Universitext. Springer, New York, 2006.

[13] J. Ernesti. *Space-time methods for acoustic waves with applications to full waveform inversion.* PhD thesis, Karlsruher Institut für Technologie (KIT), 2018.

[14] R. D. Falgout, S. Friedhoff, T. V. Kolev, S. P. MacLachlan, and J. B. Schroder. Parallel time integration with multigrid. *SIAM J. Sci. Comput.*, 36 (6): C635–C661, 2014.

[15] M. J. Gander. 50 years of time parallel time integration. In T. Carraro, M. Geiger, S. Körkel, and R. Rannacher, editors, *Multiple shooting and time domain decomposition*. Springer, 2015.

[16] M. J. Gander and M. Neumüller. Analysis of a new space-time parallel multigrid algorithm for parabolic problems. *SIAM J. Sci. Comput.*, 38 (4): A2173–A2208, 2016.

[17] M. J. Gander and S. Vandewalle. Analysis of the parareal time-parallel time-integration method. *SIAM J. Sci. Comput.*, 29 (2): 556–578, 2007.

[18] M. J. Gander, L. Halpern, and F. Nataf. Optimal Schwarz waveform relaxation for the one dimensional wave equation. *SIAM J. Numer. Anal.*, 41 (5): 1643–1681, 2003.

[19] J. S. Hesthaven and T. Warburton. *Nodal discontinuous Galerkin methods*. Springer, 2008.

[20] M. Hochbruck, T. Pazur, A. Schulz, E. Thawinan, and C. Wieners. Efficient time integration for discontinuous Galerkin approximations of linear wave equations. *ZAMM*, 95 (3): 237–259, 2015.

[21] S. Kestler, K. Steih, and K. Urban. An efficient space-time adaptive wavelet Galerkin method for time-periodic parabolic partial differential equations. *Math. Comp.*, 85 (299): 1309–1333, 2016.

[22] U. Köcher and M. Bause. Variational space-time methods for the wave equation. *J. Sci. Comput.*, 61 (2): 424–453, 2014.

[23] F. Kretzschmar, A. Moiola, I. Perugia, and S. M. Schnepp. A priori error analysis of space-time

Trefftz discontinuous Galerkin methods for wave problems. *IMA J. Numer. Anal.*, 1599–1635, 2015.
[24] U. Langer, S. E. Moore, and M. Neumüller. Space-time isogeometric analysis of parabolic evolution problems. *Comp. Meth. Appl. Mech. Engrg.*, 306: 342–363, 2016.
[25] J.–L. Lions, Y. Maday, and G. Turinici. A "parareal" in time discretization of PDE's. *C. R. Acad. Sci. Paris, Ser. I*, 332 (7): 661–668, 2001.
[26] D. Maurer and C. Wieners. A parallel block LU decomposition method for distributed finite element matrices. *Parallel Comput.*, 37 (12): 742–758, 2011.
[27] D. Maurer and C. Wieners. A scalable parallel factorization of finite element matrices with distributed Schur complements. *Numer. Linear Algebra Appl.*, 23 (5): 848–864, 2016.
[28] M. Neumüller and O. Steinbach. Refinement of flexible space-time finite element meshes and discontinuous Galerkin methods. *Comput. Vis. Sci.*, 14 (5): 189–205, 2011.
[29] M. Neumüller and O. Steinbach. A DG space-time domain decomposition method. In *Domain Decomposition Methods in Science and Engineering XX*, pages 623–630. Springer, 2013.
[30] N. C. Nguyen, J. Peraire, and B. High-order implicit hybridizable discontinuous Galerkin methods for acoustics and elastodynamics. *J. Comput. Phys.*, 230 (10): 3695–3718, 2011.
[31] S. Petersen, C. Farhat, and R. Tezaur. A space-time discontinuous Galerkin method for the solution of the wave equation in the time domain. *Int. J. Numer. Meth. Engng.*, 78 (3): 275–295, 2009.
[32] M. Renardy and R. C. Rogers. *An introduction to partial differential equations. 2nd ed.* New York, Springer, 2004.
[33] S. Rhebergen, B. Cockburn, and J. J. W. van der Vegt. A space-time discontinuous Galerkin method for the incompressible Navier–Stokes equations. *J. Comput. Phys.*, 233: 339–358, 2013.
[34] C. Schwab and R. Stevenson. Space-time adaptive wavelet methods for parabolic evolution problems. *Math. Comp.*, 78 (267): 1293–1318, 2009.
[35] R. Speck, D. Ruprecht, M. Emmett, M. Bolten, and R. Krause. A space-time parallel solver for the three-dimensional heat equation. In M. Bader, A. Bode, H.-J. Bungartz, M. Gerndt, G. R. Joubert, and F. J. Peters, editors, *Parallel computing: Accelerating computational science and engineering (CSE)*, volume 25 of *Advances in parallel computing*, pages 263–272. IOS Press, 2014.
[36] O. Steinbach. Space-time finite element methods for parabolic problems. *Comput. Meth. Appl. Math.*, 15 (4): 551–566, 2015.
[37] O. Steinbach and H. Yang. Comparison of algebraic multigrid methods for an adaptive space-time finite-element discretization of the heat equation in 3d and 4d. *Numer. Linear Algebra Appl.*, 25 (3): e2143, 2018.
[38] J. J. W. van der Vegt and S. Rhebergen. *hp*-multigrid as smoother algorithm for higher order discontinuous Galerkin discretizations of advection dominated flows: Part I. Multilevel analysis. *J. Comput. Phys.*, 231 (22): 7537–7563, 2012.
[39] D. Wang, R. Tezaur, and C. Farhat. A hybrid discontinuous in space and time Galerkin method for wave propagation problems. *Int. J. Numer. Meth. Engng.*, 99 (4): 263–289, 2014.

Johannes Ernesti and Christian Wieners

3 A space-time discontinuous Petrov–Galerkin method for acoustic waves

Abstract: We apply the discontinuous Petrov–Galerkin (DPG) method to linear acoustic waves in space and time using the framework of first-order Friedrichs systems. Based on results for operators and semigroups of hyperbolic systems, we show that the ideal DPG method is well-posed. The main task is to avoid the explicit use of traces, which are difficult to define in Hilbert spaces with respect to the graph norm of the space-time differential operator. Then, the practical DPG method is analyzed by constructing a Fortin operator numerically.

For our numerical experiments, we introduce a simplified DPG method with discontinuous ansatz functions on the faces of the space-time skeleton, where the error is bounded by an equivalent conforming DPG method. Examples for a plane-wave configuration confirms the numerical analysis, and the computation of a diffraction pattern illustrates a first step to applications.

Keywords: Discontinuous Petrov–Galerkin method, space-time discretizations, semigroups, variational space-time Hilbert spaces

MSC 2010: 65M60, 65M15, 65M55

1 Introduction

The discontinuous Petrov–Galerkin method (DPG), introduced by Demkowicz et al., provides a very flexible framework to construct and to analyze stable finite element discretizations for general linear first-order systems, see [5] for an overview and many references.

The main idea of the DPG method is to introduce a substructuring, and to use discontinuous approximations in the subdomains and traces on the skeleton. This is combined with discontinuous test functions, so that the discrete solution can be obtained by a symmetric linear system for the skeleton values.

The DPG method can be introduced as a minimal residual method, which allows for an equivalent saddle-point formulation. So the main objective for the numerical

Acknowledgement: This work was supported by the German Research Foundation (DFG) by CRC 1173. We thank Luca Hornung, Dietmar Gallistl, Jay Gopalakrishnan, and Paulina Sepúlveda for helpful discussions on first drafts of the manuscript.

Johannes Ernesti, Christian Wieners, Institut für Angewandte und Numerische Mathematik, Karlsruher Institut für Technologie, 76128 Karlsruhe, Germany, e-mails: Johannes.Ernesti@kit.edu, Christian.Wieners@kit.edu

https://doi.org/10.1515/9783110548488-003

analysis of the DPG method is to provide the corresponding inf-sup stability. This involves two steps. Firstly, in the ideal DPG method, stability has to be provided with respect to the dual norm of the residual, that is, by testing with a full Hilbert space. Secondly, the explicit construction of a Fortin operator allows analyzing the practical DPG method with discrete test functions.

Here, we discuss the application of the DPG method to linear first-order systems in space and time, where we consider, as reference example, linear acoustic waves. The required stability for the space-time operator is obtained within a semigroup approach, which also provides an estimate of Poincaré type. To establish a suitable Hilbert space setting for the closure of the operator in space and time, we use the results in [19, Section 12] for semigroups, in [20, Section 4.5] for polar sets, and the framework for symmetric Friedrichs systems in [10].

Then, following the DPG analysis in [6, 14], we show that the ideal and the practical DPG method only rely on the boundary operator, using integration by parts, without explicit reference to traces, see also [10]. Then, the construction of a Fortin operator follows the approach in [17].

In our realization of the DPG method, we use a simplified approach. Since the traces of space-time cells are different for faces in time and in space, conforming ansatz spaces on the skeleton may require nodal points on faces, edges, and vertices. It turns out that discontinuous ansatz functions on the faces of the space-time skeleton are easier to construct and yield optimal convergence rates in simple test scenarios. This variational crime is analyzed with respect to a discrete norm by comparing the simplified method with an equivalent conforming DPG method.

The method is implemented within the parallel finite element software system [21]. We test the full space-time approach by computing the diffraction pattern of a double-slit experiment, which demonstrates the advantages of a method, which is simultaneously parallel in space *and* time.

2 Linear acoustic waves

We consider the first-order system for linear acoustics,

$$\kappa^{-1}\partial_t p + \nabla \cdot \mathbf{v} = 0, \tag{2.1a}$$

$$\rho \partial_t \mathbf{v} + \nabla p = \mathbf{0}, \tag{2.1b}$$

in the space-time cylinder

$$Q = \Omega \times (0, T) \subset \mathbb{R}^d \times \mathbb{R},$$

depending on a density distribution $\rho > 0$ and bulk modulus $\kappa > 0$ (see [8, Section 2] for more details on this model and the relation to elastic waves). For simplicity of the presentation, we set $\rho = 1$, and $\kappa = 1$.

The corresponding first-order differential operator is given by

$$L(p, \mathbf{v}) = (\partial_t p + \nabla \cdot \mathbf{v}, \partial_t \mathbf{v} + \nabla p).$$

Now we want to establish an analytic setting for a unique solution of

$$L(p, \mathbf{v}) = (f, \mathbf{g}) \tag{2.2}$$

(subject to initial and boundary conditions), which depends continuously on the data.

2.1 The semigroup setting

We consider the ODE

$$\partial_t(p, \mathbf{v}) = A(p, \mathbf{v}) + (f, \mathbf{g}), \quad A(p, \mathbf{v}) = (-\nabla \cdot \mathbf{v}, -\nabla p),$$

where the operator A is associated with a dense domain

$$\mathcal{D}(A) \subset \mathrm{L}_2(\Omega; \mathbb{R} \times \mathbb{R}^d).$$

Here, we choose $\mathcal{D}(A) = \mathrm{H}^1_0(\Omega) \times \mathrm{H}(\mathrm{div}, \Omega)$, including homogeneous Dirichlet boundary conditions for the pressure on $\partial\Omega$.

We show that the operator A with domain $\mathcal{D}(A)$ generates a semigroup. Therefore, we check the requirements of the Lumer–Phillips theorem. In the first step, we show that $\mathrm{id} - A$ is surjective. For $(f, \mathbf{g}) \in \mathrm{L}_2(\Omega; \mathbb{R} \times \mathbb{R}^d)$, we define $p \in \mathrm{H}^1_0(\Omega)$, solving

$$(\nabla p, \nabla q)_\Omega + (p, q)_\Omega = (f, q)_\Omega + (\mathbf{g}, \nabla q)_\Omega, \quad q \in \mathrm{H}^1_0(\Omega),$$

and then we define $\mathbf{v} = \mathbf{g} - \nabla p$. We observe

$$(\mathbf{v}, -\nabla q)_\Omega = (f, q)_\Omega - (p, q)_\Omega, \quad q \in \mathrm{C}^1_\mathrm{c}(\Omega);$$

that is, $\mathbf{v} \in \mathrm{H}(\mathrm{div}, \Omega)$ and $\nabla \cdot \mathbf{v} = f - p$, so that together $(p, \mathbf{v}) - A(p, \mathbf{v}) = (f, \mathbf{g})$. This gives surjectivity. Moreover, we have

$$\big(A(p, \mathbf{v}), (p, \mathbf{v})\big)_\Omega = 0, \quad (p, \mathbf{v}) \in \mathcal{D}(A). \tag{2.3}$$

Thus, the operator A generates a semigroup [19, Theorem 12.22] (see also [15, Section 2.2] and [16] for the application to general linear wave equations).

2.2 Duality, adjoint operators, and the Hilbert adjoint

In the next section, many arguments will rely on duality. For this purpose, we introduce the Hilbert adjoint A^{ad} of the operator A with domain $\mathcal{D}(A^{\mathrm{ad}})$, see [19, Section 8.4.2].

The adjoint operator is defined in the domain

$$\mathcal{D}(A^{\mathrm{ad}}) = \{(q, \boldsymbol{w}) \in \mathrm{L}_2(\Omega; \mathbb{R} \times \mathbb{R}^d) \colon (f, \boldsymbol{g}) \in \mathrm{L}_2(\Omega; \mathbb{R} \times \mathbb{R}^d) \text{ exists}$$
$$\text{such that } ((f, \boldsymbol{g}), (p, \boldsymbol{v}))_\Omega = ((q, \boldsymbol{w}), A(p, \boldsymbol{v}))_\Omega \text{ for } (p, \boldsymbol{v}) \in \mathcal{D}(A)\}.$$

For the acoustic wave equation, we have $\mathcal{D}(A^{\mathrm{ad}}) = \mathrm{H}_0^1(\Omega) \times \mathrm{H}(\mathrm{div}, \Omega) = \mathcal{D}(A)$. Then, for $(q, \boldsymbol{w}) \in \mathcal{D}(A^{\mathrm{ad}})$, we define $A^{\mathrm{ad}}(q, \boldsymbol{w}) \in \mathrm{L}_2(\Omega; \mathbb{R} \times \mathbb{R}^d)$ by

$$(A^{\mathrm{ad}}(q, \boldsymbol{w}), (p, \boldsymbol{v}))_\Omega = ((q, \boldsymbol{w}), A(p, \boldsymbol{v}))_\Omega, \quad (p, \boldsymbol{v}) \in \mathcal{D}(A).$$

Since $\mathcal{D}(A) \subset \mathrm{L}_2(\Omega; \mathbb{R} \times \mathbb{R}^d)$ is dense, the operator A^{ad} is well-defined.

Correspondingly, for the space-time operator $L = \partial_t - A$, the formal adjoint of the differential operator is given by $L^{\mathrm{ad}} = -\partial_t - A^{\mathrm{ad}}$, and we obtain in $Q = \Omega \times (0, T)$

$$(L^{\mathrm{ad}}(q, \boldsymbol{w}), (p, \boldsymbol{v}))_Q = ((q, \boldsymbol{w}), L(p, \boldsymbol{v}))_Q, \quad (p, \boldsymbol{v}), (q, \boldsymbol{w}) \in \mathrm{C}_c^1(Q; \mathbb{R} \times \mathbb{R}^d).$$

In our application, the adjoint problem describes a wave equation backward in time.

In the next section, we will define suitable domains for the operators L and L^{ad}, extending the domains $\mathcal{D}(A)$ and $\mathcal{D}(A^{\mathrm{ad}})$ in $\mathrm{L}_2(\Omega; \mathbb{R} \times \mathbb{R}^d)$ to domains of the space-time operators in $\mathrm{L}_2(Q; \mathbb{R} \times \mathbb{R}^d)$, so that L^{ad} is the Hilbert adjoint of L in this setting.

2.3 Polar sets

Below, we use polar sets in Hilbert spaces X. Let X' be the topological dual of X. For $Y \subset X$ and $Z \subset X'$, the corresponding annihilator or polar sets are given by

$$Y^\perp = \{\ell \in X' \colon \langle \ell, \boldsymbol{y} \rangle = 0, \boldsymbol{y} \in Y\},$$
$${}^\perp Z = \{\boldsymbol{z} \in X \colon \langle \ell, \boldsymbol{z} \rangle = 0, \ell \in Z\},$$

see [20, Section 4.5]. In particular, ${}^\perp(Y^\perp)$ is the closure of Y in X, see [20, Theorem 4.7].

3 A variational space-time setting

We consider the ODE

$$\partial_t \boldsymbol{y} = A\boldsymbol{y} + \boldsymbol{b} \quad \text{in } [0, T], \quad \boldsymbol{y}(0) = \boldsymbol{0}, \tag{3.1}$$

where A is an operator with a dense domain $\mathcal{D}(A)$ in $Y = \mathrm{L}_2(\Omega; \mathbb{R}^m)$. We assume that the operator A generates a semigroup. Then, for all

$$\boldsymbol{b} \in \mathrm{C}^0([0, T]; \mathcal{D}(A))$$

a solution $\boldsymbol{y} \in \mathrm{C}^1([0,T];Y) \cap \mathrm{C}^0([0,T];\mathcal{D}(A))$ of (3.1) exists and is of the form

$$\boldsymbol{y}(t) = \int_0^t \exp((t-s)A)\boldsymbol{b}(s)\,\mathrm{d}s. \tag{3.2}$$

This extends to right-hand sides in

$$\mathrm{W}^{1,1}((0,T);Y) = \Big\{ v \in \mathrm{L}_1((0,T);Y) : \exists f_v \in \mathrm{L}_1((0,T);Y) : \int_0^T \varphi f_v \,\mathrm{d}t = -\int_0^T \varphi' v \,\mathrm{d}t \in Y, \varphi \in \mathrm{C}^1_\mathrm{c}(0,T) \Big\}$$

[19, Theorem 12.16], [1, Definition II.5.7]. Equation (3.2) directly implies

$$\|\boldsymbol{y}(t)\|_\Omega \le \int_0^t \|\exp((t-s)A)\|_\Omega \|\boldsymbol{b}(s)\|_\Omega \,\mathrm{d}s.$$

In case of hyperbolic operators satisfying (2.3), we have $\|\exp(A)\|_\Omega = 1$, see, for example, [19, Theorem 12.22]. Then,

$$\|\boldsymbol{y}(t)\|_\Omega \le \int_0^t \|\boldsymbol{b}(s)\|_\Omega \,\mathrm{d}s,$$

and integration in time yields

$$\begin{aligned}\|\boldsymbol{y}\|_{\Omega\times(0,T)} &\le \left(\int_0^T \left(\int_0^t \|\boldsymbol{b}(s)\|_\Omega \,\mathrm{d}s \right)^2 \mathrm{d}t \right)^{1/2} \le \left(\int_0^T t\,\|\boldsymbol{b}\|^2_{\Omega\times(0,t)} \mathrm{d}t \right)^{1/2} \\ &\le \left(\int_0^T t\,\mathrm{d}t \|\boldsymbol{b}\|^2_{\Omega\times(0,T)} \right)^{1/2} = \frac{T}{\sqrt{2}} \|\boldsymbol{b}\|_{\Omega\times(0,T)}. \end{aligned} \tag{3.3}$$

The ODE solution (3.1) belongs to the Banach space

$$\mathcal{V} = \{\boldsymbol{y} \in \mathrm{C}^1([0,T];Y) \cap \mathrm{C}^0([0,T];\mathcal{D}(A)) : \boldsymbol{y}(0) = \boldsymbol{0}\},$$

and inserting the operator $L = \partial_t - A$, we obtain for all $\boldsymbol{b} \in \mathrm{W}^{1,1}((0,T);Y)$ a solution $\boldsymbol{y} \in \mathcal{V}$ with $L\boldsymbol{y} = \boldsymbol{b}$, see [19, Theorem 12.16]. Note that L is not a closed operator in $\mathcal{V}$.

Since $\mathrm{W}^{1,1}((0,T);Y)$ is dense in $\mathrm{L}_2((0,T);Y)$, we obtain the following result:

Lemma 3.1. *$L(\mathcal{V})$ is dense in $\mathrm{L}_2((0,T);Y)$.*

In [6], a corresponding density result is obtained for the linear Schrödinger equation (see also [14] for acoustic waves).

In our application also, the adjoint operator A^{ad} generates a semigroup, so that this result transfers to the adjoint problem, given by the ODE backward in time

$$-\partial_t \boldsymbol{z} = A^{\mathrm{ad}}\boldsymbol{z} + \boldsymbol{c} \quad \text{in } [0,T], \quad \boldsymbol{z}(T) = \mathbf{0}. \tag{3.4}$$

Thus, for $\boldsymbol{c} \in \mathrm{W}^{1,1}((0,T);Y)$, the solution of $L^{\mathrm{ad}}\boldsymbol{z} = \boldsymbol{c}$ is given by

$$\boldsymbol{z}(t) = \int_t^T \exp((s-t)A^{\mathrm{ad}})\boldsymbol{c}(s)\,\mathrm{d}s.$$

Defining

$$\mathcal{V}^{\mathrm{ad}} = \{\boldsymbol{z} \in \mathrm{C}^1([0,T];Y) \cap \mathrm{C}^0([0,T];\mathcal{D}(A^{\mathrm{ad}})) : \boldsymbol{z}(T) = \mathbf{0}\}$$

shows that $L^{\mathrm{ad}}(\mathcal{V}^{\mathrm{ad}})$ is dense in $\mathrm{L}_2((0,T);Y)$, and we have

$$(L^{\mathrm{ad}}(q,\boldsymbol{w}),(p,\boldsymbol{v}))_Q = ((q,\boldsymbol{w}),L(p,\boldsymbol{v}))_Q, \quad (p,\boldsymbol{v}) \in \mathcal{V}, (q,\boldsymbol{w}) \in \mathcal{V}^{\mathrm{ad}}.$$

3.1 A Hilbert space setting for the space-time operator

In $W = \mathrm{L}_2((0,T);Y) = \mathrm{L}_2(Q;\mathbb{R}^m)$, we define the Hilbert space:

$$\begin{aligned} \mathrm{H}(L,Q) &= \{\boldsymbol{y} \in W : L\boldsymbol{y} \in W\} \\ &= \{\boldsymbol{y} \in W : \boldsymbol{b} \in W \text{ exists such that} \\ &\qquad (\boldsymbol{b},\boldsymbol{z})_Q = (\boldsymbol{y}, L^{\mathrm{ad}}\boldsymbol{z})_Q \text{ for } \boldsymbol{z} \in \mathrm{C}^1_{\mathrm{c}}(Q;\mathbb{R}^m)\} \end{aligned} \tag{3.5}$$

with respect to the graph norm

$$\|\boldsymbol{y}\|_{Q,L} = \sqrt{\|\boldsymbol{y}\|_Q^2 + \|L\boldsymbol{y}\|_Q^2}, \quad \boldsymbol{y} \in \mathrm{H}(L,Q).$$

Analogously, we define $\mathrm{H}(L^{\mathrm{ad}},Q) = \{\boldsymbol{y} \in W : L^{\mathrm{ad}}\boldsymbol{y} \in W\}$, and $\mathrm{H}(L^{\mathrm{ad}},Q)'$ denotes its dual space. We define the operator $D \in \mathcal{L}(\mathrm{H}(L,Q),\mathrm{H}(L^{\mathrm{ad}},Q)')$ by

$$\langle D\boldsymbol{y},\boldsymbol{z}\rangle = (L\boldsymbol{y},\boldsymbol{z})_Q - (\boldsymbol{y},L^{\mathrm{ad}}\boldsymbol{z})_Q, \quad \boldsymbol{y} \in \mathrm{H}(L,Q), \boldsymbol{z} \in \mathrm{H}(L^{\mathrm{ad}},Q),$$

and the kernel of D is denoted by

$$\mathcal{N}(D) = \{\boldsymbol{y} \in \mathrm{H}(L,Q) : D\boldsymbol{y} = \mathbf{0}\}.$$

By definition of the adjoint operator L^{ad}, we have $C^1_{\mathrm{c}}(Q;\mathbb{R}^m) \subset \mathcal{N}(D)$. Thus, the operator D describes traces obtained using integration by parts in abstract form.

Let $\mathrm{H}_0(L,Q) \subset \mathrm{H}(L,Q)$ be the closure of $C^1_c(Q;\mathbb{R}^m) \subset \mathcal{N}(D)$. Then, also $\mathrm{H}_0(L,Q) \subset \mathcal{N}(D)$.

In fact, we can establish equality. The proof is based on duality using the operator $D^{\mathrm{ad}} \in \mathcal{L}(\mathrm{H}(L^{\mathrm{ad}},Q),\mathrm{H}(L,Q)')$ with

$$\langle D^{\mathrm{ad}}\boldsymbol{z},\boldsymbol{y}\rangle = (L^{\mathrm{ad}}\boldsymbol{z},\boldsymbol{y})_Q - (\boldsymbol{z},L\boldsymbol{y})_Q;$$

that is, $-D^{\mathrm{ad}}$ is the transposed operator of D.

Theorem 3.1. *We have*

$$\mathrm{H}_0(L,Q) = \mathcal{N}(D).$$

Proof. We only have to show $\mathcal{N}(D) \subset \mathrm{H}_0(L,Q)$. Provided we have established $\mathrm{C}^1_c(Q;\mathbb{R}^m)^\perp \subset D^{\mathrm{ad}}(\mathrm{H}(L^{\mathrm{ad}},Q))$, the assertion follows from

$$\begin{aligned}\mathcal{N}(D) &= \{\boldsymbol{y} \in \mathrm{H}(L,Q) : \langle D\boldsymbol{y},\boldsymbol{z}\rangle = 0 = \langle D^{\mathrm{ad}}\boldsymbol{z},\boldsymbol{y}\rangle \text{ for } \boldsymbol{z} \in \mathrm{H}(L^{\mathrm{ad}},Q)\}\\ &= {}^\perp D^{\mathrm{ad}}(\mathrm{H}(L^{\mathrm{ad}},Q)) \subset {}^\perp(\mathrm{C}^1_c(Q;\mathbb{R}^m)^\perp) = \mathrm{H}_0(L,Q).\end{aligned}$$

The proof uses the technique in [10, Lemma 2.4], see also [4, Lemma 2.2] and [22, Lemma 1]. For a given functional $\ell \in \mathrm{C}^1_c(Q;\mathbb{R}^m)^\perp \subset \mathrm{H}(L,Q)'$ we construct $\boldsymbol{z} \in \mathrm{H}(L^{\mathrm{ad}},Q)$ with $D^{\mathrm{ad}}\boldsymbol{z} = \ell$. Therefore, we define $\boldsymbol{y} \in \mathrm{H}(L,Q)$ solving

$$(L\boldsymbol{y},L\boldsymbol{\phi})_Q + (\boldsymbol{y},\boldsymbol{\phi})_Q = -\langle \ell,\boldsymbol{\phi}\rangle, \quad \boldsymbol{\phi} \in \mathrm{H}(L,Q). \tag{3.6}$$

Then, since $\langle \ell,\boldsymbol{w}\rangle = 0$ for test functions $\boldsymbol{w} \in \mathrm{C}^1_c(Q;\mathbb{R}^m)$, we observe

$$(\boldsymbol{y},\boldsymbol{\phi})_Q = -(L\boldsymbol{y},L\boldsymbol{\phi})_Q, \quad \boldsymbol{\phi} \in \mathrm{C}^1_c(Q;\mathbb{R}^m).$$

Inserting $\boldsymbol{z} = L\boldsymbol{y}$ and using the definition of $\mathrm{H}(L^{\mathrm{ad}},Q)$, we observe $\boldsymbol{z} \in \mathrm{H}(L^{\mathrm{ad}},Q)$ and $L^{\mathrm{ad}}\boldsymbol{z} = -\boldsymbol{y}$. From (3.6) we now obtain

$$\begin{aligned}\langle D^{\mathrm{ad}}\boldsymbol{z},\boldsymbol{\phi}\rangle &= (L^{\mathrm{ad}}\boldsymbol{z},\boldsymbol{\phi})_Q - (\boldsymbol{z},L\boldsymbol{\phi})_Q\\ &= -(\boldsymbol{y},\boldsymbol{\phi})_Q - (L\boldsymbol{y},L\boldsymbol{\phi})_Q = \langle \ell,\boldsymbol{\phi}\rangle, \quad \boldsymbol{\phi} \in \mathrm{H}(L,Q),\end{aligned}$$

i. e., $D^{\mathrm{ad}}\boldsymbol{z} = \ell$. □

3.2 The closure of the space-time operator $(L,\mathcal{V})$

We assume that $C_L > 0$ exists such that

$$\|\boldsymbol{y}\|_Q \le C_L\|L\boldsymbol{y}\|_Q, \quad \boldsymbol{y} \in \mathcal{V}. \tag{3.7}$$

In case of linear hyperbolic operators, this is obtained from (3.3) with $C_L = \frac{1}{\sqrt{2}}T$, see also [18, Theorem 3.1], [7, Lemma 1], and [23, Lemma 6].

In particular, L is injective on $\mathcal{V}$. Now we define

$$V = {}^{\perp}(\mathcal{V}^{\perp}) \subset \mathrm{H}(L, Q).$$

That is, V is the closure of $\mathcal{V}$ in $\mathrm{H}(L, Q)$ with respect to the graph norm [20, Theorem 4.7]. The estimate (3.7) also holds for the closure; that is,

$$\|\boldsymbol{y}\|_Q \le C_L \|L\boldsymbol{y}\|_Q, \quad \boldsymbol{y} \in V. \tag{3.8}$$

Theorem 3.2. *$L \in \mathcal{L}(V, W)$ is a bijection.*

This is a general result for operators: If L satisfies (3.7) and $L(\mathcal{V}) \subset W$ is dense, then L extends to a bijection in the closure $V = {}^{\perp}(\mathcal{V})^{\perp}$.

Proof. From (3.8), we observe that L is injective, and since $V \subset \mathrm{H}(L, Q)$ is closed, $L(V) \subset W$ has closed range. This is shown as follows: For any sequence $(\boldsymbol{y}_n)_n \subset V$ with $\lim L\boldsymbol{y}_n = \boldsymbol{b} \in W$, we have

$$\|\boldsymbol{y}_n - \boldsymbol{y}_k\|_Q + \|L\boldsymbol{y}_n - L\boldsymbol{y}_k\|_Q \le (C_L + 1)\|L\boldsymbol{y}_n - L\boldsymbol{y}_k\|_Q \to 0,$$

so that $(\boldsymbol{y}_n)_n$ is a Cauchy sequence in V, and since $V \subset \mathrm{H}(L, Q)$ is closed, $\boldsymbol{y} = \lim \boldsymbol{y}_n \in V$ with $L\boldsymbol{y} = \boldsymbol{b}$ exists. Since $L(\mathcal{V}) \subset W$ is dense (Lemma 3.1), we obtain $L(V) = W$. □

The estimate (3.7) transfers to the adjoint operator. That is, we have for $\boldsymbol{z} \in \mathcal{V}^{\mathrm{ad}}$

$$\|\boldsymbol{z}\|_Q = \sup_{\boldsymbol{b} \in W} \frac{(\boldsymbol{z}, \boldsymbol{b})_Q}{\|\boldsymbol{b}\|_Q} = \sup_{\boldsymbol{y} \in \mathcal{V}} \frac{(\boldsymbol{z}, L\boldsymbol{y})_Q}{\|L\boldsymbol{y}\|_Q} = \sup_{\boldsymbol{y} \in \mathcal{V}} \frac{(L^{\mathrm{ad}}\boldsymbol{z}, \boldsymbol{y})_Q}{\|L\boldsymbol{y}\|_Q} \le C_L \|L^{\mathrm{ad}}\boldsymbol{z}\|_Q,$$

again using the density of $L(\mathcal{V})$ in W, and using

$$\langle D^{\mathrm{ad}}\boldsymbol{z}, \boldsymbol{y}\rangle = (L^{\mathrm{ad}}\boldsymbol{z}, \boldsymbol{y})_Q - (\boldsymbol{z}, L\boldsymbol{y})_Q = 0, \quad \boldsymbol{y} \in \mathcal{V}, \boldsymbol{z} \in \mathcal{V}^{\mathrm{ad}}, \tag{3.9}$$

which holds by construction of $\mathcal{V}$ and $\mathcal{V}^{\mathrm{ad}}$. Defining $V^{\mathrm{ad}} = {}^{\perp}(\mathcal{V}^{\mathrm{ad}})^{\perp} \subset \mathrm{H}(L^{\mathrm{ad}}, Q)$, the estimate corresponding to (3.8) holds also for the closure of the adjoint; that is,

$$\|\boldsymbol{z}\|_Q \le C_L \|L^{\mathrm{ad}}\boldsymbol{z}\|_Q, \quad \boldsymbol{z} \in V^{\mathrm{ad}}. \tag{3.10}$$

Theorem 3.3. *We have*

$$\begin{aligned} V &= {}^{\perp}D^{\mathrm{ad}}(\mathcal{V}^{\mathrm{ad}}) \\ &= \{\boldsymbol{y} \in \mathrm{H}(L, Q) \colon (L\boldsymbol{y}, \boldsymbol{z})_Q = (\boldsymbol{y}, L^{\mathrm{ad}}\boldsymbol{z})_Q \text{ for all } \boldsymbol{z} \in \mathcal{V}^{\mathrm{ad}}\}. \end{aligned}$$

In particular, this shows that the operator L with domain V is the Hilbert adjoint of the operator L^{ad} with domain V^{ad}.

Proof. We have $\mathcal{V} \subset {}^{\perp}D^{\mathrm{ad}}(\mathcal{V}^{\mathrm{ad}})$ by (3.9), so that $V \subset {}^{\perp}D^{\mathrm{ad}}(\mathcal{V}^{\mathrm{ad}})$, since ${}^{\perp}D^{\mathrm{ad}}(\mathcal{V}^{\mathrm{ad}})$ is closed in $\mathrm{H}(L,Q)$.

Now, for $\boldsymbol{w} \in {}^{\perp}D^{\mathrm{ad}}(\mathcal{V}^{\mathrm{ad}}) \subset \mathrm{H}(L,Q)$, set $\boldsymbol{b} = L\boldsymbol{w}$, and let $\boldsymbol{y} \in V$ be the unique solution of $L\boldsymbol{y} = \boldsymbol{b}$ (see Theorem 3.2), yielding $L(\boldsymbol{y}-\boldsymbol{w}) = \mathbf{0}$. Since $\boldsymbol{y} \in V \subset {}^{\perp}D^{\mathrm{ad}}(\mathcal{V}^{\mathrm{ad}})$, we have $\boldsymbol{y}-\boldsymbol{w} \in {}^{\perp}D^{\mathrm{ad}}(\mathcal{V}^{\mathrm{ad}})$, and we obtain for all $\boldsymbol{z} \in \mathcal{V}^{\mathrm{ad}}$

$$0 = \langle D^{\mathrm{ad}}\boldsymbol{z}, \boldsymbol{y}-\boldsymbol{w}\rangle = (L^{\mathrm{ad}}\boldsymbol{z}, \boldsymbol{y}-\boldsymbol{w})_Q - (\boldsymbol{z}, L(\boldsymbol{y}-\boldsymbol{w}))_Q = (L^{\mathrm{ad}}\boldsymbol{z}, \boldsymbol{y}-\boldsymbol{w})_Q.$$

Since $L^{\mathrm{ad}}(\mathcal{V}^{\mathrm{ad}}) \subset W$ is dense, we obtain $\boldsymbol{w} = \boldsymbol{y} \in V$. □

Remark 3.1. By the definition of V, Theorem 3.3 directly extends to

$$V = {}^{\perp}D^{\mathrm{ad}}(V^{\mathrm{ad}}).$$

4 Space-time substructuring

For a decomposition $Q_h = \bigcup_{R\in\mathcal{R}_h} R$ into open disjoint space-time cells R, we consider the corresponding discontinuous space $\mathrm{H}(L,Q_h) = \prod_R \mathrm{H}(L,R)$.

Introducing local operators $D_R \in \mathcal{L}(\mathrm{H}(L,R), \mathrm{H}(L^{\mathrm{ad}},R)')$, defined by

$$\langle D_R\boldsymbol{y}_R, \boldsymbol{z}_R\rangle = (L\boldsymbol{y}_R, \boldsymbol{z}_R)_R - (\boldsymbol{y}_R, L^{\mathrm{ad}}\boldsymbol{z}_R)_R, \quad \boldsymbol{y}_R \in \mathrm{H}(L,R), \boldsymbol{z}_R \in \mathrm{H}(L^{\mathrm{ad}},R),$$

we extend the operator D to $D_h \in \mathcal{L}(\mathrm{H}(L,Q_h), \mathrm{H}(L^{\mathrm{ad}},Q_h)')$ by

$$\langle D_h\boldsymbol{y}, \boldsymbol{z}\rangle = \sum_R \langle D_R\boldsymbol{y}_R, \boldsymbol{z}_R\rangle$$

with $\boldsymbol{y}_R = \boldsymbol{y}_{|R}$ and $\boldsymbol{z}_R = \boldsymbol{z}_{|R}$. In particular, we obtain

$$\langle D\boldsymbol{y}, \boldsymbol{z}\rangle = \sum_R ((L\boldsymbol{y})_{|R}, \boldsymbol{z}_{|R})_R - (\boldsymbol{y}_{|R}, (L^{\mathrm{ad}}\boldsymbol{z})_{|R})_R = \langle D_h\boldsymbol{y}, \boldsymbol{z}\rangle \tag{4.1}$$

for conforming functions $\boldsymbol{y} \in \mathrm{H}(L,Q)$ and $\boldsymbol{z} \in \mathrm{H}(L^{\mathrm{ad}},Q)$.

Analogously, we define $D_h^{\mathrm{ad}} \in \mathcal{L}(\mathrm{H}(L^{\mathrm{ad}},Q_h), \mathrm{H}(L,Q_h)')$.

Lemma 4.1. *We have*

$$\begin{aligned} V &= {}^{\perp}D_h^{\mathrm{ad}}(V^{\mathrm{ad}}) \\ &= \{\boldsymbol{y} \in \mathrm{H}(L,Q_h) : \langle D_h\boldsymbol{y}, \boldsymbol{z}\rangle = 0 \textit{ for all } \boldsymbol{z} \in V^{\mathrm{ad}}\}. \end{aligned}$$

Proof. It is sufficient to show ${}^{\perp}D_h^{\mathrm{ad}}(V^{\mathrm{ad}}) \subset \mathrm{H}(L,Q)$. Then, (4.1) yields the assertion by ${}^{\perp}D_h^{\mathrm{ad}}(V^{\mathrm{ad}}) \cap \mathrm{H}(L,Q) = {}^{\perp}D^{\mathrm{ad}}(V^{\mathrm{ad}}) = V$, see Theorem 3.3.

For $\boldsymbol{y} \in {}^{\perp}D_h^{\mathrm{ad}}(V^{\mathrm{ad}}) \subset \mathrm{H}(L,Q_h)$ and $\boldsymbol{b} = L\boldsymbol{y} \in W$, we have $\langle D_h\boldsymbol{y}, \boldsymbol{z}\rangle = 0$ for $\boldsymbol{z} \in \mathrm{C}_c^1(Q,\mathbb{R}^m) \subset V^{\mathrm{ad}}$. Thus, we obtain

$$(\boldsymbol{b}, \boldsymbol{z})_Q = (L\boldsymbol{y}, \boldsymbol{z})_{Q_h} = (\boldsymbol{y}, L^{\mathrm{ad}}\boldsymbol{z})_{Q_h} = (\boldsymbol{y}, L^{\mathrm{ad}}\boldsymbol{z})_Q, \quad \boldsymbol{z} \in \mathrm{C}_c^1(Q,\mathbb{R}^m),$$

so that, indeed, $\boldsymbol{y} \in \mathrm{H}(L,Q)$ by definition (3.5). □

Lemma 4.2. *We have*

$$\mathrm{H}_0(L, Q_h) = \mathcal{N}(D_h).$$

Proof. We have $\mathrm{H}_0(L, R) = \mathcal{N}(D_R)$, see Theorem 3.1. Thus, the assertion follows from

$$\mathrm{H}_0(L, Q_h) = \prod_R \mathrm{H}_0(L, R) = \prod_R \mathcal{N}(D_R) = \mathcal{N}(D_h). \qquad \square$$

This shows that the operator D_h is well-defined on the quotient space (associated with the quotient norm) denoted by

$$\hat{\mathrm{H}}(L, Q_h) = \mathrm{H}(L, Q_h)/\mathrm{H}_0(L, Q_h), \quad \|\hat{\mathbf{y}}\|_{L,\partial Q_h} = \inf_{\hat{\mathbf{y}}=\mathbf{y}+\mathrm{H}_0(L,Q_h)} \|\mathbf{y}\|_{L,Q_h}.$$

That is, $\hat{D}_h \in \mathcal{L}(\hat{\mathrm{H}}(L, Q_h), \mathrm{H}(L^{\mathrm{ad}}, Q_h)')$ is well-defined with

$$\langle \hat{D}_h \hat{\mathbf{y}}, \boldsymbol{z} \rangle = \langle D_h \mathbf{y}, \boldsymbol{z} \rangle, \quad \hat{\mathbf{y}} = \mathbf{y} + \mathrm{H}_0(L, Q_h).$$

By construction, $\hat{D}_h$ is injective; that is, $\mathcal{N}(\hat{D}_h) = \{\mathbf{0}\}$.

With respect to the substructuring, we represent the solution in $W \times \hat{\mathrm{H}}(L, Q_h)$ as follows: For given $\boldsymbol{b} \in W$, let $\mathbf{y} \in V$ be the unique solution of $L\mathbf{y} = \boldsymbol{b}$, and define $\hat{\mathbf{y}} = \mathbf{y} + \mathrm{H}_0(L, Q_h) \in \hat{\mathrm{H}}(L, Q_h)$. Then, inserting $\hat{D}_h$ yields

$$(\boldsymbol{b}, \boldsymbol{z})_Q = (L\mathbf{y}, \boldsymbol{z})_Q = (\mathbf{y}, L^{\mathrm{ad}}\boldsymbol{z})_{Q_h} + \langle \hat{D}_h \hat{\mathbf{y}}, \boldsymbol{z} \rangle, \quad \boldsymbol{z} \in \mathrm{H}(L^{\mathrm{ad}}, Q_h).$$

For the corresponding Petrov–Galerkin method in $W \times \hat{\mathrm{H}}(L, Q_h)$, we define the operator

$$B_h \in \mathcal{L}(W \times \hat{\mathrm{H}}(L, Q_h), \mathrm{H}(L^{\mathrm{ad}}, Q_h)'), \quad \langle B_h(\mathbf{y}, \hat{\mathbf{y}}), \boldsymbol{z} \rangle = (\mathbf{y}, L^{\mathrm{ad}}\boldsymbol{z})_Q + \langle \hat{D}_h \hat{\mathbf{y}}, \boldsymbol{z} \rangle.$$

As a result, the pair $(\mathbf{y}, \hat{\mathbf{y}})$ also fulfills the equation

$$\langle B_h(\mathbf{y}, \hat{\mathbf{y}}), \boldsymbol{z} \rangle = (\boldsymbol{b}, \boldsymbol{z})_Q, \quad \boldsymbol{z} \in \mathrm{H}(L^{\mathrm{ad}}, Q_h),$$

for test functions $\boldsymbol{z} \in \mathrm{H}(L^{\mathrm{ad}}, Q_h)$.

The norm in $W \times \hat{\mathrm{H}}(L, Q_h)$ is denoted by

$$\|(\mathbf{y}, \hat{\mathbf{y}})\|_{Q;L,\partial Q_h} = \sqrt{\|\mathbf{y}\|_Q^2 + \|\hat{\mathbf{y}}\|_{L,\partial Q_h}^2}.$$

Now, we show that B_h is invertible in $W \times \hat{V}$ with $\hat{V} = V/\mathrm{H}_0(L, Q_h) \subset \hat{\mathrm{H}}(L, Q_h)$.

Theorem 4.1. *We have for* $(\mathbf{y}, \hat{\mathbf{y}}) \in W \times \hat{V}$

$$\sup_{\boldsymbol{z} \in \mathrm{H}(L^{\mathrm{ad}}, Q_h)} \frac{\langle B_h(\mathbf{y}, \hat{\mathbf{y}}), \boldsymbol{z} \rangle}{\|\boldsymbol{z}\|_{L^{\mathrm{ad}}, Q_h}} \geq \frac{1}{\sqrt{4C_L^2 + 2}} \|(\mathbf{y}, \hat{\mathbf{y}})\|_{Q;L,\partial Q_h}. \tag{4.2}$$

Proof. In the first step, we establish

$$\sup_{(\mathbf{y},\hat{\mathbf{y}})\in W\times\hat{V}} \frac{\langle B_h(\mathbf{y},\hat{\mathbf{y}}),\mathbf{z}\rangle}{\|(\mathbf{y},\hat{\mathbf{y}})\|_{Q;L,\partial Q_h}} \geq \frac{1}{\sqrt{4C_L^2+2}}\|\mathbf{z}\|_{L^{\mathrm{ad}},Q}, \quad \mathbf{z}\in \mathrm{H}(L^{\mathrm{ad}},Q_h). \tag{4.3}$$

For given $\mathbf{z}\in \mathrm{H}(L^{\mathrm{ad}},Q_h)\subset W$, we find a unique function $\mathbf{y}_0\in V$, solving $L\mathbf{y}_0=\mathbf{z}$, and we set $\hat{\mathbf{y}}_0=\mathbf{y}_0+\mathrm{H}_0(L,Q_h)\in\hat{V}$. Then,

$$\langle B_h(\mathbf{y}_0,\hat{\mathbf{y}}_0),\mathbf{z}\rangle = (\mathbf{y}_0,L^{\mathrm{ad}}\mathbf{z})_{Q_h} + \langle \hat{D}_h\hat{\mathbf{y}}_0,\mathbf{z}\rangle = (L\mathbf{y}_0,\mathbf{z})_Q = \|\mathbf{z}\|_Q^2,$$

and inserting (3.8) yields

$$\begin{aligned}\|(\mathbf{y}_0,\hat{\mathbf{y}}_0)\|^2_{Q;L,\partial Q_h} &= \|\mathbf{y}_0\|_Q^2 + \|\hat{\mathbf{y}}_0\|^2_{L,\partial Q}\\ &\leq \|\mathbf{y}_0\|_Q^2 + \|\mathbf{y}_0\|^2_{L,Q} = 2\|\mathbf{y}_0\|_Q^2 + \|L\mathbf{y}_0\|_Q^2\\ &\leq (2C_L^2+1)\|L\mathbf{y}_0\|_Q^2 = (2C_L^2+1)\|\mathbf{z}\|_Q^2,\end{aligned}$$

so that we obtain

$$\begin{aligned}\sup_{(\mathbf{y},\hat{\mathbf{y}})\in W\times\hat{V}} \frac{\langle B_h(\mathbf{y},\hat{\mathbf{y}}),\mathbf{z}\rangle}{\|(\mathbf{y},\hat{\mathbf{y}})\|_{Q;L,\partial Q_h}} &\geq \frac{\langle B_h(\mathbf{y}_0,\hat{\mathbf{y}}_0),\mathbf{z}\rangle}{\|(\mathbf{y}_0,\hat{\mathbf{y}}_0)\|_{Q;L,\partial Q_h}} = \frac{\|\mathbf{z}\|_Q^2}{\|(\mathbf{y}_0,\hat{\mathbf{y}}_0)\|_{Q;L,\partial Q_h}}\\ &\geq \frac{1}{\sqrt{2C_L^2+1}}\|\mathbf{z}\|_Q.\end{aligned}$$

Then, choosing $(\mathbf{y},\hat{\mathbf{y}})=(L^{\mathrm{ad}}\mathbf{z},\mathbf{0})$ yields

$$\sup_{(\mathbf{y},\hat{\mathbf{y}})\in W\times\hat{V}} \frac{\langle B_h(\mathbf{y},\hat{\mathbf{y}}),\mathbf{z}\rangle}{\|(\mathbf{y},\hat{\mathbf{y}})\|_{Q;L,\partial Q_h}} \geq \frac{\langle B_h(L^{\mathrm{ad}}\mathbf{z},\mathbf{0}),\mathbf{z}\rangle}{\|(L^{\mathrm{ad}}\mathbf{z},\mathbf{0})\|_{Q;L,\partial Q_h}} = \|L^{\mathrm{ad}}\mathbf{z}\|_Q.$$

Now, (4.3) follows from $\|\mathbf{z}\|^2_{L^{\mathrm{ad}},Q}\leq 2\max\{\|\mathbf{z}\|_Q^2,\|L^{\mathrm{ad}}\mathbf{z}\|_Q^2\}$; that is,

$$\max\{\|\mathbf{z}\|_Q,\|L^{\mathrm{ad}}\mathbf{z}\|_Q\} \geq \frac{1}{\sqrt{2}}\|\mathbf{z}\|_{L^{\mathrm{ad}},Q}.$$

In the second step, we show that the operator B_h is injective in $W\times\hat{V}$; then, (4.2) is obtained by duality [2, Lemma 4.4.2].

Therefore, we consider $(\mathbf{y},\hat{\mathbf{y}})\in W\times\hat{V}$ with

$$\langle B_h(\mathbf{y},\hat{\mathbf{y}}),\mathbf{z}\rangle = 0, \quad \mathbf{z}\in\mathrm{H}(L^{\mathrm{ad}},Q_h).$$

This yields

$$0 = \langle B_h(\mathbf{y},\hat{\mathbf{y}}),\mathbf{z}\rangle = (\mathbf{y},L^{\mathrm{ad}}\mathbf{z})_{Q_h}, \quad \mathbf{z}\in\mathrm{C}^1_{\mathrm{c}}(Q_h,\mathbb{R}^m);$$

that is, $\boldsymbol{y} \in \mathrm{H}(L, Q_h)$, and $L\boldsymbol{y} = \mathbf{0}$. Thus, from $\hat{\boldsymbol{y}} \in \hat{V}$ and $\langle \hat{D}_h \hat{\boldsymbol{y}}, \boldsymbol{z} \rangle = 0$ for $\boldsymbol{z} \in V^{\mathrm{ad}}$, we conclude for all $\boldsymbol{z} \in V^{\mathrm{ad}}$:

$$\begin{aligned} 0 &= \langle B_h(\boldsymbol{y}, \hat{\boldsymbol{y}}), \boldsymbol{z} \rangle - (L\boldsymbol{y}, \boldsymbol{z})_{Q_h} = (\boldsymbol{y}, L^{\mathrm{ad}}\boldsymbol{z})_{Q_h} + \langle \hat{D}_h \hat{\boldsymbol{y}}, \boldsymbol{z} \rangle - (L\boldsymbol{y}, \boldsymbol{z})_{Q_h} \\ &= \langle \hat{D}_h \hat{\boldsymbol{y}}, \boldsymbol{z} \rangle - \langle D_h \boldsymbol{y}, \boldsymbol{z} \rangle = -\langle D_h \boldsymbol{y}, \boldsymbol{z} \rangle, \end{aligned}$$

which shows $\boldsymbol{y} \in V$, see Lemma 4.1. Together with $L\boldsymbol{y} = \mathbf{0}$ and (3.8), this implies $\boldsymbol{y} = \mathbf{0}$. Thus,

$$0 = \langle B_h(\boldsymbol{y}, \hat{\boldsymbol{y}}), \boldsymbol{z} \rangle = \langle \hat{D}_h \hat{\boldsymbol{y}}, \boldsymbol{z} \rangle, \quad \boldsymbol{z} \in \mathrm{H}(L^{\mathrm{ad}}, Q_h),$$

which implies $\hat{\boldsymbol{y}} = \mathbf{0}$, see Lemma 4.2. □

Theorem 4.1 provides stability for the *ideal DPG method* with discrete approximations in $W \times \hat{V}$, and for test functions the continuous space $\mathrm{H}(L^{\mathrm{ad}}, Q_h)$. Using a discrete test space yields the *practical DPG method.*

5 The DPG method

Now we select a global conforming discrete ansatz space $\hat{V}_h \subset \hat{V}$ on the skeleton and local ansatz and test spaces $W_{R,h} \subset \mathrm{L}_2(R, \mathbb{R}^m)$, and $Z_{R,h} \subset \mathrm{H}(L^{\mathrm{ad}}, R)$. We set $\hat{V}_{R,h} = \hat{V}_{h|\partial R}$, $W_h = \prod W_{R,h}$ and $Z_h = \prod Z_{R,h}$.

To verify discrete inf-sup stability, we construct a suitable local Fortin operator $\Pi_{R,h} \in \mathcal{L}(\mathrm{H}(L^{\mathrm{ad}}, R), Z_{R,h})$ in every space-time cell R following the approach presented in [17, Section 3.1.4], see also the construction in [9, Theorem 1]. This yields a mesh-dependent estimate. Then we show by a scaling argument that it is sufficient to construct a local Fortin operator on a reference cell, so that, finally, a mesh-independent a priori bound for the DPG approximation is obtained.

5.1 The local construction of the Fortin operator

We define $B_R \in \mathcal{L}(\mathrm{L}_2(R, \mathbb{R}^m) \times \hat{\mathrm{H}}(L, R), \mathrm{H}(L^{\mathrm{ad}}, R)')$ by

$$\langle B_R(\boldsymbol{y}_R, \hat{\boldsymbol{y}}_R), \boldsymbol{z}_R \rangle = (\boldsymbol{y}_R, L^{\mathrm{ad}}\boldsymbol{z}_R)_R + \langle \hat{D}_R \hat{\boldsymbol{y}}_R, \boldsymbol{z}_R \rangle.$$

We assume that for given $\hat{V}_{R,h}$ and $W_{R,h}$, the local test spaces $Z_{R,h}$ are selected so that for all $\boldsymbol{z}_R \in \mathrm{H}(L^{\mathrm{ad}}, R)$ the affine space

$$\begin{aligned} \mathcal{N}(\boldsymbol{z}_R) = \{ \boldsymbol{z}_{R,h} \in Z_{R,h} \colon & \langle B_R(\boldsymbol{y}_{R,h}, \hat{\boldsymbol{y}}_{R,h}), \boldsymbol{z}_{R,h} \rangle = \langle B_R(\boldsymbol{y}_{R,h}, \hat{\boldsymbol{y}}_{R,h}), \boldsymbol{z}_R \rangle, \\ & (\boldsymbol{y}_{R,h}, \hat{\boldsymbol{y}}_{R,h}) \in W_{R,h} \times \hat{V}_{R,h} \} \end{aligned}$$

is not empty for all $R \in \mathcal{R}$. Then, a Fortin operator with $\Pi_{R,h}\boldsymbol{z}_R \in \mathcal{N}(\boldsymbol{z}_R)$ exists. For the scaling argument below, we require the stability estimate

$$|\Pi_{R,h}\boldsymbol{z}_R|_{L^{\mathrm{ad}},R} \le |\boldsymbol{z}_R|_{L^{\mathrm{ad}},R}$$

with respect to the seminorm $|\boldsymbol{z}_R|_{L^{\mathrm{ad}},R} = \|L^{\mathrm{ad}}\boldsymbol{z}_R\|_R$. This can easily be achieved by extending $W_{R,h}$ to $W_{R,h}^{\mathrm{ext}} \supset W_{R,h} + L^{\mathrm{ad}}Z_{R,h}$, since the orthogonality

$$0 = \langle B_R(\boldsymbol{y}_{R,h}, \mathbf{0}), \boldsymbol{z}_{R,h} - \boldsymbol{z}_R\rangle = (\boldsymbol{y}_{R,h}, L^{\mathrm{ad}}(\boldsymbol{z}_{R,h} - \boldsymbol{z}_R))_R, \quad \boldsymbol{y}_{R,h} \in W_{R,h}^{\mathrm{ext}}$$

implies $|\boldsymbol{z}_{R,h}|_{L^{\mathrm{ad}},R} \le |\boldsymbol{z}_R|_{L^{\mathrm{ad}},R}$. We assume that also $\mathcal{N}^{\mathrm{ext}}(\boldsymbol{z}_R) \subset \mathcal{N}(\boldsymbol{z}_R)$, obtained by testing with the larger space $W_{R,h}^{\mathrm{ext}} \supset W_{R,h}$, is not empty.

To compute a bound for the norm of $\Pi_{R,h}$ numerically, we assume that extensions $V_{R,h} \subset \mathrm{H}(L, R)$ of $\hat{V}_{R,h}$ exists with $\dim V_{R,h} = \dim \hat{V}_{R,h}$, so that for every trace function $\hat{\boldsymbol{y}}_{R,h} \in \hat{V}_{R,h}$ a unique extension $\bar{\boldsymbol{y}}_{R,h} \in V_{R,h}$ exists, which can be locally evaluated in R, and which satisfies

$$\langle D_R\bar{\boldsymbol{y}}_{R,h}, \boldsymbol{z}_R\rangle = \langle \hat{D}_R\hat{\boldsymbol{y}}_{R,h}, \boldsymbol{z}_R\rangle, \quad \boldsymbol{z}_R \in \mathrm{H}(L^{\mathrm{ad}}, R); \tag{5.1}$$

that is, $\hat{\boldsymbol{y}}_{R,h} = \bar{\boldsymbol{y}}_{R,h} + \mathrm{H}_0(L, Q_h)$. This defines a well-defined bijection

$$\hat{I}_{R,h}\colon V_{R,h} \to \hat{V}_{R,h}$$

such that $\hat{\boldsymbol{y}}_{R,h} = \hat{I}_{R,h}\bar{\boldsymbol{y}}_{R,h}$ satisfies (5.1).

The minimizer $\boldsymbol{z}_{R,h} = \Pi_{R,h}\boldsymbol{z}_R \in \mathcal{N}^{\mathrm{ext}}(\boldsymbol{z}_R)$ with respect to the norm in $\mathrm{H}(L^{\mathrm{ad}}, R)$ can be computed by a discrete linear saddle-point problem as follows: We define the discrete operators:

$$\begin{aligned} B_{R,h} &\in \mathcal{L}(W_{R,h}^{\mathrm{ext}} \times V_{R,h}, Z_{R,h}'),\\ C_{R,h} &\in \mathcal{L}(Z_{R,h}, Z_{R,h}'),\\ G_{R,h} &\in \mathcal{L}(W_{R,h}^{\mathrm{ext}} \times V_{R,h}, (W_{R,h}^{\mathrm{ext}} \times V_{R,h})') \end{aligned}$$

by

$$\begin{aligned} \langle B_{R,h}(\boldsymbol{y}_{R,h}, \bar{\boldsymbol{y}}_{R,h}), \boldsymbol{z}_{R,h}\rangle &= \langle B_R(\boldsymbol{y}_{R,h}, \hat{I}_{R,h}\bar{\boldsymbol{y}}_{R,h}), \boldsymbol{z}_{R,h}\rangle,\\ \langle C_{R,h}\boldsymbol{z}_{R,h}, \boldsymbol{\psi}_{R,h}\rangle &= (\boldsymbol{z}_{R,h}, \boldsymbol{\psi}_{R,h})_{L^{\mathrm{ad}},R},\\ \langle G_{R,h}(\boldsymbol{y}_{R,h}, \bar{\boldsymbol{y}}_{R,h}), (\boldsymbol{\phi}_{R,h}, \bar{\boldsymbol{\phi}}_{R,h})\rangle &= (\boldsymbol{y}_{R,h}, \boldsymbol{\phi}_{R,h})_R + (\bar{\boldsymbol{y}}_{R,h}, \bar{\boldsymbol{\phi}}_{R,h})_{L,R}, \end{aligned}$$

and the embedding $E_{R,h} \in \mathcal{L}(W_{R,h}^{\mathrm{ext}} \times V_{R,h}, W \times \hat{\mathrm{H}}(L, R))$. Then, $\boldsymbol{z}_{R,h} = \Pi_{R,h}\boldsymbol{z}_R$ solves the discrete saddle-point problem

$$C_{R,h}\boldsymbol{z}_{R,h} + B_{R,h}(\boldsymbol{y}_{R,h}, \bar{\boldsymbol{y}}_{R,h}) = \mathbf{0}, \tag{5.2a}$$

$$B_{R,h}'\boldsymbol{z}_{R,h} = E_{R,h}'B_R'\boldsymbol{z}_R, \tag{5.2b}$$

where $(\boldsymbol{y}_{R,h}, \bar{\boldsymbol{y}}_{R,h}) \in W_{R,h}^{\mathrm{ext}} \times V_{R,h}$ is the Lagrange multiplier.

Remark 5.1. Inf-sup stability requires that B_h is injective in $W_h \times \hat{V}_h$, but locally we cannot expect that $B_{R,h}$ is injective, since $B_{R,h}(\mathbf{y}_{R,h}, \mathbf{y}_{R,h}) = \mathbf{0}$ for all functions $\mathbf{y}_{R,h} \in V_{R,h} \cap W_{R,h} \cap \mathcal{N}(L)$.

On the other hand, since we assume that $\mathcal{N}^{\text{ext}}(\mathbf{z}_R)$ is not empty for all $\mathbf{z}_R$, the equation (5.2b) has always a solution, and since $C_{R,h}$ is positive definite, $\mathbf{z}_{R,h} = \Pi_{R,h}\mathbf{z}_R$ is the unique solution of the optimization problem. The Lagrange parameter $(\mathbf{y}_{R,h}, \bar{\mathbf{y}}_{R,h})$ is only unique up to $\mathcal{N}(B_{R,h})$.

Inserting $\mathbf{z}_{R,h} = -C_{R,h}^{-1}B_{R,h}(\mathbf{y}_{R,h}, \bar{\mathbf{y}}_{R,h})$ yields

$$S_{R,h}(\mathbf{y}_{R,h}, \bar{\mathbf{y}}_{R,h}) = -E_{R,h}'B_R'\mathbf{z}_R$$

with the Schur complement operator

$$S_{R,h} = B_{R,h}'C_{R,h}^{-1}B_{R,h} \in \mathcal{L}\big(W_{R,h}^{\text{ext}} \times V_{R,h}, (W_{R,h}^{\text{ext}} \times V_{R,h})'\big).$$

Inserting the pseudoinverse (with respect to the inner product in $W_{R,h}^{\text{ext}} \times V_{R,h}$)

$$S_{R,h}^{+} = \lim_{\delta\to 0}\big(S_{R,h}G_{R,h}^{-1}S_{R,h} + \delta G_{R,h}\big)^{-1}S_{R,h}G_{R,h}^{-1},$$

satisfying $S_{R,h}^{+}S_{R,h}S_{R,h}^{+} = S_{R,h}^{+}$, defines

$$\Pi_{R,h} = C_{R,h}^{-1}B_{R,h}S_{R,h}^{+}E_{R,h}'B_R'. \tag{5.3}$$

We compute $\alpha_{R,h} > 0$ such that

$$\langle \ell_{R,h}, S_{R,h}^{+}\ell_{R,h}\rangle \le \alpha_{R,h}\langle \ell_{R,h}, G_{R,h}^{-1}\ell_{R,h}\rangle, \quad \ell_{R,h} \in (W_{R,h}^{\text{ext}} \times V_{R,h})'. \tag{5.4}$$

That is, we determine the largest eigenvalue of a finite dimensional symmetric generalized eigenvalue problem. For given $\mathbf{z}_R \in \mathrm{H}(L^{\text{ad}}, R)$, we select the discrete functional $\ell_{R,h} = E_{R,h}'B_R'\mathbf{z}_R$, and the norm of the Fortin operator is estimated by

$$\begin{aligned}
\|\Pi_{R,h}\mathbf{z}_R\|_{L^{\text{ad}},R}^2 &= \langle B_{R,h}S_{R,h}^{+}\ell_{R,h}, C_{R,h}^{-1}B_{R,h}S_{R,h}^{+}\ell_{R,h}\rangle \\
&= \langle \ell_{R,h}, S_{R,h}^{+}S_{R,h}S_{R,h}^{+}\ell_{R,h}\rangle \\
&= \langle \ell_{R,h}, S_{R,h}^{+}\ell_{R,h}\rangle \\
&\le \alpha_{R,h}\langle \ell_{R,h}, G_{R,h}^{-1}\ell_{R,h}\rangle \\
&\le 2\alpha_{R,h}\|\mathbf{z}_R\|_{L^{\text{ad}},R}^2
\end{aligned}$$

using

$$\sqrt{\langle \ell_{R,h}, G_{R,h}^{-1}\ell_{R,h}\rangle} = \sup_{(\mathbf{y}_{R,h},\bar{\mathbf{y}}_{R,h})\in W_{R,h}^{\text{ext}}\times V_{R,h}} \frac{\langle \ell_{R,h}, (\mathbf{y}_{R,h}, \bar{\mathbf{y}}_{R,h})\rangle}{\sqrt{\langle G_{R,h}(\mathbf{y}_{R,h}, \bar{\mathbf{y}}_{R,h}), (\mathbf{y}_{R,h}, \bar{\mathbf{y}}_{R,h})\rangle}}$$

$$= \sup_{(\boldsymbol{y}_{R,h},\bar{\boldsymbol{y}}_{R,h})\in W^{\text{ext}}_{R,h}\times V_{R,h}} \frac{\langle E'_{R,h}B'_R\boldsymbol{z}_R,(\boldsymbol{y}_{R,h},\bar{\boldsymbol{y}}_{R,h})\rangle}{\|(\boldsymbol{y}_{R,h},\bar{\boldsymbol{y}}_{R,h})\|_{R;L,R}}$$

$$= \sup_{(\boldsymbol{y}_{R,h},\bar{\boldsymbol{y}}_{R,h})\in W^{\text{ext}}_{R,h}\times V_{R,h}} \frac{\langle B_RE_{R,h}(\boldsymbol{y}_{R,h},\bar{\boldsymbol{y}}_{R,h}),\boldsymbol{z}_R\rangle}{\|(\boldsymbol{y}_{R,h},\bar{\boldsymbol{y}}_{R,h})\|_{R;L,R}} \le \sqrt{2}\|\boldsymbol{z}_R\|_{L^{\text{ad}},R},$$

with

$$\langle G_{R,h}(\boldsymbol{y}_{R,h},\bar{\boldsymbol{y}}_{R,h}),(\boldsymbol{y}_{R,h},\bar{\boldsymbol{y}}_{R,h})\rangle = \|(\boldsymbol{y}_{R,h},\bar{\boldsymbol{y}}_{R,h})\|^2_{R;L,R},$$

and

$$\begin{aligned}
&\langle B_R(\boldsymbol{y}_{R,h},\hat{I}_{R,h}\bar{\boldsymbol{y}}_{R,h}),\boldsymbol{z}_R\rangle\\
&= (\boldsymbol{y}_{R,h},L^{\text{ad}}\boldsymbol{z}_R)_R + (L\bar{\boldsymbol{y}}_{R,h},\boldsymbol{z}_R)_R - (\bar{\boldsymbol{y}}_{R,h},L^{\text{ad}}\boldsymbol{z}_R)_R\\
&\le \|\boldsymbol{y}_{R,h}\|_R\|L^{\text{ad}}\boldsymbol{z}_R\|_R + \|L\bar{\boldsymbol{y}}_{R,h}\|_R\|\boldsymbol{z}_R\|_R + \|\bar{\boldsymbol{y}}_{R,h}\|_R\|L^{\text{ad}}\boldsymbol{z}_R\|_R\\
&\le \sqrt{\|\boldsymbol{y}_{R,h}\|^2_R + \|L\bar{\boldsymbol{y}}_{R,h}\|^2_R + \|\bar{\boldsymbol{y}}_{R,h}\|^2_R}\sqrt{2\|L^{\text{ad}}\boldsymbol{z}_R\|^2_R + \|\boldsymbol{z}_R\|^2_R}\\
&\le \sqrt{2}\|(\boldsymbol{y}_{R,h},\bar{\boldsymbol{y}}_{R,h})\|_{R;L,R}\|\boldsymbol{z}_R\|_{L^{\text{ad}},R}.
\end{aligned}$$

The construction is completely local, so that it extends to

$$\|\Pi_h\boldsymbol{z}\|_{L^{\text{ad}},Q_h} \le \sqrt{2\alpha_h}\|\boldsymbol{z}\|_{L^{\text{ad}},Q_h}, \quad \boldsymbol{z}\in \mathrm{H}(L^{\text{ad}},Q_h) \tag{5.5}$$

with $\alpha_h = \max \alpha_{R,h}$. This gives discrete inf-sup stability

$$\sup_{\boldsymbol{z}_h\in Z_h} \frac{\langle B_h(\boldsymbol{y}_h,\hat{\boldsymbol{y}}_h),\boldsymbol{z}_h\rangle}{\|\boldsymbol{z}_h\|_{L^{\text{ad}},Q_h}} \ge \beta_h\|(\boldsymbol{y}_h,\hat{\boldsymbol{y}}_h)\|_{Q;L,\partial Q_h} \tag{5.6}$$

for $(\boldsymbol{y}_h,\hat{\boldsymbol{y}}_h)\in W_h\times\hat{V}_h$ with $\beta_h = \frac{1}{2\sqrt{\alpha_{R,h}}\sqrt{2C_L^2+1}}$ [3, Proposition II.2.8].

5.2 A scaling argument

Numerically, we observe that the eigenvalue estimate (5.4) is mesh-dependent. Thus, we compute $\alpha_0 = \alpha_{R_0,h_0}$ on a reference element $R_0 = (0,h_0)^d\times(0,h_0/c)$ for the speed of sound $c>0$, and we analyze the transformation $\varphi_R\colon R_0\to R$. For simplicity, we only discuss a scaling of the form $\varphi_R(\boldsymbol{x},t) = (\boldsymbol{x}_R,t_R)+(h/h_0)(\boldsymbol{x},t)$ with $R = (\boldsymbol{x}_R,t_R)+(0,h)^d\times(0,h/c)$.

Let Π_{R_0,h_0} be a local Fortin operator on the reference cell R_0. For the seminorm $|\boldsymbol{z}_R|_{L^{\text{ad}},R} = \|L^{\text{ad}}\boldsymbol{z}_R\|_R$ and the operator B_R, we assume the scaling properties

$$\begin{aligned}
h^{-d+1}\,|\boldsymbol{z}_R|^2_{L^{\text{ad}},R} &= h_0^{-d+1}\,|\boldsymbol{z}_R\circ\varphi_R|^2_{L^{\text{ad}},R_0},\\
h^{-d}\,\langle B_R(\boldsymbol{y}_{R,h},\hat{\boldsymbol{y}}_{R,h}),\boldsymbol{z}_R\rangle &= h_0^{-d}\,\langle B_{R_0}(\boldsymbol{y}_{R,h}\circ\varphi_R,\hat{\boldsymbol{y}}_{R,h}\circ\varphi_R),\boldsymbol{z}_R\circ\varphi_R\rangle.
\end{aligned}$$

This holds for acoustic waves with constant coefficients. Then, the transformation

$$\Pi_{R,h}\boldsymbol{z}_R = \big(\Pi_{R_0,h_0}(\boldsymbol{z}_R \circ \varphi_R)\big) \circ \varphi_R^{-1}$$

defines a local Fortin operator in R. By scaling, we obtain for $h \le h_0$

$$\begin{aligned}
h^{-d-1}\,\|\Pi_{R,h}\boldsymbol{z}_R\|_R^2 &= h_0^{-d-1}\,\|(\Pi_{R,h}\boldsymbol{z}_R)\circ\varphi_R\|_{R_0}^2\\
&= h_0^{-d-1}\,\|\Pi_{R_0,h_0}(\boldsymbol{z}_R\circ\varphi_R)\|_{R_0}^2\\
&\le h_0^{-d-1}\,\|\Pi_{R_0,h_0}(\boldsymbol{z}_R\circ\varphi_R)\|_{L^{\mathrm{ad}},R_0}^2\\
&\le h_0^{-d-1}\,\|\Pi_{R_0,h_0}\|_{L^{\mathrm{ad}},R_0}^2\,\|\boldsymbol{z}_R\circ\varphi_R\|_{L^{\mathrm{ad}},R_0}^2,\\
h_0^{-d-1}\,\|\boldsymbol{z}_R\circ\varphi_R\|_{L^{\mathrm{ad}},R_0}^2 &= h^{-d-1}\,\|\boldsymbol{z}_R\|_R^2 + h_0^{-2}\,h^{-d+1}\,|\boldsymbol{z}_R|_{L^{\mathrm{ad}},R}^2\\
&\le h^{-d-1}\,\|\boldsymbol{z}_R\|_{L^{\mathrm{ad}},R}^2,\\
h^{-d+1}\,|\Pi_{R,h}\boldsymbol{z}_R|_{L^{\mathrm{ad}},R}^2 &= h_0^{-d+1}\,|(\Pi_{R,h}\boldsymbol{z}_R)\circ\varphi_R|_{L^{\mathrm{ad}},R_0}^2\\
&= h_0^{-d+1}\,|\Pi_{R_0,h_0}(\boldsymbol{z}_R\circ\varphi_R)|_{L^{\mathrm{ad}},R_0}^2\\
&\le h_0^{-d+1}\,|\boldsymbol{z}_R\circ\varphi_R|_{L^{\mathrm{ad}},R_0}^2 = h^{-d+1}\,|\boldsymbol{z}_R|_{L^{\mathrm{ad}},R}^2,
\end{aligned}$$

which yields together

$$\|\Pi_{R,h}\boldsymbol{z}_R\|_{L^{\mathrm{ad}},R} \le \sqrt{1+\|\Pi_{R_0,h_0}\|_{L^{\mathrm{ad}},R_0}^2}\,\|\boldsymbol{z}_R\|_{L^{\mathrm{ad}},R}.$$

For simple meshes, this results into the computable inf-sup constant

$$\beta_h = \frac{1}{\sqrt{1+2\alpha_{R_0,h_0}}\,\sqrt{4C_L^2+2}}.$$

5.3 An a priori error estimate for the practical DPG method

The discrete DPG solution $(\boldsymbol{y}_h, \hat{\boldsymbol{y}}_h) \in W_h \times \hat{V}_h$ is obtained by minimizing the residual $B_h(\boldsymbol{y}_h, \hat{\boldsymbol{y}}_h) - \boldsymbol{b}$ in Z_h'. That is, by minimizing the functional

$$\Psi_h(\boldsymbol{y}_h, \hat{\boldsymbol{y}}_h) = \sup_{\boldsymbol{z}_h \in Z_h} \frac{\langle B_h(\boldsymbol{y}_h, \hat{\boldsymbol{y}}_h), \boldsymbol{z}_h\rangle - (\boldsymbol{b}, \boldsymbol{z}_h)_Q}{\|\boldsymbol{z}_h\|_{L^{\mathrm{ad}},Q}}.$$

The unique minimizer is the Petrov–Galerkin solution obtained by testing with the optimal test space $Z_h^{\mathrm{opt}} = C_h^{-1} B_h(W_h \times \hat{V}_h)$; that is,

$$\big\langle B_h(\boldsymbol{y}_h, \hat{\boldsymbol{y}}_h), \boldsymbol{z}_h\big\rangle = (\boldsymbol{b}, \boldsymbol{z}_h)_Q, \quad \boldsymbol{z}_h \in Z_h^{\mathrm{opt}}. \tag{5.7}$$

Since B_h is continuous, and since we assume that $\mathcal{N}(\boldsymbol{z}_R) \neq \{\boldsymbol{0}\}$ for all $\boldsymbol{z}_R \in \mathrm{H}(L^{\mathrm{ad}}, R)$ (so that a computable, but, in general, mesh dependent, inf-sup constant exists, as discussed in Section 5.1), Petrov–Galerkin estimates apply [24, Theorem 2]. In simple cases where the scaling argument applies, this yields a mesh-independent estimate for α_h, and thus for the inf-sup constant β_h.

In our experiments, the assumption $\mathcal{N}(\boldsymbol{z}_R) \neq \{\boldsymbol{0}\}$ can be achieved easily by choosing polynomials of higher order in the local test spaces $Z_{R,h}$ than in the local ansatz spaces $W_{R,h}$, $V_{R,h}$.

Theorem 5.1. *Let $\boldsymbol{y} \in V$ be the solution of $L\boldsymbol{y} = \boldsymbol{b}$, and set $\hat{\boldsymbol{y}} = \boldsymbol{y} + \mathrm{H}_0(L, Q_h) \in \hat{V}$. If a Fortin operator can be constructed and bounded by (5.5), a unique Petrov–Galerkin approximation $(\boldsymbol{y}_h, \hat{\boldsymbol{y}}_h) \in W_h \times \hat{V}_h$ of (5.7) exists and satisfies the a priori error estimate*

$$\|(\boldsymbol{y} - \boldsymbol{y}_h, \hat{\boldsymbol{y}} - \hat{\boldsymbol{y}}_h)\|_{Q;L,\partial Q_h} \le \frac{\sqrt{2}}{\beta_h} \inf_{(\boldsymbol{\phi}_h, \hat{\boldsymbol{\phi}}_h) \in W_h \times \hat{V}_h} \|(\boldsymbol{y} - \boldsymbol{\phi}_h, \hat{\boldsymbol{y}} - \hat{\boldsymbol{\phi}}_h)\|_{Q;L,\partial Q_h}.$$

6 The simplified DPG method

For the realization of the practical DPG method, it is advantageous to use traces on the skeleton ∂Q_h. This process depends on the application, and is now described for linear acoustic waves. For space-time tensor-product decompositions with space-time cells $R = K \times (a, b) \subset \Omega \times (0, T)$, we define a trace mapping $I_{\partial R}$ to $\mathrm{L}_2(\partial R; \mathbb{R} \times \mathbb{R}^d)$ by

$$I_{\partial R}(p_R, \boldsymbol{v}_R) = \begin{cases} (p_R, \boldsymbol{v}_R)_{|K \times \{t\}} & \text{for traces at time } t \in \{a, b\}, \\ (p_R, (\boldsymbol{v}_R \cdot \boldsymbol{n}_F)\boldsymbol{n}_F)_{|F \times (a,b)} & \text{in space with } F \subset \partial K \end{cases}$$

for all sufficiently smooth functions $(p_R, \boldsymbol{v}_R)$. We define local trace bilinear forms

$$\begin{aligned} &\gamma_R((\tilde{p}_R, \tilde{\boldsymbol{v}}_R), (q_R, \boldsymbol{w}_R)) \\ &\quad = ((\tilde{p}_R, \tilde{\boldsymbol{v}}_R), (q_R, \boldsymbol{w}_R))_{K \times \{b\}} - ((\tilde{p}_R, \tilde{\boldsymbol{v}}_R), (q_R, \boldsymbol{w}_R))_{K \times \{a\}} \\ &\qquad + \sum_{F \subset \partial K} (\tilde{p}_R, \boldsymbol{w}_R \cdot \boldsymbol{n}_K)_{F \times (a,b)} + (\tilde{\boldsymbol{v}}_R \cdot \boldsymbol{n}_K, q_R)_{F \times (a,b)} \end{aligned}$$

for $(\tilde{p}_R, \tilde{\boldsymbol{v}}_R) \in \mathrm{L}_2(\partial R; \mathbb{R} \times \mathbb{R}^d)$ and $(q_R, \boldsymbol{w}_R) \in \mathrm{H}(L^{\mathrm{ad}}, R)$ sufficiently smooth with $I_{\partial R}(q_R, \boldsymbol{w}_R) \in \mathrm{L}_2(\partial R; \mathbb{R} \times \mathbb{R}^d)$, and we define

$$\begin{aligned} b_h(((p, \boldsymbol{v}), (\tilde{p}, \tilde{\boldsymbol{v}})), (q, \boldsymbol{w})) = -(p, \partial_t q + \nabla \cdot \boldsymbol{w})_{Q_h} - (\boldsymbol{v}, \partial_t \boldsymbol{w} + \nabla q)_{Q_h} \\ + \gamma_h((\tilde{p}, \tilde{\boldsymbol{v}}), (q, \boldsymbol{w})) \end{aligned}$$

for $(p, \boldsymbol{v}) \in \mathrm{L}_2(Q; \mathbb{R} \times \mathbb{R}^d)$, $(\tilde{p}, \tilde{\boldsymbol{v}}) \in \mathrm{L}_2(\partial Q_h; \mathbb{R} \times \mathbb{R}^d)$, and for $(q, \boldsymbol{w}) \in \mathrm{H}(L^{\mathrm{ad}}, Q_h)$ with traces in L_2, where

$$\gamma_h((\tilde{p}, \tilde{\boldsymbol{v}}), (q, \boldsymbol{w})) = \sum_R \gamma_R((\tilde{p}_R, \tilde{\boldsymbol{v}}_R), (q_R, \boldsymbol{w}_R)).$$

By construction, we observe

$$\gamma_R\big(I_{\partial R}(p_R,\mathbf{v}_R),(q_R,\mathbf{w}_R)\big) = \big\langle D_R(p_R,\mathbf{v}_R),(q_R,\mathbf{w}_R)\big\rangle$$

for $(p_R,\mathbf{v}_R) \in \mathrm{H}(L,R)$ and $(q_R,\mathbf{w}_R) \in \mathrm{H}(L^{\mathrm{ad}},R)$ with traces in L_2, and

$$b_h\big(((p,\mathbf{v}),I_{\partial Q_h}(\tilde{p},\tilde{\mathbf{v}})),(q,\mathbf{w})\big) = \big\langle B_h((p,\mathbf{v}),\hat{I}_h(\tilde{p},\tilde{\mathbf{v}})),(q,\mathbf{w})\big\rangle$$

for $(p,\mathbf{v}) \in \mathrm{L}_2(Q;\mathbb{R}\times\mathbb{R}^d)$, and for $(\tilde{p},\tilde{\mathbf{v}}) \in \mathrm{H}(L,Q_h)$, and $(q,\mathbf{w}) \in \mathrm{H}(L^{\mathrm{ad}},Q_h)$ with traces in L_2.

Thus, in the realization of the DPG method, we can replace the operator B_h by the bilinear form $b_h(\cdot,\cdot)$, so that it is sufficient to represent $\hat{V}_h$ by its trace values on ∂Q_h.

In the simplified DPG method, we select independently polynomial ansatz spaces for the traces on every space-time face of the skeleton ∂Q_h; that is, we choose a discontinuous space

$$\tilde{V}_h = \prod_{K\times\{a\}\subset\partial Q_h} V_{K\times\{a\},h} \times \prod_{F\times(a,b)\subset\partial Q_h} V_{F\times(a,b),h} \subset \mathrm{L}_2(\partial Q_h;\mathbb{R}\times\mathbb{R}^d).$$

The representation of Neumann traces for $(\tilde{p}_h,\tilde{\mathbf{v}}_h) \in \tilde{V}_h$ requires to select an orientation $\mathbf{n}_F \in \{\pm\mathbf{n}_K\}$. Then, $\tilde{\mathbf{v}}_{h|F\times(a,b)} = \tilde{v}_h\mathbf{n}_F$ with $\tilde{v}_h \in \mathrm{L}_2(F\times(a,b))$. In the case where $\tilde{V}_h$ is the trace of a conforming subspace $V_h \subset V$, i. e., $\tilde{V}_h = I_{\partial Q_h}V_h$, the simplified method coincides with a conforming DPG method. In general, the skeleton space $\tilde{V}_h$ may be nonconforming. Then, we assume a weaker condition, which is described in what follows.

To obtain a well-defined method and to provide an a priori error analysis, we assume that a conforming reconstruction $V_h \subset V$ of $\tilde{V}_h$ exists such that for given $(\tilde{p}_h,\tilde{\mathbf{v}}_h) \in \tilde{V}_h$, the reconstruction $(\bar{p}_h,\bar{\mathbf{v}}_h) \in V_h$ is uniquely defined by

$$\gamma_R\big((\bar{p}_{R,h},\bar{\mathbf{v}}_{R,h}),(q_{R,h},\mathbf{w}_{R,h})\big) = \gamma_R\big(I_{\partial R,h}(\tilde{p}_h,\tilde{\mathbf{v}}_h),(q_{R,h},\mathbf{w}_{R,h})\big) \tag{6.1}$$

for all $(q_{R,h},\mathbf{w}_{R,h}) \in Z_{R,h}$ and all space-time cells R. In particular, this implies $\dim V_h = \dim\tilde{V}_h$. Note that the traces in V_h only coincide with functions $\tilde{V}_h$ when tested with the finite dimensional space $Z_{R,h}$.

Then, by construction, the simplified method with ansatz space $W_h \times \tilde{V}_h$ and test space Z_h yields the same discrete linear system as the practical method with $\tilde{V}_h$ replaced by $\hat{V}_h = V_h/\mathrm{H}_0(L,Q_h)$. For the error analysis, we introduce the discrete norm

$$\|(\tilde{p}_h,\tilde{\mathbf{v}}_h)\|_{Z_h'} = \sup_{(q_h,\mathbf{w}_h)\in Z_h} \frac{\gamma_h((\tilde{p}_h,\tilde{\mathbf{v}}_h),(q_h,\mathbf{w}_h))}{\|(q_h,\mathbf{w}_h)\|_{L^{\mathrm{ad}},Q_h}}, \qquad (\tilde{p}_h,\tilde{\mathbf{v}}_h) \in \tilde{V}_h. \tag{6.2}$$

This extends to a (mesh-dependent) seminorm in $\mathrm{L}_2(\partial Q_h;\mathbb{R}\times\mathbb{R}^d)$, and we observe for $(p,\mathbf{v}) \in V$ with trace $(\tilde{p},\tilde{\mathbf{v}}) = I_{\partial Q_h}(p,\mathbf{v}) \in \mathrm{L}_2(\partial Q_h;\mathbb{R}\times\mathbb{R}^d)$

$$\begin{aligned}
\|(\tilde p,\tilde{\mathbf v})\|_{Z_h'} &= \sup_{(q_h,\mathbf w_h)\in Z_h} \frac{\gamma_h((\tilde p,\tilde{\mathbf v}),(q_h,\mathbf w_h))}{\|(q_h,\mathbf w_h)\|_{L^{\mathrm{ad}},Q_h}}\\
&= \sup_{(q_h,\mathbf w_h)\in Z_h} \inf_{(p_0,\mathbf v_0)\in \mathrm H_0(L,Q_h)} \frac{\langle D_h(p+p_0,\mathbf v+\mathbf v_0),(q_h,\mathbf w_h)\rangle}{\|(q_h,\mathbf w_h)\|_{L^{\mathrm{ad}},Q_h}}\\
&\le \sup_{(q,\mathbf w)\in \mathrm H(L^{\mathrm{ad}},Q_h)} \inf_{(p_0,\mathbf v_0)\in \mathrm H_0(L,Q_h)} \frac{\langle D_h(p+p_0,\mathbf v+\mathbf v_0),(q,\mathbf w)\rangle}{\|(q,\mathbf w)\|_{L^{\mathrm{ad}},Q_h}}\\
&\le \inf_{(p_0,\mathbf v_0)\in \mathrm H_0(L,Q_h)} \|(p+p_0,\mathbf v+\mathbf v_0)\|_{L,Q_h} = \|(\hat p,\hat{\mathbf v})\|_{L,\partial Q_h},
\end{aligned} \tag{6.3}$$

with $(\hat p,\hat{\mathbf v}) = (p,\mathbf v) + \mathrm H_0(L,Q_h) \in \hat V$.

With respect to the seminorm (6.2), we can transfer the result in Theorem 5.1 to the simplified DPG method.

Theorem 6.1. *Assume that a conforming reconstruction $V_h \subset V$ of $\tilde V_h$ exists, satisfying* (6.1) *and* $\dim V_h = \dim \tilde V_h$.

(a) *If a Fortin operator can be constructed and bounded by* (5.5), *a unique Petrov–Galerkin approximation* $((p_h,\mathbf v_h),(\tilde p_h,\tilde{\mathbf v}_h)) \in W_h \times \tilde V_h$ *exists, solving*

$$b_h(((p_h,\mathbf v_h),(\tilde p_h,\tilde{\mathbf v}_h)),(q_h,\mathbf w_h)) = ((f,\mathbf g),(q_h,\mathbf w_h))_Q, \quad (q_h,\mathbf w_h) \in Z_h^{\mathrm{opt}}.$$

(b) *Let $(p,\mathbf v) \in V$ be the solution of* (2.2), *and assume that $(p,\mathbf v)$ is sufficiently regular with traces $(\tilde p,\tilde{\mathbf v}) = I_{\partial Q_h}(p,\mathbf v) \in \mathrm L_2(\partial Q_h;\mathbb R\times\mathbb R^d)$. Then, the error can be bounded by*

$$\begin{aligned}
&\|((p,\mathbf v)-(p_h,\mathbf v_h)),((\tilde p,\tilde{\mathbf v})-(\tilde p_h,\tilde{\mathbf v}_h))\|_{W\times Z_h'}\\
&\quad\le (1+\sqrt2\beta_h^{-1}) \inf_{((\phi_h,\boldsymbol\psi_h),(\tilde\phi_h,\tilde{\boldsymbol\psi}_h))\in W_h\times\tilde V_h} \|((p,\mathbf v)-(\phi_h,\boldsymbol\psi_h)),((\tilde p,\tilde{\mathbf v})-(\tilde\phi_h,\tilde{\boldsymbol\psi}_h))\|_{W\times Z_h'}.
\end{aligned}$$

Proof. For the conforming reconstruction $V_h \subset V$ of $\tilde V_h$ and $\hat V_h = V_h/\mathrm H_0(L,Q_h)$, the practical DPG method applies, providing an error estimate in $W_h \times \hat V_h$ by Theorem 5.1.

Let $(p,\mathbf v) \in V$ be the solution of (2.2), let

$$(\tilde p,\tilde{\mathbf v}) = I_{\partial Q_h}(p,\mathbf v) \in \mathrm L_2(\partial Q_h;\mathbb R\times\mathbb R^d)$$

its trace, and set $(\hat p,\hat{\mathbf v}) = (p,\mathbf v) + \mathrm H_0(L,Q_h) \in \hat V$.

For the discrete solution $((p_h,\mathbf v_h),(\tilde p_h,\tilde{\mathbf v}_h)) \in W_h \times \tilde V_h$ let $(\bar p_h,\bar{\mathbf v}_h) \in V_h$ be the conforming reconstruction of $(\tilde p_h,\tilde{\mathbf v}_h)$, and set

$$(\hat p_h,\hat{\mathbf v}_h) = (\bar p_h,\bar{\mathbf v}_h) + \mathrm H_0(L,Q_h) \in \hat V_h.$$

Then, (6.1) and (6.3) yield

$$\|(\tilde p_h,\tilde{\mathbf v}_h)\|_{Z_h'} = \|I_{\partial R,h}(\bar p_h,\bar{\mathbf v}_h)\|_{Z_h'} \le \|(\hat p_h,\hat{\mathbf v}_h)\|_{L,\partial Q_h}.$$

Now, for some $((\phi_h,\boldsymbol\psi_h),(\tilde\phi_h,\tilde{\boldsymbol\psi}_h)) \in W_h \times \tilde V_h$ let $(\bar\phi_h,\bar{\boldsymbol\psi}_h) \in V_h$ be the conforming reconstruction of $(\tilde\phi_h,\tilde\psi_h)$, and set $(\hat\phi_h,\hat{\boldsymbol\psi}_h) = (\bar\phi_h,\bar{\boldsymbol\psi}_h) + \mathrm H_0(L,Q_h)$. Then

$$
\begin{aligned}
&\beta_h \|((p_h, \mathbf{v}_h) - (\phi_h, \boldsymbol{\psi}_h), (\hat{p}_h, \hat{\mathbf{v}}_h) - (\hat{\phi}_h, \hat{\boldsymbol{\psi}}_h))\|_{Q;L,\partial Q_h} \\
&\quad \le \sup_{(q_h, \mathbf{w}_h)\in Z_h} \frac{\langle B_h((p_h, \mathbf{v}_h) - (\phi_h, \boldsymbol{\psi}_h), (\hat{p}_h, \hat{\mathbf{v}}_h) - (\hat{\phi}_h, \hat{\boldsymbol{\psi}}_h)), (q_h, \mathbf{w}_h)\rangle}{\|(q_h, \mathbf{w}_h)\|_{L^{\mathrm{ad}}, Q_h}} \\
&\quad \le \|(p, \mathbf{v}) - (\phi_h, \boldsymbol{\psi}_h)\|_W + \sup_{(q_h, \mathbf{w}_h)\in Z_h} \frac{\langle \hat{D}_h((\hat{p}, \hat{\mathbf{v}}) - (\hat{\phi}_h, \hat{\boldsymbol{\psi}}_h)), (q_h, \mathbf{w}_h)\rangle}{\|(q_h, \mathbf{w}_h)\|_{L^{\mathrm{ad}}, Q_h}} \\
&\quad = \|(p, \mathbf{v}) - (\phi_h, \boldsymbol{\psi}_h)\|_W + \sup_{(q_h, \mathbf{w}_h)\in Z_h} \frac{\gamma_h((\tilde{p}, \tilde{\mathbf{v}}) - (\tilde{\phi}_h, \tilde{\boldsymbol{\psi}}_h), (q_h, \mathbf{w}_h))}{\|(q_h, \mathbf{w}_h)\|_{L^{\mathrm{ad}}, Q_h}} \\
&\quad = \|(p, \mathbf{v}) - (\phi_h, \boldsymbol{\psi}_h)\|_W + \|(\tilde{p}, \tilde{\mathbf{v}}) - (\tilde{\phi}_h, \tilde{\boldsymbol{\psi}}_h)\|_{Z_h'} \\
&\quad \le \sqrt{2} \|((p, \mathbf{v}) - (\phi_h, \boldsymbol{\psi}_h), (\tilde{p}, \tilde{\mathbf{v}}) - (\tilde{\phi}_h, \tilde{\boldsymbol{\psi}}_h))\|_{W\times Z_h'},
\end{aligned}
$$

using discrete inf-sup stability (5.6). Together with

$$
\begin{aligned}
\|(\tilde{\phi}_h, \tilde{\boldsymbol{\psi}}_h) - (\tilde{p}_h, \tilde{\mathbf{v}}_h)\|_{Z_h'} &= \|I_{\partial Q_h}(\hat{\phi}_h, \hat{\boldsymbol{\psi}}_h) - I_{\partial Q_h}(\hat{p}_h, \hat{\mathbf{v}}_h)\|_{Z_h'} \\
&\le \|(\hat{\phi}_h, \hat{\boldsymbol{\psi}}_h) - (\hat{p}_h, \hat{\mathbf{v}}_h)\|_{L,\partial Q_h},
\end{aligned}
$$

we finally obtain

$$
\begin{aligned}
&\|((p, \mathbf{v}) - (p_h, \mathbf{v}_h), (\tilde{p}, \tilde{\mathbf{v}}) - (\tilde{p}_h, \tilde{\mathbf{v}}_h))\|_{W\times Z_h'} \\
&\quad \le \|((p, \mathbf{v}) - (\phi_h, \boldsymbol{\psi}_h), (\tilde{p}, \tilde{\mathbf{v}}) - (\tilde{\phi}_h, \tilde{\boldsymbol{\psi}}_h))\|_{W\times Z_h'} \\
&\qquad + \|((\phi_h, \boldsymbol{\psi}_h) - (p_h, \mathbf{v}_h), (\tilde{\phi}_h, \tilde{\boldsymbol{\psi}}_h) - (\tilde{p}_h, \tilde{\mathbf{v}}_h))\|_{W\times Z_h'} \\
&\quad \le \|((p, \mathbf{v}) - (\phi_h, \boldsymbol{\psi}_h), (\tilde{p}, \tilde{\mathbf{v}}) - (\tilde{\phi}_h, \tilde{\boldsymbol{\psi}}_h))\|_{W\times Z_h'} \\
&\qquad + \|((\phi_h, \boldsymbol{\psi}_h) - (p_h, \mathbf{v}_h), (\hat{\phi}_h, \hat{\boldsymbol{\psi}}_h) - (\hat{p}_h, \hat{\mathbf{v}}_h))\|_{Q;L,\partial Q_h} \\
&\quad \le \left(1 + \frac{\sqrt{2}}{\beta_h}\right) \|((p, \mathbf{v}) - (\phi_h, \boldsymbol{\psi}_h), (\tilde{p}, \tilde{\mathbf{v}}) - (\tilde{\phi}_h, \tilde{\boldsymbol{\psi}}_h))\|_{W\times Z_h'}. \qquad \square
\end{aligned}
$$

The reconstruction space V_h is completely virtual; it is not required for the realization of the simplified DPG solution. On the other hand, one needs an explicit representation of V_h for the estimate of the discrete inf-sup constant as described in the previous section.

For the numerical solution, the discrete Petrov–Galerkin equation is reduced to a positive definite Schur complement problem for $(\tilde{p}_h, \tilde{\mathbf{v}}_h)$; see [22, Lemma 9] for explicit estimates for the Schur complement, depending on β_h and C_L.

7 Numerical experiments

To evaluate the efficiency of the simplified space-time DPG method, we consider two examples in two space dimensions; that is, $d = 2$. In the first numerical test, we use a configuration, where the exact solution is known, so that we can compare the approximation results with the a priori estimate in Theorem 6.1. The second test illustrates the application to a double-slit experiment. Also see [11, Section 5] for a more detailed evaluation of this space-time DPG method.

For both examples, we use the discrete spaces

$$W_{R,h} = \mathbb{Q}_2(R) \times \mathbb{Q}_2(R)^2, \quad Z_{R,h} = \mathbb{Q}_4(R) \times \mathbb{Q}_4(R)^2$$

with tensor product polynomial spaces $\mathbb{Q}_2(R) = \mathbb{P}_2 \otimes \mathbb{P}_2 \otimes \mathbb{P}_2$ in the space-time cells $R = K \times (a, b)$, and on the skeleton ∂Q_h

$$V_{K\times\{t\},h} = \mathbb{Q}_2(K) \times \mathbb{Q}_2(K)^2 \text{ for faces in time, and}$$
$$V_{F\times(a,b),h} = \mathbb{Q}_2(F \times (a,b)) \times \mathbb{Q}_2(F \times (a,b))\boldsymbol{n}_F \text{ for faces in space.}$$

7.1 The construction of the Fortin operator

In case of conforming trace approximations $\tilde{V}_h$ and simple meshes, it is sufficient to construct the Fortin operator in a reference element R_0, and then the estimates for the Fortin operator in $R \subset Q_h$ follow from the scaling argument in Section 5.2.

In the nonconforming case, a conforming reconstruction $V_h \subset V$ with (6.1) has to be computed. Therefore, we compute a minimum energy extension of trace functions in $\tilde{V}_{R,h}$. On each cell R, we select a basis $\{(\tilde{p}_1, \tilde{\boldsymbol{v}}_1), \dots, (\tilde{p}_N, \tilde{\boldsymbol{v}}_N)\}$ of $\tilde{V}_{R,h}$ and an extension space $V_{R,h} \subset \mathrm{H}(L, R)$. Then we compute $(\bar{p}_1, \bar{\boldsymbol{v}}_1), \dots, (\bar{p}_N, \bar{\boldsymbol{v}}_N) \in V_{R,h}$ by solving the minimization problem

$$\min_{(\bar{p}_n, \bar{\boldsymbol{v}}_n) \in V_{R,h}(\tilde{p}_n, \tilde{\boldsymbol{v}}_n)} \|(\bar{p}_n, \bar{\boldsymbol{v}}_n)\|_{L,R}$$

in the affine space

$$V_{R,h}(\tilde{p}_n, \tilde{\boldsymbol{v}}_n) = \big\{ (\bar{p}_n, \bar{\boldsymbol{v}}_n) \in V_{R,h} \colon \gamma_R\big((\bar{p}_n, \bar{\boldsymbol{v}}_n) - (\tilde{p}_n, \tilde{\boldsymbol{v}}_n), (q_{R,h}, \boldsymbol{w}_{R,h})\big) = 0 \text{ for } (q_{R,h}, \boldsymbol{w}_{R,h}) \in Z_{R,h} \big\},$$

see Fig. 3.1 for an illustration. The resulting estimates for the Fortin operator for different polynomial degrees are listed in Tab. 3.1.

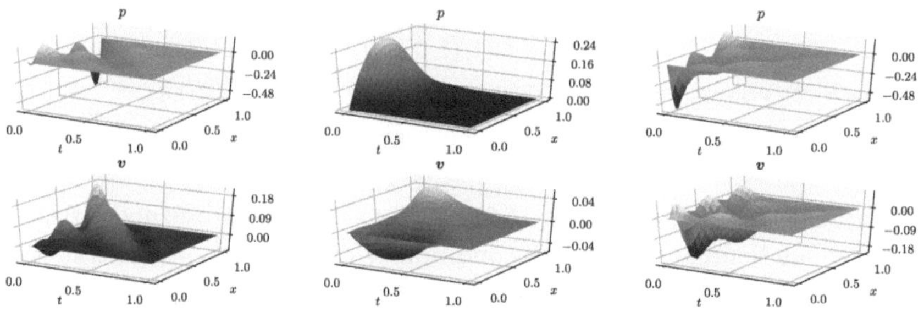

Figure 3.1: Conforming reconstructions in $V_{R,h} = \mathbb{Q}_6(R) \times \mathbb{Q}_6(R)$ for $d = 1$ of the trace space $\tilde{V}_K = \mathbb{P}_2 \times \mathbb{P}_2$ on a face $K \subset \partial R$, and test space $Z_{R,h} = \mathbb{Q}_4(R)^2$. We show the extensions $\bar{p}_n$ and $\bar{\boldsymbol{v}}_n$ for the three nodal basis functions in $\mathbb{P}_2$.

Table 3.1: We present two upper bounds for $\|\Pi_{R,h}\|_{L^{\mathrm{ad}},R}$ in two space dimensions with $R = (0, a_1 h_k) \times (0, a_2 h_k) \times (0, c h_k)$ and $a_1 \approx a_2 \approx c \approx 1$. Left: Numerical norm estimates with ansatz space $W_{R,h} = \mathbb{Q}_p(R)^3$, test space $Z_{R,h} = \mathbb{Q}_{p+2}(R)^3$, and extension space $\mathbb{Q}_{p+4}(R)^3 \supset \tilde{V}_{R,h}$. The estimates depend on the mesh size $h_k = 2^k$ and the polynomial degree p. Right: Numerical estimate on the reference cell R_0 with $W^{\mathrm{ext}}_{R,h} = \mathbb{Q}_{p+1}(R)^3$. This yields an inf-sup constant independent of h by the scaling argument in Section 5.2.

$\|\Pi_{R,h}\|_{L^{\mathrm{ad}},R}$	h_0	h_1	h_2	h_3	h_0
$p = 0$	2.067	2.161	2.18	2.19	2.98
$p = 1$	12.039	18.817	32.87	62.68	34.85
$p = 2$	35.861	64.140	116.78	239.71	144.78

Computational setup

The numerical calculations were performed on a parallel computer consisting of two nodes, each of which features AMD Opteron(TM) 6274 processors with 64 cores and 512 GB memory. We used a GMRES method preconditioned by a local symmetric Gauß–Seidel iteration on every parallel subdomain to solve the linear systems. This performs reasonably well for our examples; it remains an open task to realize an efficient multigrid preconditioner for this application.

7.2 A plane-wave solution

We consider a rectangular domain $\Omega = (-2, 2) \times (0, 1)$ with a nonhomogeneous material distribution for density ρ and bulk modulus κ:

$$(\rho(x_1, x_2), \kappa(x_1, x_2)) = \begin{cases} (1, 1), & x_1 < 0, \\ (0.5, 2), & x_1 \in (0, 1), \\ (7/5, 5/7), & x_1 > 1, \end{cases}$$

so that the system (2.1) has a plane-wave solution with amplitude $a(\cdot)$:

$$(p(x_1, x_2, t), \mathbf{v}(x_1, x_2, t)) = \begin{cases} a(x_1 - t)(1, (1, 0)), & x_1 < 0, \\ a(0.5x_1 - t)(1, (1, 0)), & x_1 \in (0, 1), \\ a(1.4x_1 - 0.9 - t)(1, (1, 0)), & x_1 > 1, \end{cases}$$

see [13, Section 3.5], where $a(s) = \cos((5s - 4.5)\pi/2)^2$ for $s \in (-1.7, 1.3)$ and $a(s) = 0$ else. We use homogenous Neumann boundary conditions $\mathbf{v} \cdot \mathbf{n} = 0$ for $x_2 = 0$ and $x_2 = 1$, and homogeneous Dirichlet boundary conditions $p = 0$ for $x_1 = \pm 2$.

Note that due to the special choice of material parameters, the analytic solution does not feature reflections on the interfaces. Figure 3.2 illustrates the evolution of this solution in the space-time cylinder.

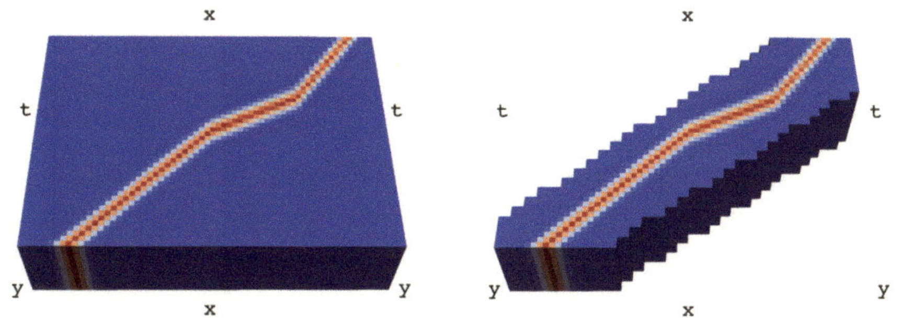

Figure 3.2: A wave front traveling from left to the right through three different materials. Here, the full space-time mesh with 3,193,344 DoFs can be truncated resulting in 1,284,984 DoFs whereas the approximation quality remains unchanged.

The convergence of the DPG method for this example is evaluated for a sequence of regular meshes, where each cuboid is subdivided in 8 congruent parts on refinement, see Tab. 3.2.

Table 3.2: The convergence of $(p_h, \mathbf{v}_h)$ in $L_2(Q; \mathbb{R} \times \mathbb{R}^2)$ on a sequence of uniformly refined meshes with $h_k = 2^{-k} h_0$, $h_0 = 1$.

DoFs:	**7,560**	**54,432**	**411,264**	**3,193,344**
e_h:	0.493784	0.244424	0.081376	0.028764
e_{2h}/e_h:		2.020197	3.003630	2.829075
$\log_2 e_{2h}/e_h$:		1.014495	1.586707	1.500330

Now we exploit the support of the solution contained in a small fraction of the space-time cylinder Q. To this end, we use the simulation on a coarse mesh to identify a superset of this support. In a second step, we truncate the space-time mesh by dropping cells, where the solution vanishes. The resulting new space-time boundaries are enhanced with zero boundary or initial conditions. As a result, we have reduced the amount of DoFs to approximate the solution, while conserving the approximation quality. See Figure 3.2 for an example of this procedure and Table 3.3 for a convergence study.

Table 3.3: The convergence of $(p, \mathbf{v})$ in the $L_2(Q)$ norm on a sequence of truncated meshes for $h_k = 2^{-k} h_0$, $h_0 = 2^{-3}$, is shown.

DoFs:	**168,534**	**1,284,984**	**10,026,720**	**79,201,152**
e_h:	0.08127	0.02902	0.00519	0.00114
e_{2h}/e_h:		2.79989	5.59012	4.51952
$\log_2 e_{2h}/e_h$:		1.485373	2.482881	2.176172

7.3 The diffraction pattern from a double slit

The second example considers a double-slit experiment, where two coherent wave fronts enter the domain through a pair of small slits. By Huygens principle, a circular wave is propagated from each of the slits, yielding a characteristic inference pattern, see Figure 3.3 for a description of the setup, and Figure 3.4 for visualizations of the solution. The boundary $\partial\Omega = \Gamma_\mathrm{N} \cup \Gamma_\mathrm{D}$ is partitioned in a Neumann part Γ_N and a Dirichlet part Γ_D, where we use $\boldsymbol{v} \cdot \boldsymbol{n} = 0$ on $\Gamma_\mathrm{N} \times (0, T)$ and $p(x, t) = \sin(2\pi\omega(x - t))$ for $(x, t) \in \Gamma_\mathrm{D} \times (0, T)$ with $\omega = 2$.

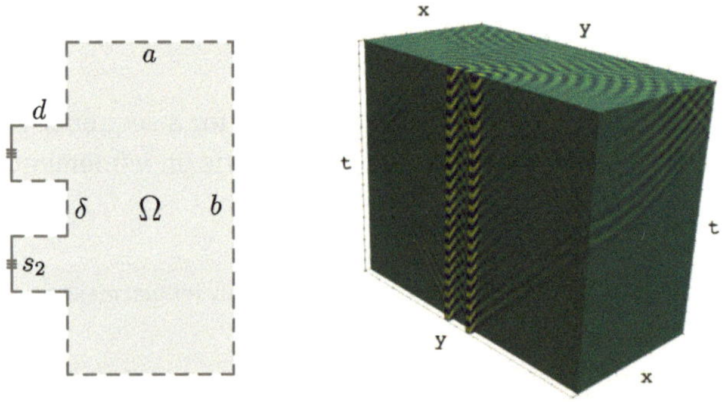

Figure 3.3: The spatial domain Ω is described on the left, where the slit dimensions are $d = s_1 = s_2 = 0.25$, their distance is $\delta = 1$, and the dimensions of the large rectangle are $a = 6$, $b = 12$. Ω is substructured using a regular mesh Ω_h of squares with side lengths 0.25. The corresponding space-time cylinder $Q = \Omega \times (0, T)$ is discretized using tensor-product elements $R = K \times (t_{n-1}, t_n)$ for each cell $K \in \Omega_h$ and $t_n = nT/N$, $n = 0, \ldots, N$, with $T = 10$ and $N = 50$. The dashed portion of $\partial\Omega$ indicates Γ_N and the remaining faces, marked by three lines, represent Γ_D. On the right, a space-time plot of the solution is given on a two times refined version of this mesh featuring 3,692,800 space-time cells and 234,210,528 face DoFs.

This example compares the results in [12, Section 8.3.2] for a space-time discontinuous Galerkin method for 2D Maxwell's equations and a similar double-slit setup. The experiment demonstrates that the space-time DPG method is able to approximate a complex wave pattern in a physically motivated example. An in-depth comparison of the space-time DPG method to other methods like discontinuous Galerkin remains as a challenge for the future.

8 Conclusion and outlook

We presented a space-time framework for acoustic waves, including an appropriate abstract trace space, and we constructed and analyzed a space-time DPG method

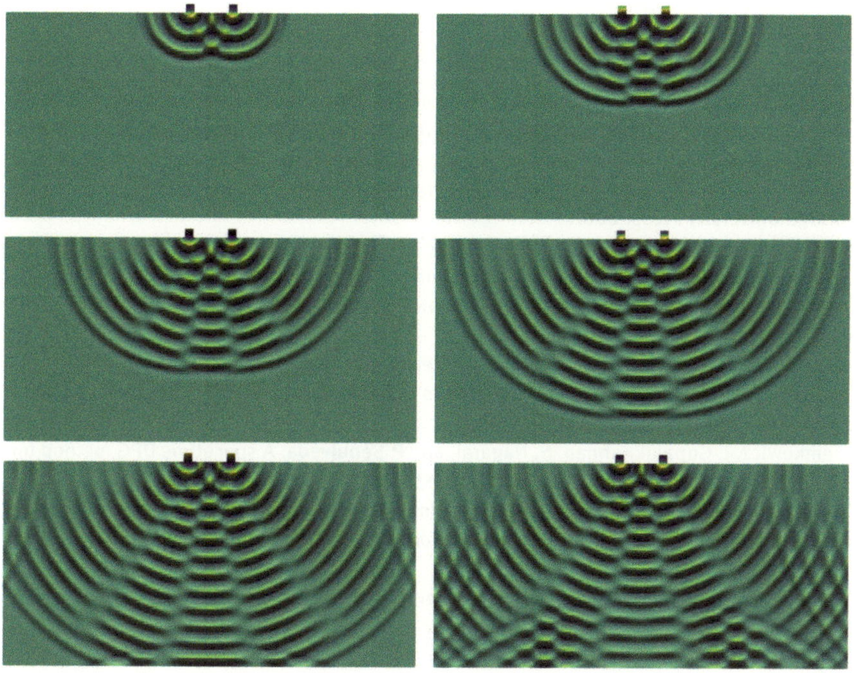

Figure 3.4: Snapshots of the pressure component at times $t = 0.6, 2.08, 3.56, 5.04, 6.52, 8$. These were obtained by slicing the space-time solution from Figure 3.3 along planes that are orthogonal to the time direction.

within this setting. Moreover, we considered a nonconforming variant with appealing properties from an implementation point of view; we also provided a numerically accessible criterion to compute a bound for the norm of the Fortin operator. We demonstrated the flexibility of the method by providing a numerical example with a truncated space-time mesh, which reduces the size of the linear system by a factor of three. Furthermore, we presented an example for a double-slit experiment as a starting point for upcoming in-depth comparisons of space-time DPG to established discretization methods.

For the presented theory, we used tools from semigroups and functional analysis, which are also applicable to other first-order systems, such as electro–magnetic or elastic waves. Moreover, we assumed a constant material distribution to keep our notation simple. A consideration of more general wave equations, taking into account spatially varying material properties, remains as a future challenge. To render this space-time discretization competitive to classical schemes, a preconditioner is necessary that scales well with respect to the mesh size and also with the number of processes used. In the future, we would like to consider multigrid algorithms, which are promising candidates in this respect, see, for example, [8, 12] for a multigrid preconditioned space-time discontinuous Galerkin method.

Bibliography

[1] F. Boyer and P. Fabrie. *Mathematical tools for the study of the incompressible Navier–Stokes equations and related models.* Volume 183 of Applied Mathematical Sciences, Springer, New York, 2013.

[2] D. Braess. *Finite elements. Theory, fast solvers, and applications in solid mechanics. 3th ed.* Cambridge University Press, 2007.

[3] F. Brezzi and M. Fortin. *Mixed and hybrid finite element methods.* Springer, 1991.

[4] C. Carstensen, L. Demkowicz, and J. Gopalakrishnan. Breaking spaces and forms for the DPG method and applications including Maxwell equations. *Comput. Math. Appl.*, 72 (3): 494–522, 2016.

[5] L. F. Demkowicz and J. Gopalakrishnan. An overview of the discontinuous Petrov–Galerkin method. In *Recent developments in discontinuous Galerkin finite element methods for partial differential equations*, pages 149–180. Springer, 2014.

[6] L. F. Demkowicz, J. Gopalakrishnan, S. Nagaraj, and P. Sepulveda. A spacetime DPG method for the Schrödinger equation. *SIAM J. Numer. Anal.*, 55 (4): 1740–1759, 2017.

[7] W. Dörfler, S. Findeisen, and C. Wieners. Space-time discontinuous Galerkin discretizations for linear first-order hyperbolic evolution systems. *Comput. Meth. Appl. Math.*, 16 (3): 409–428, 2016.

[8] W. Dörfler, S. Findeisen, C. Wieners, and D. Ziegler. Parallel adaptive discontinuous Galerkin discretizations in space and time for linear elastic and acoustic waves. In U. Langer and O. Steinbach, editors, *Space-time methods. Applications to partial differential equations*, Volume 25 of Radon Series on Computational and Applied Mathematics, pages 61–88, de Gruyter, Berlin, 2019.

[9] A. Ern and J.-L. Guermond. A converse to Fortin's lemma in Banach spaces. *Comptes Rendus Mathematique*, 354 (11): 1092–1095, 2016.

[10] A. Ern, J.-L. Guermond, and G. Caplain. An intrinsic criterion for the bijectivity of Hilbert operators related to Friedrichs systems. *Comm. Partial Differential Equations*, 32 (1–3): 317–341, 2007.

[11] J. Ernesti. *Space-time methods for acoustic waves with applications to full waveform inversion.* PhD thesis, Karlsruher Institut für Technologie (KIT), 2018.

[12] S. Findeisen. *A parallel and adaptive space-time method for Maxwell's equations.* PhD thesis, Karlsruher Institut für Technologie (KIT), 2016.

[13] J.-P. Fouque, J. Garnier, G. Papanicolaou, and K. Solna. *Wave propagation and time reversal in randomly layered media.* Volume 56 of Stochastic Modelling and Applied Probability, Springer, New York, 2007.

[14] J. Gopalakrishnan and P. Sepulveda. A space-time DPG method for the wave equation in multiple dimensions. In U. Langer and O. Steinbach, editors, *Space-time methods. Applications to partial differential equations.* Volume 25 of Radon Series on Computational and Applied Mathematics, pages 117–139, de Gruyter, Berlin, 2019.

[15] M. Hochbruck, T. Pazur, A. Schulz, E. Thawinan, and C. Wieners. Efficient time integration for discontinuous Galerkin approximations of linear wave equations. *ZAMM*, 95 (3): 237–259, 2015.

[16] A. Kirsch and A. Rieder. Inverse problems for abstract evolution equations with applications in electrodynamics and elasticity. *Inverse Problems*, 32 (8): 085001, 2016.

[17] S. Nagaraj, S. Petrides, and L. F. Demkowicz. Construction of DPG Fortin operators for second order problems. *Comput. Math. Appl.*, 74 (8): 1964–1980, 2017.

[18] R. Racke. *Lectures on nonlinear evolution equations.* Birkhäuser, 1992. 2nd ed., 2015.

[19] M. Renardy and R. C. Rogers. *An introduction to partial differential equations.* Springer, 2006.

[20] W. Rudin. *Functional analysis*. International Series in Pure and Applied Mathematics. McGraw–Hill, Inc., New York, 1991.
[21] C. Wieners. A geometric data structure for parallel finite elements and the application to multigrid methods with block smoothing. *Comput. Visual. Sci.*, 13: 161–175, 2010.
[22] C. Wieners. The skeleton reduction for substructuring methods. In *Numerical mathematics and advanced applications ENUMATH 2015*. Volume 112 of Lecture Notes in Computational Science and Engineering, pages 133–142. Springer, 2016.
[23] C. Wieners and B. Wohlmuth. Robust operator estimates and the application to substructuring methods for first-order systems. *ESAIM Math. Model. Numer. Anal.*, 48 (5): 1473–1494, 2014.
[24] J. Xu and L. Zikatanov. Some observations on Babuška and Brezzi theories. *Numer. Math.*, 94 (1): 195–202, 2003.

Jay Gopalakrishnan and Paulina Sepúlveda

4 A space-time DPG method for the wave equation in multiple dimensions

Abstract: A space-time discontinuous Petrov–Galerkin (DPG) method for the linear wave equation is presented. This method is based on a weak formulation that uses a broken graph space. The wellposedness of this formulation is established using a previously presented abstract framework. One of the main tasks in the verification of the conditions of this framework is proving a density result. This is done in detail for a simple domain in arbitrary dimensions. The DPG method based on the weak formulation is then studied theoretically and numerically. Error estimates and numerical results are presented for triangular, rectangular, tetrahedral, and hexahedral meshes of the space-time domain. The potential for using the built-in error estimator of the DPG method for an adaptive mesh refinement strategy in two and three dimensions is also presented.

Keywords: time-dependent, wave equation, hyperbolic, discontinuous Petrov–Galerkin, finite element method

MSC 2010: 35L05, 58J45, 65M60

1 Introduction

This is a study on the feasibility of the discontinuous Petrov–Galerkin (DPG) method [5, 6] for the space-time wave equation. We follow the approach laid out in our earlier study of the DPG method for the space-time Schrödinger equation [7].

Currently, the most widely used numerical techniques for transient problems are time-stepping schemes (based on the method of lines approach). However, there has been increasing interest recently in direct space-time discretizations (where time is viewed as just another coordinate). Some reasons for investigating these approaches include their potential for performing natural space-time adaptivity, possibility to ob-

Acknowledgement: This work was partly supported by AFOSR grant FA9550–17–1–0090. Numerical studies were partially facilitated by the Portland Institute of Sciences (PICS) established under NSF grant DMS–1624776. Paulina Sepúlveda has received funding from the Spanish Ministry of Economy and Competitiveness with reference MTM2016–76329–R (AEI/FEDER, EU) and BCAM "Severo Ochoa" accreditation of excellence SEV–2013–0323 and SEV–2017–0718.

Jay Gopalakrishnan, Portland State University, PO Box 751, Portland, OR 97207-0751, USA, e-mail: gjay@pdx.edu
Paulina Sepúlveda, Basque Center for Applied Mathematics, Mazarredo 14, Bilbao, E48009, Spain, e-mail: psepulveda@bcamath.org

https://doi.org/10.1515/9783110548488-004

tain convergence even under limited space-time regularity, exploitation of parallelism without causality constraints, and treatment of moving boundaries (see, for example, [7, 16, 17, 19, 20]). The analysis and implementation of 4D finite element discretizations is already underway [14, 20], hence our interest in obtaining a well-posed formulation in arbitrary dimensions.

Since the DPG method has a built-in error estimator and exhibits good pre-asymptotic mesh-independent stability properties (see [3]), it is natural to consider its extension to space-time problems. Applications of the DPG method for space-time problems have already been computationally studied in [11] for the transient parabolic partial differential equations and [10] for the time-dependent convection-diffusion equation. We also note that a scheme that combines DPG spatial discretization with backward Euler time stepping for the heat equation has been analyzed in [12].

In contrast to these works, here we consider the transient acoustic wave system in arbitrary dimensions. One contribution of this work is a proof of the wellposedness of the ultraweak DPG formulation for the space-time wave problem in a nonstandard Hilbert space, without developing a trace theory for this function space. By using the abstract theory developed in [7], the proof reduces to verification of some conditions. This verification proceeds by proving a density result. The presented proof only applies for a multidimensional hyper-rectangle.

We also present, both practically and theoretically, how the built-in DPG error estimator is useful for space-time adaptive refinement in two and three dimensions using conforming meshes of simplices. We also show that depending on how the interfacial variables are treated, one may end up with a discrete DPG system that has a nontrivial kernel for some alignment of mesh facets, a difficulty that we have not previously encountered in any other DPG example. We then provide practical solutions for solving for the DPG wave approximations despite the null space. The solutions computed using these techniques were observed to converge at the optimal rate.

In Section 2, we introduce the model wave problem and put it into the abstract variational setting of [7]. In Section 3, we introduce a broken weak formulation (upon which the DPG method is based) and prove its wellposedness subject to a density condition. In Section 4, we give a proof of the density condition for a simple multidimensional domain. In Section 5, we prove error estimates for the ideal DPG method for solutions with enough regularity. Finally, numerical experiments and implementation techniques are presented in Section 6.

2 The transient wave problem

Let Ω_0 be a spatial domain in $\mathbb{R}^d$, with Lipschitz boundary $\partial\Omega_0$, and let $\Omega = \Omega_0 \times (0, T)$ be the space-time domain, where $T > 0$ represents the final time. We consider the first-order system for the wave equation given by

$$\partial_t q - c\,\mathrm{grad}_x\, \mu = g, \tag{2.1a}$$
$$\partial_t \mu - c\,\mathrm{div}_x\, q = f, \tag{2.1b}$$

where $f \in L^2(\Omega)$ and $g \in L^2(\Omega)^d$ and $c > 0$ is the constant wave speed. Here, the differential operators div_x and grad_x represent the (distributional) divergence and gradient operators that differentiate only along the spatial components (x). We add homogeneous initial and boundary conditions:

$$\mu|_{t=0} = 0, \quad q|_{t=0} = 0, \quad \mu|_{\partial\Omega_0\times(0,T)} = 0. \tag{2.1c}$$

Here, q represents the velocity and μ the pressure. We now cast this problem in the framework of the abstract setting in [7, Appendix A].

2.1 The formal wave operator

Formally, the wave operator generated by the above system may be considered as a first-order distributional derivative operator. Namely, set $A : L^2(\Omega)^{d+1} \to \mathcal{D}'(\Omega)^{d+1}$ by

$$Au = \begin{pmatrix} \partial_t u_q - c\,\mathrm{grad}_x\, u_\mu \\ \partial_t u_\mu - c\,\mathrm{div}_x\, u_q \end{pmatrix}, \tag{2.2}$$

where u in $L^2(\Omega)^{d+1}$ is block partitioned into

$$u = \begin{pmatrix} u_q \\ u_\mu \end{pmatrix}, \quad u_q \in L^2(\Omega)^d, \quad u_\mu \in L^2(\Omega). \tag{2.3}$$

Next, we introduce the space

$$W(\Omega) = \{u \in L^2(\Omega)^{d+1} : Au \in L^2(\Omega)^{d+1}\}.$$

By $W(K)$, we mean the similarly defined space on an open subset K of Ω, but when considering this space with domain Ω, we abbreviate $W(\Omega) = W$. Hereon, we denote by $(\cdot,\cdot)$ and $\|\cdot\|$ the $L^2(\Omega)^{d+1}$ inner product and norm, respectively, and $\mathcal{D}(\Omega)^{d+1}$ and $\mathcal{D}'(\Omega)^{d+1}$ is the space of infinitely differentiable vector functions with compact support in Ω, and its dual space, respectively. It is well known that the space $W(\Omega)$ is a Hilbert space when endowed with the graph norm $\|u\|_W = (\|u\|^2 + \|Au\|^2)^{1/2}$ (see [7, Lemma A.1.]). The formal adjoint of A is the distributional differentiation operator $-A$, and it satisfies

$$(Aw, \tilde{w}) = -(w, A\tilde{w}) \quad \text{for all } w, \tilde{w} \in \mathcal{D}(\Omega)^{d+1}.$$

Define the operator $D : W \to W'$ by

$$\langle Du, v\rangle_W = (Au, v)_\Omega + (u, Av)_\Omega \quad \text{for all } u, v \in W. \tag{2.4}$$

Here, W' is the dual space of W, and $\langle\cdot,\cdot\rangle_W$ represents the duality pairing of a functional in W' with an element of W. For smooth functions $u, v \in \mathcal{D}(\bar{\Omega})^{d+1}$, integration by parts shows that

$$\langle Du, v\rangle_W = \int_{\partial\Omega} u_q \cdot (n_t v_q - c n_x v_\mu) + u_\mu(n_t v_\mu - c n_x \cdot v_q). \tag{2.5}$$

Here and throughout, $n = (n_x^\top, n_t)^\top$ represents the unit outward normal vector of Ω in $\mathbb{R}^{d+1}$, and functions in $L^2(\Omega)^{d+1}$, like the u and v above, are block partitioned as in (2.3).

2.2 The unbounded wave operator

To consider the boundary and initial conditions, we now proceed as suggested in [7, Appendix A], to define an unbounded operator with a domain that takes these conditions into account. Below, by abusing the notation, we shall denote this unbounded operator also by A.

First, let us partition the space-time boundary $\partial\Omega$ into

$$\Gamma_0 = \Omega_0 \times \{0\}, \quad \Gamma_T = \Omega_0 \times \{T\}, \quad \Gamma_b = \partial\Omega_0 \times [0, T].$$

We define the following sets of smooth functions:

$$\mathcal{V} = \{u \in \mathcal{D}(\bar{\Omega})^{d+1} : u|_{\Gamma_0} = 0, u_\mu|_{\Gamma_b} = 0\}, \tag{2.6}$$

$$\mathcal{V}^* = \{v \in \mathcal{D}(\bar{\Omega})^{d+1} : v|_{\Gamma_T} = 0, v_\mu|_{\Gamma_b} = 0\}. \tag{2.7}$$

Next, let $A : \operatorname{dom}(A) \subset L^2(\Omega)^{d+1} \to L^2(\Omega)^{d+1}$ be the unbounded operator in $L^2(\Omega)^{d+1}$ defined by the right hand side of (2.2) with

$$\operatorname{dom}(A) = \{u \in W : \langle Du, v\rangle_W = 0 \text{ for all } v \in \mathcal{V}^*\}. \tag{2.8}$$

From (2.5), we see that the set of smooth functions $\mathcal{D}(\Omega)^{d+1}$ is contained in $\operatorname{dom}(A)$. Hence, A is a densely defined operator in $L^2(\Omega)^{d+1}$. Therefore, it has a uniquely defined adjoint A^*, which is again an unbounded operator. The adjoint A^* equals the distributional derivative operator $-A$ when applied to $\operatorname{dom}(A^*)$. This domain is prescribed as in standard functional analysis [2] by

$$\operatorname{dom}(A^*) = \{\tilde{w} \in L^2(\Omega)^{d+1} : \exists \ell \in L^2(\Omega)^{d+1} \text{ such that } (Au, \tilde{w}) = (u, \ell) \\ \text{for all } u \in \operatorname{dom}(A)\}.$$

By definition, $\operatorname{dom}(A)$ is a subset of $W(\Omega)$. When this subset is given the topology of $W(\Omega)$, we obtain a closed subset of $W(\Omega)$, which we call V; that is, V and $\operatorname{dom}(A)$ coincide as sets or vector spaces, but not as topological spaces. Note that $\operatorname{dom}(A^*)$ is also a subset of W, since, for any $\tilde{w} \in \operatorname{dom}(A^*)$, the distribution $-A\tilde{w}$ is in $L^2(\Omega)^{d+1}$.

When $\mathrm{dom}(A^*)$ is given the topology of W, it will be denoted by V^*. Observe that since V is closed, A is a closed operator. For any $S \subset W$ subspace, the right annihilator of S, denoted by ${}^{\perp}S$, is defined by

$$^{\perp}S = \{w \in W : \langle s', w\rangle_W = 0 \text{ for all } s' \in S\}. \tag{2.9}$$

The definition of $\mathrm{dom}(A^*)$, when written in terms of D reveals that

$$V^* = {}^{\perp}D(V). \tag{2.10}$$

Thus, V^* is also a closed subset of W.

The next observation is that from the definitions of V and the operator D (namely (2.8) and (2.4)); it immediately follows that $\mathcal{V} \subset V$. Note also that if $v^* \in \mathcal{V}^*$, then $(Au, v^*) = -(u, Av^*) + \langle Du, v^*\rangle_W = -(u, Av^*)$ for all $u \in V$, since $\langle Du, v^*\rangle_W = 0$ by the definition of V. Therefore, v^* is in V^*. To summarize these observations, we have introduced $\mathcal{V}, \mathcal{V}^*, V$, and V^*, satisfying

$$\mathcal{V} \subset V \quad \text{and} \quad \mathcal{V}^* \subset V^*. \tag{2.11}$$

These are the abstract ingredients in the framework of [7, Appendix A] applied to the wave problem.

3 The broken weak formulation

Following the settings of [4] and [7, Appendix], we partition the space-time Lipschitz domain Ω into a mesh Ω_h of finitely many open elements K (for example, $(d+1)$–simplices or $(d+1)$–hyperrectangles), such that $\bar{\Omega} = \bigcup_{K\in\Omega_h} \bar{K}$, where $h = \max_{K\in\Omega_h} \mathrm{diam}(K)$. The DPG method is based on a variational formulation in a "broken" analogue of W, which we call W_h, defined below.

We let A_h be the wave operator applied element by element; that is,

$$(A_h w)|_K = A(w|_K), \quad w \in W(K), \quad K \in \Omega_h.$$

Let $W_h = \{w \in L^2(\Omega)^{d+1} : A_h w \in L^2(\Omega)^{d+1}\}$. The operator $D_h : W_h \to W_h'$ is defined by

$$\langle D_h w, v\rangle_{W_h} = (A_h w, v)_\Omega + (w, A_h v)_\Omega$$

for all $w, v \in W_h$, where $\langle\cdot,\cdot\rangle_{W_h}$ denotes the duality pairing in W_h in accordance with our previous notation. Below we abbreviate $\langle\cdot,\cdot\rangle_{W_h}$ to $\langle\cdot,\cdot\rangle_h$. Let $D_{h,V} : V \to W_h'$ denote the restriction of D_h to V; that is, $D_{h,V} = D_h|_V$. The range of $D_{h,V}$, denoted by Q, is made into a complete space by the quotient norm

$$\|\rho\|_Q = \inf_{v\in D_{h,V}^{-1}(\{\rho\})} \|v\|_W, \quad \rho \in Q \equiv \mathrm{ran}(D_{h,V}). \tag{3.1}$$

Define the bilinear form on $(L^2(\Omega)^{d+1} \times Q) \times W_h$ by

$$b((v,\rho),w) = -(v,A_h w)_\Omega + \langle \rho, w\rangle_h.$$

The "broken" variational formulation for the wave problem now reads as follows: Given any F in the dual space W_h', find $u \in L^2(\Omega)^{d+1}$ and $\lambda \in Q$, such that

$$b((u,\lambda),w) = F(w) \quad \text{for all } w \in W_h. \tag{3.2}$$

Critical to the success of any numerical approximation of this formulation, in particular, the DPG approximation, is its wellposedness. By [7, Theorem A.5], this formulation is well-posed, provided we verify

$$V = {}^{\perp}D(V^*), \tag{3.3}$$

$$A : V \to L^2(\Omega)^{d+1} \text{ is a bijection.} \tag{3.4}$$

Therefore, our next focus is on proving (3.3) and (3.4). Recall from (2.11) that $\mathcal{V}$ and $\mathcal{V}^*$ are subspaces of smooth functions within V and V^*. We now show that (3.3) and (3.4) follow if these are dense subspaces.

Theorem 3.1. *Suppose*

$$\mathcal{V} \textit{ is dense in } V \textit{ and } \mathcal{V}^* \textit{ is dense } V^*. \tag{3.5}$$

Then (3.3) *and* (3.4) *hold. Consequently, the broken weak formulation* (3.2) *is well-posed.*

Proof. In view of the continuity of D, (2.11), and the assumption that $\mathcal{V}^*$ is dense in V^*, the condition (3.3) now immediately follows.

Next, we will prove that

$$\|u\| \le 2T\|Au\|, \quad \text{for all } u \in \mathcal{V}, \tag{3.6a}$$

$$\|v\| \le 2T\|A^* v\|, \quad \text{for all } v \in \mathcal{V}^*. \tag{3.6b}$$

These inequalities follow by well-known energy arguments, as shown in [8, Lemma 3]. We briefly include the proof for completeness. Let $v \in \mathcal{V}^*$. Then

$$\begin{aligned}
\|v\|^2 &= \int_0^T \Big(\int_{\Omega_0} |v(x,t)|^2\,dx \Big) dt = 2 \int_0^T \int_T^t \int_{\Omega_0} \partial_s v(x,s)\cdot v(x,s)\,dx\,ds\,dt \\
&= 2 \int_0^T \int_t^T \int_{\Omega_0} v(x,s)\cdot A^* v(x,s)\,dx\,ds\,dt - 2c \int_0^T \int_t^T \int_{\partial\Omega_0} (v_q\cdot n_x) v_\mu\,ds_x\,ds\,dt \\
&\le 2T\|v\|\|A^* v\|.
\end{aligned}$$

The inequality for $\mathcal{V}$ is similarly proved by using its boundary conditions instead of those of $\mathcal{V}^*$.

Using the density assumptions, we conclude that (3.6) implies

$$\|u\| \leq 2T\|Au\| \quad \text{for all } u \in V \text{ and} \tag{3.7a}$$

$$\|v\| \leq 2T\|A^*v\| \quad \text{for all } v \in V^*. \tag{3.7b}$$

The inequality (3.7a) and the closed range theorem for closed operators imply that $A : \mathrm{dom}(A) = V \to L^2(\Omega)^{d+1}$ is injective and has closed range. Moreover, its adjoint A^* is injective (on its domain) by (3.7b), so the range of A must be all of $L^2(\Omega)^{d+1}$ (see, for example, [2, Corollary 2.18]). Hence, A is a bijection; that is, condition (3.4) holds. Finally, since we have verified both (3.3) and (3.4), applying [7, Theorem A.5], the wellposedness follows. □

Note that the wellposedness result of Theorem 3.1, in particular, implies that

$$\beta = \inf_{0 \neq (v,\rho) \in L^2(\Omega)^{d+1} \times Q} \sup_{0 \neq w \in W_h} \frac{b((v,\rho),w)}{\|(v,\rho)\|_{L^2(\Omega)^{d+1} \times Q} \|w\|_{W_h}} > 0. \tag{3.8}$$

4 Verification of the density condition

In this section, we verify (3.5) for a simple domain, namely a hyperrectangle (or an orthotope). Accordingly, throughout this section, we fix $\Omega = \Omega_0 \times (0, T)$ and

$$\Omega_0 = \prod_{i=1}^{d} (0, a_i),$$

for some $a_i > 0$. Whereas density of smooth functions in general graph spaces can be proved by standard Sobolev space techniques [1], to obtain the density of smooth functions with boundary conditions (like those in $\mathcal{V}$) we need more arguments. The proof that follows has some similarities with the proof of [7, Theorem 3.1], an analogous density result for the one-dimensional Schrödinger operator. The main differences from [7] in the proof below include the consideration of multiple spatial dimensions, and the construction of extension operators for vector functions in the wave graph space by combining even and odd reflections appropriately.

Theorem 4.1. *On the above Ω, $\mathcal{V}^*$ is dense in V^*, and $\mathcal{V}$ is dense in V.*

Proof. We shall only prove that $\mathcal{V}$ is dense in V, since the proof of the density of $\mathcal{V}^*$ in V^* is similar. We divide the proof into three main steps:

Step 1. Extension: In this step, we will extend a function in V using spatial reflections to a domain, which has larger spatial extent than Ω (see Figure 4.1).

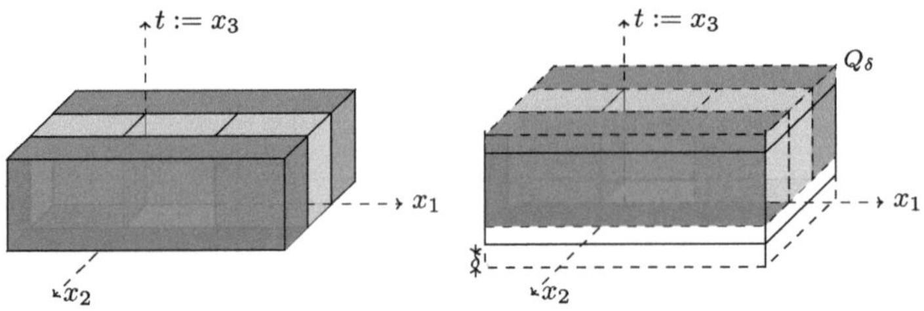

Figure 4.1: *Left:* Extended domains Q_1 and Q_2 when $\Omega \subseteq \mathbb{R}^3$. *Right:* Translation by δ in the t direction.

Let e_i denote the standard unit basis vectors in $\mathbb{R}^{d+1}$ and $y \in \mathbb{R}^{d+1}$ arbitrary. The following operations

$$R_{i,-}y = y - 2y_i e_i, \quad R_{i,+}y = y + 2(a_i - y_i)e_i$$

perform reflections of the coordinate vector y about $y_i = 0$ and $y_i = a_i$, for $i = 1, \ldots, d$. We set $Q_0 \equiv \overline{\Omega}$ and then define extended domains Q_i in a recursive way, starting from $i = 1$ through $i = d$ as follows:

$$Q_{i,-} = R_{i,-}^{-1} Q_{i-1}, \quad Q_{i,+} = R_{i,+}^{-1} Q_{i-1}, \quad Q_i = Q_{i,-} \cup Q_{i-1} \cup Q_{i,+}.$$

The final extended domain is $Q \equiv Q_d$.

Next, we introduce even and odd extensions (in the x_i–direction) of scalar functions. Namely, let $G_{i,e}, G_{i,o} : L^2(Q_{i-1}) \to L^2(Q_i)$ be defined by

$$G_{i,e}f(x,t) = \begin{cases} f(R_{i,-}(x,t)) & \text{if } (x,t) \in Q_{i,-}, \\ f(R_{i,+}(x,t)) & \text{if } (x,t) \in Q_{i,+}, \\ f(x,t) & \text{if } (x,t) \in Q_{i-1}, \end{cases} \tag{4.1}$$

and

$$G_{i,o}f(x,t) = \begin{cases} -f(R_{i,-}(x,t)) & \text{if } (x,t) \in Q_{i,-}, \\ -f(R_{i,+}(x,t)) & \text{if } (x,t) \in Q_{i,+}, \\ f(x,t) & \text{if } (x,t) \in Q_{i-1}. \end{cases} \tag{4.2}$$

In the case of a vector function $v \in L^2(Q_{i-1})^{d+1}$, we define $G_i v(x,t)$ to be the extended vector function obtained by extending (in the ith direction) all the components of v using the odd scalar extension, except the ith component, which is extended using the even scalar extension. In other words, for any $i = 1, \ldots, d$, we define $G_i : L^2(Q_{i-1})^{d+1} \to L^2(Q_i)^{d+1}$ by

$$G_i v = (G_{i,e} v_i) e_i + \sum_{j \neq i} (G_{i,o} v_j) e_j, \tag{4.3}$$

where the sum runs over all $j = 1, \ldots, d+1$, except i. Let $E_k = G_k \circ G_{k-1} \circ \cdots \circ G_1$. The cumulative extension over all spatial directions is thus obtained using $E = E_d$. It extends functions in $\overline{\Omega}$ to Q.

By changing the variable of integration, we obtain

$$(G_{i,o}f, g)_{Q_i} = (f, G'_{i,o}g)_{Q_{i-1}}, \quad \text{for all } f \in L^2(Q_{i-1}), \quad g \in L^2(Q_i),$$
$$(G_{i,e}f, g)_{Q_i} = (f, G'_{i,e}g)_{Q_{i-1}}, \quad \text{for all } f \in L^2(Q_{i-1}), \quad g \in L^2(Q_i),$$

where the "folding" operators $G'_{i,e/o} : L^2(Q_i) \to L^2(Q_{i-1})$ that go the reverse direction of the extension operators, are defined by

$$G'_{i,o}g(x,t) = g(x,t) - g(R_{i,-}^{-1}(x,t)) - g(R_{i,+}^{-1}(x,t)), \tag{4.4}$$
$$G'_{i,e}g(x,t) = g(x,t) + g(R_{i,-}^{-1}(x,t)) + g(R_{i,+}^{-1}(x,t)). \tag{4.5}$$

These scalar folding operators combine to form an analogue for vector functions as in (4.3), namely

$$G'_i w = (G'_{i,e}w_i)e_i + \sum_{j \neq i} (G'_{i,o}w_j)e_j.$$

It satisfies $(G_i v, w)_{Q_i} = (v, G'_i w)_{Q_{i-1}}$ for all $v \in L^2(Q_{i-1})^{d+1}$, $w \in L^2(Q_i)^{d+1}$, and for each i from 1 to d. Let $E'_k = G'_k \circ G'_{k+1} \cdots \circ G'_d$. Then $E' = E'_1$ folds functions in Q to $\overline{\Omega}$ and is the adjoint of the extension E in the following sense:

$$(Eu, w)_Q = (u, E'w)_\Omega, \quad \text{for all } u \in L^2(\Omega)^{d+1}, \quad w \in L^2(Q)^{d+1}. \tag{4.6}$$

We want to prove that Ev is in $W(Q)$ for any $v \in V$. Note that if $w \in L^2(\Omega)^{d+1}$, then Ew in $L^2(Q)^{d+1}$, because each G_i maps L^2 functions into L^2 per (4.3). Therefore, to prove Ev is in $W(Q)$, it only remains to prove that $A(Ev)$ is in $L^2(Q)^{d+1}$. Let $\varphi \in \mathcal{D}(Q)^{d+1}$ (where we abuse the notation and write $\mathcal{D}(Q)$ for $\mathcal{D}(Q^0)$ whenever Q^0 is the interior of Q). Using (4.6), the action of the distribution AEv on φ equals

$$\langle AEv, \varphi \rangle_{\mathcal{D}(Q)^{d+1}} = -(Ev, A\varphi)_Q = -(v, E'A\varphi)_\Omega. \tag{4.7}$$

To analyze the last term, first observe that by the chain rule applied to a smooth scalar function ϕ on Q_1,

$$\partial_t(G'_{i,o}\phi) = G'_{i,o}\partial_t\phi, \quad \partial_i(G'_{i,o}\phi) = G'_{i,e}(\partial_i\phi), \quad \partial_j(G'_{i,o}\phi) = G'_{i,o}(\partial_j\phi),$$
$$\partial_t(G'_{i,e}\phi) = G'_{i,e}\partial_t\phi, \quad \partial_i(G'_{i,e}\phi) = G'_{i,o}(\partial_i\phi), \quad \partial_j(G'_{i,e}\phi) = G'_{i,e}(\partial_j\phi),$$

for all $j \neq i$. Combining these appropriately for smooth vector function ψ on Q_i, and considering the constant wave speed $c > 0$, we find that

$$\partial_t(G'_i\psi) = G'_i\partial_t\psi, \quad c\nabla_x G'_{i,o}\psi_\mu = G'_i(c\nabla_x\psi_\mu), \quad c\,\mathrm{div}_x\, G'_i\psi_q = G'_{i,o}(c\,\mathrm{div}_x\,\psi_q).$$

Thus, for any $\varphi \in \mathcal{D}(Q)^{d+1}$, we have $E'_i A\varphi = AE'_i\varphi$ for all $i = 1, \ldots, d$, and, in particular,

$$E'A\varphi = AE'\varphi. \tag{4.8}$$

Returning to (4.7) and using (4.8) and (2.4),

$$\langle AEv, \varphi\rangle_{\mathcal{D}(Q)^{d+1}} = (Av, E'\varphi)_\Omega - \langle Dv, E'\varphi\rangle_{W(\Omega)}. \tag{4.9}$$

We shall now show that the last term above vanishes. Since v is in V, the last term will vanish by the definition of V, provided $E'\varphi$ is in $\mathcal{V}^*$. To prove that $E'\varphi$ is in $\mathcal{V}^*$, we only need to verify that $E'\varphi$ satisfies the boundary conditions in (2.7). Since φ is compactly supported in Q, we obviously have $(E'\varphi)|_{\Gamma_T} = 0$, as E' only involves spatial folding.

Next, we claim that $[E'\varphi]_\mu|_{\Gamma_b} = 0$ also. To see this, let Γ^j denote the two facets of ∂Q_j, where x_j is constant and γ^j denote the two facets of ∂Q_{j-1}, where x_j is constant. The value of $G'_{d,o}\varphi_\mu(x,t)$ for any (x,t) in γ^{d-1} is the sum of the three terms in (4.4), two of which cancel each other, and one of which vanishes, because $\varphi_\mu|_{\Gamma^d} = 0$. Thus, $\varphi_\mu|_{\Gamma^d} = 0$ implies $(G'_{d,o}\varphi_\mu)|_{\partial Q_{d-1}} = 0$ (where we have also used the fact that φ_μ vanishes on the remainder $\partial Q_{d-1} \setminus \gamma^{d-1}$). The same argument can now be repeated to get that $(G'_{d,o}\varphi_\mu)|_{\Gamma^{d-1}} = 0$ implies $(G_{d-1,o}(G_{d,o}\varphi_\mu))|_{\partial Q_{d-2}} = 0$. Continuing similarly, we obtain that $[E'\varphi]_\mu = G'_{1,o} \circ G'_{2,o} \circ \cdots \circ G'_{d,o}\varphi_\mu$ vanishes on $\partial Q_0 = \Gamma_b$. Thus, the last term in (4.9) is zero and, by (4.6), we conclude that

$$\langle AEv, \varphi\rangle_{\mathcal{D}(Q)^{d+1}} = (EAv, \varphi)_Q \tag{4.10}$$

for all φ in $\mathcal{D}(Q)^{d+1}$.

By virtue of (4.10), we have proved that for any $v \in V$, AEv is in $L^2(Q)^{d+1}$, AEv coincides with EAv, and Ev is in $W(Q)$.

Step 2. Translation: In this step, we will translate up the previously obtained extension in time coordinate. This will give us room to mollify in the next step. Such a translation step is standard in many density proofs (see, for example, [1]).

Let $v \in V$, and let $\tilde{E}v$ denote the extension of Ev by zero to $\mathbb{R}^{d+1}$; that is, $\tilde{E}v$ equals Ev in Q and it is zero elsewhere. Denote by τ_δ the translation operator in the t-direction by $\delta > 0$. That is, $(\tau_\delta w)(x,t) = w(x, t-\delta)$ for scalar or vector functions w. It is well known [2] that

$$\lim_{\delta\to 0} \|\tau_\delta g - g\|_{L^2(\mathbb{R}^{d+1})} = 0, \quad \forall g \in L^2(\mathbb{R}^{d+1}). \tag{4.11}$$

Let $Q_\delta = \prod_{i=1}^{i=d}(-a_i, 2a_i) \times (-\delta, T+\delta)$, and let H_δ be the restriction from $\mathbb{R}^{d+1}$ to Q_δ. We will now show that

$$AH_\delta\tau_\delta\tilde{E}v = H_\delta\tau_\delta\tilde{E}Av. \tag{4.12}$$

By a change of variable,

$$(\tau_\delta\tilde{E}w, \tilde{w})_{Q_\delta} = (Ew, \tau_{-\delta}\tilde{w})_Q \tag{4.13}$$

for all $w \in L^2(\Omega)^{d+1}$ and $\tilde{w} \in L^2(Q_\delta)^{d+1}$. Note that $\tau_\delta \tilde{E} v|_\Omega \in L^2(\Omega)^{d+1}$. The distribution $AH_\delta\tau_\delta\tilde{E}v$ applied to a smooth function $\varphi \in \mathcal{D}(Q_\delta)^{d+1}$ equals

$$\langle AH_\delta\tau_\delta\tilde{E}v, \varphi\rangle_{\mathcal{D}(Q_\delta)^{d+1}} = -(\tau_\delta\tilde{E}v, A\varphi)_{Q_\delta} = -(Ev, A\tau_{-\delta}\varphi)_Q$$

due to (4.13), and the fact that $\tau_{-\delta}A\varphi = A\tau_{-\delta}\varphi$. Using also (4.6) and (4.8),

$$\begin{aligned}\langle AH_\delta\tau_\delta\tilde{E}v, \varphi\rangle_{\mathcal{D}(Q_\delta)^{d+1}} &= -(v, E'A\tau_{-\delta}\varphi)_\Omega = -(v, AE'\tau_{-\delta}\varphi)_\Omega \\ &= (Av, E'\tau_{-\delta}\varphi)_\Omega - \langle Dv, E'\tau_{-\delta}\varphi\rangle_W.\end{aligned}$$

Note that $E'\tau_{-\delta}\varphi$ satisfies all the boundary conditions required for it to be in $\mathcal{V}^*$. Hence, the last term in the above display is zero. We therefore conclude that

$$\langle AH_\delta\tau_\delta\tilde{E}v, \varphi\rangle_{\mathcal{D}(Q_\delta)^{d+1}} = (\tau_\delta EAv, \varphi)_{Q_\delta},$$

which proves (4.12). In particular, $H_\delta\tau_\delta\tilde{E}v \in W(Q_\delta)$ whenever $v \in V$.

Step 3. Mollification: In this step, we finish the proof by considering a $v \in V$ and mollifying the time-translated extension $\tau_\delta\tilde{E}v$ constructed above.

To recall the standard symmetric mollifier, let $\rho_\varepsilon \in \mathcal{D}(\mathbb{R}^{d+1})$ for each $\varepsilon > 0$ be defined by

$$\rho_\varepsilon(x, t) = \varepsilon^{-(d+1)}\rho_1(\varepsilon^{-1}x, \varepsilon^{-1}t),$$

where

$$\rho_1(x, t) = \begin{cases} k\exp(-\frac{1}{1-|x|^2-t^2}) & \text{if } |x|^2 + t^2 < 1, \\ 0 & \text{if } |x|^2 + t^2 \geq 1, \end{cases}$$

and k is a constant chosen, so that $\int_{\mathbb{R}^{d+1}} \rho_1 = 1$. Here, $|\cdot|$ is the Euclidean norm in $\mathbb{R}^d$. Let $\rho_\varepsilon * v$ denote the function obtained by component-wise convolution. That is, $[\rho_\varepsilon * v]_j = [v]_j * \rho_\varepsilon$ for all j-components. Then $\rho_\varepsilon * v$ is an infinitely smooth vector function that satisfies

$$\lim_{\varepsilon\to 0} \|v - \rho_\varepsilon * v\|_{\mathbb{R}^{d+1}} = 0, \quad \forall v \in L^2(\mathbb{R}^{d+1})^{d+1}. \tag{4.14}$$

Consider any

$$0 < \delta < \min_{1\leq i\leq d}(a_i/2, T/2),$$

and define two functions $v_\varepsilon = \rho_\varepsilon * \tau_\delta\tilde{E}v$ and $a_\varepsilon = \rho_\varepsilon * \tau_\delta\tilde{E}Av$. Note that $Av_\varepsilon = a_\varepsilon$ on Ω whenever $\varepsilon < \delta/2$, thanks to (4.12).

We now proceed to show that

$$\lim_{\varepsilon\to 0} \|v_{\varepsilon|\Omega} - v\|_W = 0. \tag{4.15}$$

Set $\delta = 3\varepsilon$, and let $\varepsilon < \min_{1\le i\le d}(a_i/2, T/2)/3$ go to zero. Note that

$$\begin{aligned}
\|Av_\varepsilon - Av\| &= \|a_\varepsilon - Av\| = \|\rho_\varepsilon * \tau_\delta \tilde{E}Av - Av\|_\Omega \\
&\le \|\rho_\varepsilon * \tau_\delta \tilde{E}Av - \tilde{E}Av\|_{\mathbb{R}^{d+1}} \\
&\le \|\rho_\varepsilon * \tau_\delta \tilde{E}Av - \tau_\delta \tilde{E}Av\|_{\mathbb{R}^{d+1}} + \|\tau_\delta \tilde{E}Av - \tilde{E}Av\|_{\mathbb{R}^{d+1}}, \\
\|v_\varepsilon - v\| &\le \|\rho_\varepsilon * \tau_\delta \tilde{E}v - \tau_\delta \tilde{E}v\|_\Omega + \|\tau_\delta \tilde{E}v - v\|_\Omega \\
&\le \|\rho_\varepsilon * \tau_\delta \tilde{E}v - \tau_\delta \tilde{E}v\|_{\mathbb{R}^{d+1}} + \|\tau_\delta \tilde{E}v - \tilde{E}v\|_{\mathbb{R}^{d+1}}.
\end{aligned}$$

Using (4.11) and (4.14), it now immediately follows that (4.15) holds.

To conclude, it suffices to prove that $v_{\varepsilon|\Omega}$ is in $\mathcal{V}$. Clearly, $\tau_\delta \tilde{E}v$ is identically zero in a neighborhood of Γ_0. Hence, we conclude that $v_\varepsilon = \rho_\varepsilon * \tau_\delta \tilde{E}v$ vanishes on Γ_0 for small enough ε. Next, let us examine the value of $[v_\varepsilon]_\mu$ at points (x,t) on Γ_b, namely

$$[v_\varepsilon]_\mu(x,t) = \int_{\mathbb{R}} \int_{\mathbb{R}^d} \rho_\varepsilon(x - x', t - t')[\tau_\delta \tilde{E}v]_\mu(x', t')\, dx'\, dt'.$$

Note that $\rho_\varepsilon(x - x', t - t')$ is a symmetric function of x' about x. The other term in the integrand, namely $[\tau_\delta \tilde{E}v]_\mu(x', t')$, is odd about every facet of Γ_b. Hence, the integral of their product vanishes whenever $(x,t) \in \Gamma_b$. Thus, $[v_\varepsilon]_{\mu|\Gamma_b} = 0$, and $v_\varepsilon \in \mathcal{V}$. □

5 The method and its error estimates

In this section, we present the approximation of the previously described broken weak formulation by the (ideal) DPG method and provide *a priori* and *a posteriori* error estimates.

5.1 The DPG method

The ideal DPG method [5] seeks u_h and λ_h in finite-dimensional subspaces $U_h \subset L^2(\Omega)^{d+1}$ and $Q_h \subset Q$, respectively, satisfying

$$b((u_h, \lambda_h), w_h) = F(w_h), \quad \text{for all } w_h \in T(U_h \times Q_h), \tag{5.1}$$

where $T : L^2(\Omega)^{d+1} \times Q \to W_h$ is such that $(T(v,\rho), w)_{W_h} = b((v,\rho), w)$ for all $w \in W_h$, and any $(v,\rho) \in L^2(\Omega)^{d+1} \times Q$. Hereon, we denote U to be $L^2(\Omega)^{d+1}$ and abbreviate the W_h inner product $(\cdot,\cdot)_{W_h}$ to simply $(\cdot,\cdot)_h$.

It is well known [6] that there is a mixed method that is equivalent to the above Petrov–Galerkin method (5.1). One of the variables in this mixed method is the error representation function $\varepsilon_h \in W_h$ defined by

$$(\varepsilon_h, w)_h = (f, w) - b((u_h, \lambda_h), w), \quad \text{for all } w \in W_h. \tag{5.2}$$

One of the two equations in the mixed formulation given below is a restatement of this defining equation for ε_h. The mixed formulation seeks $\varepsilon_h \in W_h$ and $(u_h, \lambda_h) \in (U_h \times Q_h)$ such that

$$\begin{aligned} (\varepsilon_h, w)_h + b((u_h, \lambda_h), w) &= F(w) && \text{for all } w \in W_h, \\ b((v, \rho), \varepsilon_h) &= 0 && \text{for all } (v, \rho) \in U_h \times Q_h. \end{aligned} \tag{5.3}$$

We think of

$$\eta = \|\varepsilon_h\|_{W_h} \equiv \left(\sum_{K \in \Omega_h} \|\varepsilon_h\|^2_{W(K)} \right)^{1/2}$$

as an *a posteriori* error estimator, because ε_h can be computed from (5.2), after u_h and λ_h has been computed. Alternately, one can view ε_h as one of the unknowns together with u_h and λ_h as in (5.3). Note that (5.2) implies

$$\eta = \sup_{w \in W_h} \frac{b((u - u_h, \lambda - \lambda_h), w)}{\|w\|_{W_h}},$$

so it immediately follows that the estimator is globally reliable and efficient, namely

$$\beta \|(u - u_h, \lambda - \lambda_h)\|_{U \times Q} \le \eta \le \|b\| \|(u - u_h, \lambda - \lambda_h)\|_{U \times Q},$$

where β is as in (3.8). In practice, we use element-wise norms of ε_h as *a posteriori* element error indicator.

To give an *a priori* error estimate with convergence rates, we need to specify all the approximation subspaces. We choose the space $Q_h \subset Q$ by first selecting a finite element space $V_h \subset V$ and then applying D_h to all functions in it, namely

$$Q_h = D_h V_h.$$

This way, we guarantee that Q_h is a subspace of Q. The definition of V_h and the finite element subspaces of U are based on the type of elements in Ω_h. We consider two cases:

Case A Ω_h is a geometrically conforming mesh of $(d+1)$-simplices:

$$V_h = \{u \in V \cap C(\bar{\Omega})^{d+1} : u|_K \in P_{p+1}(K)^{d+1} \text{ for all } K \in \Omega_h\}, \tag{5.4a}$$

$$U_h = \{u \in L^2(\Omega)^{d+1} : u|_K \in P_p(K)^{d+1} \text{ for all } K \in \Omega_h\}, \tag{5.4b}$$

where $P_p(K)$ is the space of polynomials of total degree $\le p$ on K.

Case B Ω_h is a geometrically conforming mesh of hyperrectangles,

$$V_h = \{u \in V \cap C(\bar{\Omega})^{d+1} : u|_K \in Q_{p+1}(K)^{d+1} \text{ for all } K \in \Omega_h\}, \tag{5.5a}$$

$$U_h = \{u \in L^2(\Omega)^{d+1} : u|_K \in Q_p(K)^{d+1} \text{ for all } K \in \Omega_h\}, \tag{5.5b}$$

where $Q_p(K)$ is the space of polynomials on K that are of degree at most p in each variable.

Since the wave operator A is a first-order differential operator, $H^1(\Omega)^{d+1} \subset W(\Omega)$. Hence, the Lagrange finite element space V_h is contained in W. The space V_h has a nodal interpolation operator $I_h : H^{s+1}(\Omega)^{d+1} \to V_h$, which is bounded for $s+1 > (d+1)/2$, and which we shall use in the proof below. We will use C to denote a generic mesh-independent constant, whose value at different occurrences may differ. Note that in the estimate of the theorem below, h is the discretization parameter in both space and time.

Theorem 5.1. *Suppose $u \in V \cap H^{s+1}(\Omega)^{d+1}$ and $\lambda = D_h u$ solve (3.2). Suppose also that U_h and V_h are set as in (5.4) or (5.5) depending on the mesh type, and $Q_h = D_h V_h$. Then, there exists a constant $C > 0$ independent of h, such that the discrete solution $u_h \in U_h$ and $\lambda_h \in Q_h$ solving (5.1) satisfy*

$$\|u - u_h\| + \|\lambda - \lambda_h\|_Q \leq C h^s |u|_{H^{s+1}(\Omega)^{d+1}} \tag{5.6}$$

for $(d-1)/2 < s \leq p + 1$.

Proof. The ideal DPG method is quasioptimal, that is, by [5, Theorem 2.2],

$$\begin{aligned} \|(u,\lambda) - (u_h, \lambda_h)\|_{U\times Q}^2 &\leq C \inf_{(v_h,\rho_h)\in U_h\times Q_h} \|(u,\lambda) - (v_h, \rho_h)\|_{U\times Q}^2 \\ &\leq C \inf_{(v_h,\rho_h)\in U_h\times Q_h} [\|u - v_h\|^2 + \|\lambda - \rho_h\|_Q^2]. \end{aligned}$$

The well-known best approximation estimates for U_h imply

$$\inf_{v_h\in U_h} \|u - v_h\| \leq C h^s |u|_{H^s(\Omega)^{d+1}}, \text{ for all } 0 < s \leq p + 1. \tag{5.7}$$

To estimate the remaining term, choose $\rho_h = D_h I_h u$. Then, since $\lambda = D_h u$, by the definition of the Q-norm in (3.1) and the Bramble–Hilbert lemma,

$$\begin{aligned} \inf_{\rho_h\in Q_h} \|\lambda - \rho_h\|_Q &\leq \|u - I_h u\|_W \\ &\leq C\|u - I_h u\|_{H^1(\Omega)^{d+1}} \leq C h^s |u|_{H^{s+1}(\Omega)^{d+1}} \end{aligned} \tag{5.8}$$

for any $u \in H^{s+1}(\Omega)$, for $(d-1)/2 < s \leq p + 1$. Thus, from (5.7) and (5.8), we have that (5.6) holds. □

6 Implementation and numerical results

We implemented the DPG discretization in the form (5.3) with the following change. Since W_h is infinite-dimensional, to get a practical method, we must replace W_h by a sufficiently rich finite-dimensional space Y_h^m. A full theoretical analysis of this practical realization of the ideal DPG method is currently open, but we will provide numerical studies showing its efficacy in this section. For some nonnegative integer m, set Y_h^m as follows:

- In Case A (see (5.4)) we set $Y_h^m = \{w \in W_h(\Omega) : w|_K \in P_m(K)^{d+1}\}$,
- In Case B (see (5.5)) we set $Y_h^m = \{w \in W_h(\Omega) : w|_K \in Q_m(K)^{d+1}\}$.

Then, we compute $e_h \in Y_h^m$, $u_h \in U_h$ and $\lambda_h \in Q_h$, satisfying

$$\begin{aligned} (e_h, w)_h + b((u_h, \lambda_h), w) &= F(w) \quad \text{for all } w \in Y_h^m, \\ b((v,\rho), e_h) &= 0 \quad \text{for all } (v,\rho) \in U_h \times Q_h. \end{aligned} \tag{6.1}$$

In our numerical experience, the choice $m = p + d + 1$ gave optimal convergence rates (as reported in detail below). This choice is motivated by the study in [15]. The choice $m = p + d$ did not give optimal convergence rates for $p > 2$ and $d = 1$. A brief report of the performance of an adaptive algorithm is also included in the $d = 1$ case. Here again, we observed marked deterioration of adaptivity if $m = p + d$ is used instead of $m = p + d + 1$ for higher degrees. Beyond these comments, we shall not describe these negative results further, but will henceforth focus solely on the $m = p + d + 1$ case. All the numerical results have been implemented using the NGSolve [18] finite element software, and the codes used for the experiments below are available in [9].

6.1 A null space

To implement (6.1), one strategy is to set $\lambda_h = D_h z_h$ for some $z_h \in V_h$ and solve

$$\begin{aligned} (e_h, w)_h + b((u_h, D_h z_h), w) &= F(w) \quad \text{for all } w \in Y_h^m, \\ b((v, D_h r), e_h) &= 0 \quad \text{for all } v \in U_h, r \in V_h. \end{aligned} \tag{6.2}$$

We can decompose V_h into interior "bubbles" in $V_h^0 = \{z \in V_h : z|_{\partial K} = 0 \text{ for all } K \in \Omega_h\}$, and a remainder $V_h^1 \equiv V_h / V_h^0$. Since $b((v, D_h V_h^0), w) = 0$, we may replace V_h by V_h^1 in (6.2) (and compute a $z_h \in V_h^1$). Let $\{y_k\}$, $\{u_i\}$, and $\{z_j\}$ denote a local finite element basis for Y_h^m, U_h and V_h^1, respectively. Using this basis, the system (6.2) with V_h, replaced by V_h^1, yields a matrix equation of the following form

$$\begin{bmatrix} \mathsf{A} & \mathsf{B} \\ \mathsf{B}^\top & 0 \end{bmatrix} \begin{bmatrix} \mathsf{e} \\ \mathsf{x} \end{bmatrix} = \begin{bmatrix} \mathsf{f} \\ 0 \end{bmatrix}, \tag{6.3}$$

where e and x are the vectors of coefficients in the basis expansion of $e_h \in Y_h^m$ and $(u_h, z_h) \in U_h \times V_h$, respectively, $\mathsf{A}_{kl} = (\mathsf{y}_l, \mathsf{y}_k)_h$, $[\mathsf{B}_0]_{ki} = \mathsf{b}((\mathsf{u}_i, 0), \mathsf{y}_k)$, $[\mathsf{B}_1]_{kj} = \mathsf{b}((0, \mathsf{D}_h \mathsf{z}_j), \mathsf{y}_k)$, and $\mathsf{B} = [\mathsf{B}_0, \mathsf{B}_1]$. In all our numerical experiments, for the above-mentioned choice of $m = p + d + 1$, we observed that the matrices A and B_0 have trivial null spaces.

However, we caution that B_1 may have a null space. This runs contrary to our experience with DPG methods on nonspacetime problems, so we expand on it. Note that (see (2.5))

$$[\mathsf{B}_1]_{kj} = b((0, D_h z_j), y_k) = \sum_{K \in \Omega_h} \int_{\partial K} \mathsf{D}_{x,t} z_j \cdot y_k,$$

where

$$\mathsf{D}_{\mathsf{x},\mathsf{t}} = \begin{bmatrix} n_t \mathrm{I}_{\mathsf{d}} & -cn_x \\ -cn_x^{\mathsf{T}} & n_t \end{bmatrix}$$

and I_{d} is the $d \times d$ identity matrix. It is immediate that on mesh facets with certain combinations of n_x and n_t, the matrix $\mathsf{D}_{\mathsf{x},\mathsf{t}}$ is singular. Then B_1 will have a nontrivial kernel.

As an example, in Figure 4.2, we show one of the z_j that is in the null space of B_1 on a triangular mesh for $p = 1$ and $c = 1$. In fact, on the mesh shown, there are 8 basis functions of V_h^1 that are in the null space of B_1, two for each diagonal edge. Recall that the wave speed is 1 for our model wave problem, so these edges align with the light cone for $d = 1$. In the case of $d = 2$ space dimensions, we continued to find a nontrivial null space for B_1 on analogous meshes.

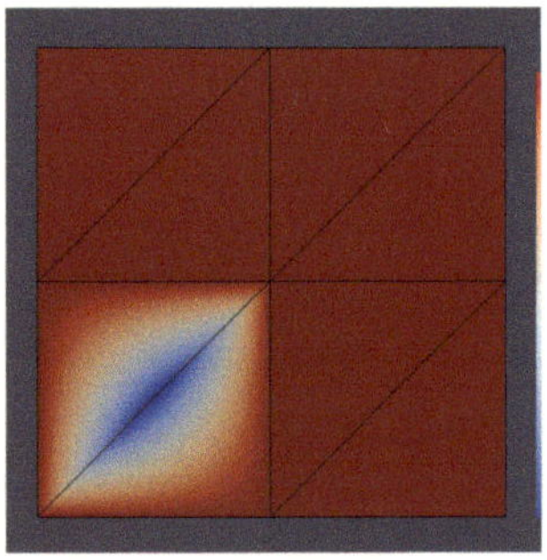

Figure 4.2: Example of a spacetime shape function z_j in the kernel.

This null space problem occurs because the interface variable λ_h is set indirectly by applying the singular operator D_h on V_h. If one could directly construct a basis for $Q_h = D_h V_h$, then one can directly implement (6.1) (instead of (6.2)). However, we do not know how to construct such a basis easily on general simplicial meshes. Hence, we continue on to describe how to solve (6.2) despite its kernel.

6.2 Techniques to solve despite the null space

Despite the above-mentioned problem, one may solve the DPG system using one of the following approaches:

6.2.1 Technique 1: remaining orthogonal to null space in conjugate gradients

The matrix system (6.3) can be solved by reducing it to its Schur complement

$$\mathsf{B}^{\mathsf{T}}\mathsf{A}^{-1}\mathsf{B}\mathsf{x} = \mathsf{B}^{\mathsf{T}}\mathsf{A}^{-1}\mathsf{f} \tag{6.4}$$

first. Let $\mathsf{C} = \mathsf{B}^\mathsf{T}\mathsf{A}^{-1}\mathsf{B}$ and $\mathsf{g} = \mathsf{B}^\mathsf{T}\mathsf{A}^{-1}\mathsf{f}$. The matrix C is symmetric and positive semidefinite. Its easy to see that

$$\ker \mathsf{C} = \ker \mathsf{B}.$$

Thus, solutions of (6.4) are defined only up to this kernel. Note, however, that since $\ker \mathsf{B} = \ker \mathsf{B}_1$ and B_0 has only the trivial kernel, the U_h-component of the DPG solution is uniquely defined independently of $\ker \mathsf{B}_1$.

One may obtain one solution of (6.4) using the conjugate gradient method, which computes its nth iterate x_n in the Krylov space

$$K_n(\mathsf{C}, \mathsf{r}_0) = \operatorname{span}\{\mathsf{C}^k \mathsf{r}^0 : k = 0, \dots, n-1\},$$

where $\mathsf{r}_0 = \mathsf{g} - \mathsf{A}\mathsf{x}_0$ is the initial residual. This iteration will converge if $K_n(\mathsf{C}, \mathsf{r}_0)$ remains (ℓ^2) orthogonal to $\ker(\mathsf{C})$ for all n. A simple prescription to guarantee this orthogonality is to choose the initial iterate $\mathsf{x}_0 = 0$. Indeed, if $\mathsf{x}_0 = 0$, then $\mathsf{r}_0 = \mathsf{g} = \mathsf{B}^\mathsf{T}\mathsf{A}^{-1}\mathsf{f}$ is in the range of B^T, which equals the orthogonal complement of $\ker \mathsf{B} = \ker \mathsf{C}$. Then for all $n \geq 1$, its obvious that $\mathsf{C}^n \mathsf{r}_0$ is also orthogonal to $\ker \mathsf{C}$. Thus, $K_n(\mathsf{C}, \mathsf{r}_0)$ is orthogonal to $\ker \mathsf{C}$.

To summarize this technique, we use the conjugate gradient algorithm to compute one solution orthogonal to $\ker(\mathsf{C})$ and extract the unique U_h-component from that solution for reporting the errors.

6.2.2 Technique 2: regularization of the linear system

Another technique to solve the singular system (6.4) approximately is regularization. First, we rewrite (6.4) in block form as

$$\begin{pmatrix} \mathsf{B}_0^\mathsf{T}\mathsf{A}^{-1}\mathsf{B}_0 & \mathsf{B}_0^\mathsf{T}\mathsf{A}^{-1}\mathsf{B}_1 \\ \mathsf{B}_1^\mathsf{T}\mathsf{A}^{-1}\mathsf{B}_0 & \mathsf{B}_1^\mathsf{T}\mathsf{A}^{-1}\mathsf{B}_1 \end{pmatrix} \mathsf{x} = \mathsf{g}.$$

Since only B_1 may have a nontrivial kernel in V_h^1, we can convert this to an invertible system by adding a small positive-definite term in V_h^1. Namely, let M be the mass matrix $\mathsf{M}_{jl} = (z_l, z_j)$. Instead of solving (6.4), we solve for

$$\begin{pmatrix} \mathsf{B}_0^\mathsf{T}\mathsf{A}^{-1}\mathsf{B}_0 & \mathsf{B}_0^\mathsf{T}\mathsf{A}^{-1}\mathsf{B}_1 \\ \mathsf{B}_1^\mathsf{T}\mathsf{A}^{-1}\mathsf{B}_0 & \mathsf{B}_1^\mathsf{T}\mathsf{A}^{-1}\mathsf{B}_1 + \alpha \mathsf{M} \end{pmatrix} \mathsf{x} = \mathsf{g}, \tag{6.5}$$

where α is a positive regularization parameter, usually set much smaller than the order of the expected discretization errors. In all our reported experiments, it was set to 10^{-9}. The regularized system (6.5) is invertible and can be solved using any direct or iterative methods.

6.3 Convergence rates in two-dimensional space-time

Let $\Omega = (0,1)^2$. We consider a problem with homogeneous boundary and initial conditions, where the exact solution to the second-order wave equation is given by $\phi(x,t) = \sin(\pi x)\sin^2(\pi t)$. Then, the exact solution for the first-order system is

$$u = \begin{pmatrix} c\pi\cos(\pi x)\sin^2(\pi t) \\ \pi\sin(\pi x)\sin(2\pi t) \end{pmatrix},$$

and the corresponding source terms are

$$g = 0, \quad f = \pi^2\sin(\pi x)(2\cos(2\pi t) + c^2\sin^2(\pi t)).$$

In each experiment, a (nonuniform) coarse triangular mesh of Ω was constructed, with element diameters not exceeding a reported mesh size h and consider $c = 1$. Successive refinements of the mesh were obtained by connecting the midpoints of the edges.

We observe in Table 4.1 that the order of convergence for u_h in the L^2 norm is $O(h^{p+1})$, in accordance with Theorem 5.1. Similarly, in Table 4.2, we observe the same convergence rates for rectangular meshes. All results in both tables were obtained using Technique 1.

Table 4.1: Convergence rates for $\|u - u_h\|$ on triangular meshes using Technique 1.

h	$p = 0$	eoc	$p = 1$	eoc	$p = 2$	eoc	$p = 3$	eoc
1/4	1.285×10^{0}	–	1.537×10^{-1}	–	2.038×10^{-2}	–	1.262×10^{-3}	–
1/8	5.638×10^{-1}	1.19	5.613×10^{-2}	1.45	4.754×10^{-3}	2.10	1.537×10^{-4}	3.04
1/16	2.207×10^{-1}	1.35	1.247×10^{-2}	2.17	5.490×10^{-4}	3.11	7.852×10^{-6}	4.29
1/32	1.021×10^{-1}	1.11	3.031×10^{-3}	2.04	6.696×10^{-5}	3.00	4.786×10^{-7}	4.04

Table 4.2: Convergence rates for $\|u - u_h\|$ on rectangular meshes using Technique 1.

h	$p = 0$	eoc	$p = 1$	eoc	$p = 2$	eoc	$p = 3$	eoc
1/4	9.723×10^{-1}	–	1.683×10^{-1}	–	6.672×10^{-3}	–	2.091×10^{-3}	–
1/8	4.736×10^{-1}	1.04	4.287×10^{-2}	1.97	8.506×10^{-4}	2.97	1.331×10^{-4}	3.97
1/16	2.329×10^{-1}	1.35	1.076×10^{-2}	1.99	1.071×10^{-4}	2.99	8.377×10^{-6}	3.99
1/32	1.159×10^{-1}	1.11	2.694×10^{-3}	2.00	1.341×10^{-5}	3.00	5.261×10^{-7}	3.99

6.4 Adaptivity

Let $\Omega = (0,1)^2$. We consider the same model problem (2.1), but now with zero sources $f = g = 0$ and the nonzero initial condition

$$\mu|_{t=0} = -\phi_0, \quad q|_{t=0} = \phi_0$$

in place of (2.1c), where $\phi_0 = \exp(-1000((x-0.5)^2))$. The boundary condition $\mu = 0$ continues to remain the same. This simulates a beam reflecting off the Dirichlet boundary.

In Figure 4.3, we display a few iterates from the standard adaptive refinement algorithm using $p = 3$ and the DPG error estimator. We started with the extremely coarse mesh shown in Figure 4.3(a), used the element-wise norms of e_h to compute the DPG element error indicator, marked elements with more than 50 % of the total indicated error, refined the marked elements (and more for conformity) by bisection, and repeated this adaptivity loop. The few iterates from the adaptivity loop (shown in Figure 4.3) show the potential of the space-time DPG method to easily capture localized features in space-time.

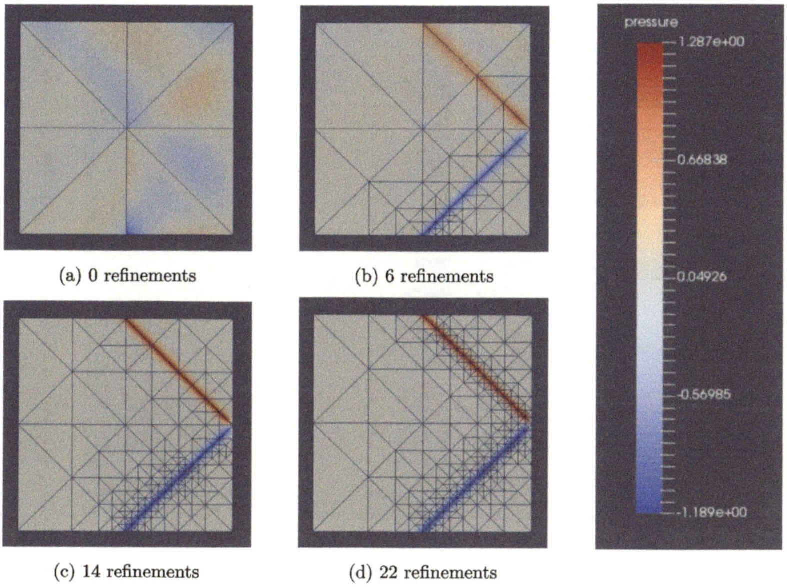

Figure 4.3: Iterates from the adaptive algorithm. Numerical pressure μ is shown for $p = 3$. Time axis is vertical.

6.5 Adaptivity with inhomogeneous materials

Consider the case when the domain consists of two regions, namely $\Omega_l = (0, 0.5) \times (0, 1.4)$, and $\Omega_r = (0.5, 1.4) \times (0, 1.4)$, and a more general first-order wave equation

$$\begin{pmatrix} \kappa_1 & 0 \\ 0 & \kappa_2 \end{pmatrix} \partial_t u - \begin{pmatrix} 0 & c \\ c & 0 \end{pmatrix} \partial_x u = 0,$$

where

$$\kappa_1 = \begin{cases} 2, & 0 < x < 1/2, \\ 1/2, & 1/2 < x < 1, \end{cases} \qquad \kappa_2 = \begin{cases} 2, & 0 < x < 1/2, \\ 1/2, & 1/2 < x < 1, \end{cases}$$

as in [13], we set $c = 1$. Here, κ_1, κ_2 are material parameters. The wave speed is given by $c/\sqrt{\kappa_1 \kappa_2}$, and jumps between 0.5 to 2. The impedance, given by κ_1/κ_2, is the same in both regions. Therefore, we expect no reflections between the regions. We set vanishing Dirichlet boundary conditions as the previous example, and

$$f = g = 0, \quad u_q(x,0) = e^{-5000((x-0.2)^2)}, \quad \text{and} \quad u_\mu(x,0) = -e^{-5000((x-0.2)^2)}.$$

We can observe the results of the adaptive algorithm in Figure 4.4.

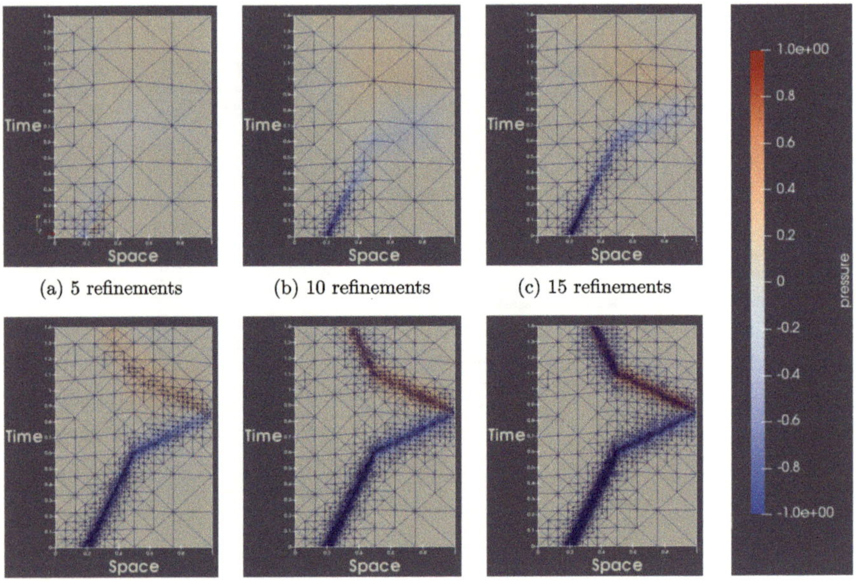

(a) 5 refinements (b) 10 refinements (c) 15 refinements
(d) 20 refinements (e) 25 refinements (f) 30 refinements

Figure 4.4: Iterate from the adaptive algorithm. Numerical pressure μ is shown for $p = 1$.

6.6 Convergence rates in three-dimensional space-time

On $\Omega = (0,1)^3$, we consider the problem, where the exact solution to the second- order wave equation is given by $\phi(x,t) = \sin(\pi x)\sin(\pi y)t^2$. This corresponds to

$$u = \begin{pmatrix} \pi\cos(\pi x)\sin(\pi y)t^2 \\ \pi\cos(\pi y)\sin(\pi x)t^2 \\ 2\sin(\pi x)\sin(\pi y)t \end{pmatrix},$$

$f = \sin(\pi x)\sin(\pi y)(2 + 2\pi^2 t^2)$ and $g = 0$.

In Table 4.3, we show the convergence rates of u_h for successively refined tetrahedral meshes, obtained using Technique 2 for $p = 0, 1, 2, 3$. Table 4.4 shows analogous results obtained for successively refined hexahedral meshes using Technique 1. In all these cases, we observe $O(h^{p+1})$ convergence rates for u_h.

Table 4.3: Convergence rates for $\|u - u_h\|$ on tetrahedral meshes obtained using Technique 2.

h	$p = 0$	eoc	$p = 1$	eoc	$p = 2$	eoc	$p = 3$	eoc
1	9.060×10^{-1}	–	4.783×10^{-1}	–	1.415×10^{-1}	–	4.395×10^{-2}	–
1/2	6.056×10^{-1}	0.58	1.392×10^{-1}	1.78	1.391×10^{-2}	3.35	3.285×10^{-3}	3.74
1/4	3.390×10^{-1}	0.84	3.351×10^{-2}	2.05	1.477×10^{-3}	3.24	1.649×10^{-4}	4.32
1/8	1.547×10^{-1}	1.13	8.955×10^{-3}	1.90	1.721×10^{-4}	3.10	9.969×10^{-6}	4.05

Table 4.4: Convergence rates for $\|u - u_h\|$ on hexahedral meshes using Technique 1.

h	$p = 0$	eoc	$p = 1$	eoc	$p = 2$	eoc	$p = 3$	eoc
1	1.115×10^{0}	–	$6.007 \cdot 10^{-1}$	–	2.883×10^{-2}	–	$3.326 \cdot 10^{-2}$	–
1/2	7.577×10^{-1}	0.56	1.512×10^{-1}	1.99	2.826×10^{-3}	3.35	2.054×10^{-3}	4.02
1/4	4.203×10^{-1}	0.85	3.859×10^{-2}	1.97	3.526×10^{-4}	3.00	1.323×10^{-4}	3.96
1/8	2.134×10^{-1}	0.98	9.692×10^{-3}	1.99	3.802×10^{-5}	3.21	9.377×10^{-6}	3.82

6.7 Adaptivity in 3D

Consider $\Omega = (0,1)^3$, and the problem, where the exact solution is given by

$$u(x,y,t) = e^{-200((x-x_0-ct)^2+(y-y_0-ct)^2)} \begin{pmatrix} 1 \\ 1 \\ -1 \end{pmatrix}.$$

Here, we have chosen $x_0 = y_0 = 0.2$. This corresponds to set $f = 0$, and

$$g = 400ce^{-200((x-x_0-ct)^2+(y-y_0-ct)^2)} \begin{pmatrix} y - y_0 - ct \\ x - x_0 - ct \end{pmatrix}.$$

After setting $c = 1/2$ and homogeneous Dirichlet boundary conditions, we observe that the adaptive scheme captures with precision the behavior of the wave propagation in Figure 4.5.

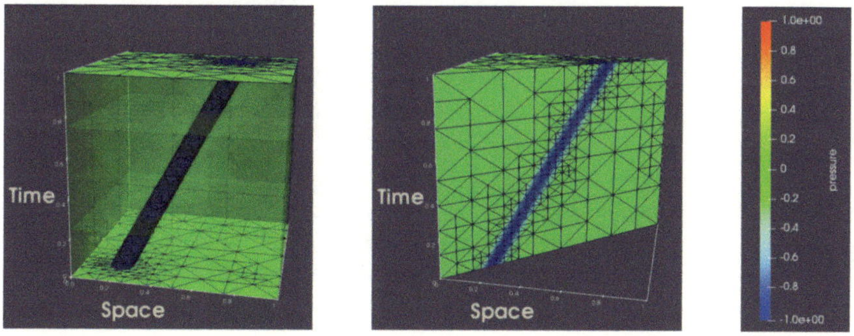

Figure 4.5: Adaptivity example in three dimensions after 10 iterations, u_μ component is shown, and $p = 1$.

6.8 Adaptivity with varying wave speed

Consider $\Omega = (-4,4)^2 \times (0,8)$ and the exact solution

$$u(x,y,t) = \begin{pmatrix} -1 \\ -1 \\ 1 \end{pmatrix} e^{-20((x-c\cos(\frac{\pi}{2}t))^2+(y+c\sin(\frac{\pi}{2}t))^2)}.$$

After setting $c = 1$, the component u_μ corresponds to a pulse propagating from the spatial coordinate $(1,0,0)$ at time $t = 0$ to $(1,0,8)$ at time $t = 8$, rotating in time and maintaining unit distance with the t-axis. We have chosen the solution, so two complete rotations from $t = 0$ to $t = 8$ are performed. The initial condition is set by

$$u(x,y,0) = \begin{pmatrix} -1 \\ -1 \\ 1 \end{pmatrix} e^{-20((x-1)^2+y^2)},$$

homogeneous Dirichlet boundary conditions are imposed, and

$$g = 40c \begin{pmatrix} x(\frac{\pi}{2}\sin(\frac{\pi}{2}t)+1)+\cos(\frac{\pi}{2}t)(\frac{\pi}{2}y-c)) \\ \sin(\frac{\pi}{2}t)(\frac{\pi}{2}x+c)+y(1+\frac{\pi}{2}\cos(\frac{\pi}{2}t)) \end{pmatrix} u_\mu(x,y,t),$$

$$f = -40c\Big(x\Big(\frac{\pi}{2}\sin\Big(\frac{\pi}{2}t\Big)+1\Big)$$
$$+y\Big(\frac{\pi}{2}\cos\Big(\frac{\pi}{2}t\Big)+1\Big) - \sqrt{2}c\cos\Big(\frac{\pi}{2}t+\frac{\pi}{4}\Big)\Big)u_\mu(x,y,t).$$

A sample of the results from the adaptive process are shown in Figure 4.6.

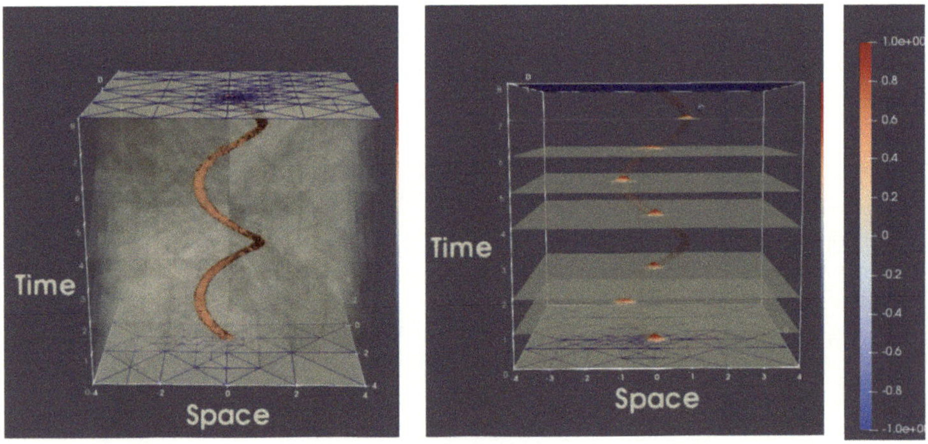

(a) Spacetime contour of the solution u_μ at an adaptive step.

(b) Slides of the solution u_μ at particular time steps.

Figure 4.6: Iterate from the adaptive algorithm. Numerical pressure μ is shown for $p = 1$.

Bibliography

[1] R. A. Adams. *Sobolev spaces*. Volume 65 of Pure and Applied Mathematics, Academic Press, New York, 1975.

[2] H. Brezis. *Functional analysis, Sobolev spaces and partial differential equations*. Universitext, Springer, 2011.

[3] C. Carstensen, L. Demkowicz, and J. Gopalakrishnan. A posteriori error control for DPG methods. *SIAM J. Numer. Anal.*, 36 (3): 1335–1353, 2014.

[4] C. Carstensen, L. Demkowicz, and J. Gopalakrishnan. Breaking spaces and forms for the DPG method and applications including Maxwell equations. *Comput. Math. Appl.*, 72: 494–522, 2016.

[5] L. Demkowicz and J. Gopalakrishnan. A class of discontinuous Petrov–Galerkin methods. Part II: Optimal test functions. *Numer. Methods Partial Differential Equations*, 27: 70–105, 2011.

[6] L. Demkowicz and J. Gopalakrishnan. An overview of the discontinuous Petrov Galerkin method. In X. Feng, O. Karakashian, and Y. Xing, editors, *Recent developments in discontinuous Galerkin finite element methods for partial differential equations: 2012 John H Barret memorial lectures*. Volume 157 of IMA Vol. Math. Appl., pages 149–180, Springer, Cham, 2014.

[7] L. Demkowicz, J. Gopalakrishnan, S. Nagaraj, and P. Sepúlveda. A space-time DPG method for the Schrödinger equation. *SIAM J. Numer. Anal.*, 55 (4): 1740–1759, 2017.

[8] W. Dörfler, S. Findeisen, and C. Wieners. Space-time discontinuous Galerkin discretizations for linear first-order hyperbolic evolution systems. *Comput. Methods Appl. Math.*, 16: 409–428, 2016.

[9] DPG Methods in NGSolve. Software hosted on GitHub, https://github.com/jayggg/DPG.

[10] T. E. Ellis, J. Chan, and L. F. Demkowicz. Robust DPG methods for the transient convection-diffusion. In G. Barrenechea, F. Brezzi, A. Cangiani, E. Georgoulis, editors, *Building bridges: Connections and challenges in modern approaches to numerical partial differential equations*. Volume 114 of Lecture Notes in Computational Science and Engineering, Springer, Cham, 2016.

[11] T. E. Ellis, L. F. Demkowicz, J. L. Chan, and R. D. Moser. Space-time DPG: Designing a method for massively parallel CFD. ICES Report 14–32, The Institute for Computational Engineering and Sciences, The University of Texas at Austin, 2014.

[12] T. Führer, N. Heuer, and J. Sen Gupta. A time-stepping DPG scheme for the heat equation. *Comput. Methods Appl. Math.*, 17 (2): 237–252, 2017.

[13] J. Gopalakrishnan, P. Monk, and P. Sepúlveda. A tent pitching scheme motivated by Friedrichs theory. *Comput. Math. Appl.*, 70 (5): 1114–1135, 2015.

[14] J. Gopalakrishnan, M. Neumüller, and P. S. Vassilevski. The auxiliary space preconditioner for the de Rham complex. *SIAM J. Numer. Anal.*, 56 (6): 3196–3218, 2018.

[15] J. Gopalakrishnan and W. Qiu. An analysis of the practical DPG method. *Math. Comput.*, 83 (286): 537–552, 2014.

[16] M. Neumüller. *Space-time methods: Fast solvers and applications*. Volume 20 of Monographic Series TU Graz: Computation in Engineering and Science, 2013.

[17] M. Neumüller, P. S. Vassilevski, and U. E. Villa. Space-time CFOSLS methods with AMGe upscaling. In C.-O. Lee, X.-C. Cai, D. E. Keyes, H. H. Kim, A. Klawonn, E.-J. Park, and O. B. Widlund, editors *Domain decomposition methods in science and engineering XXIII*, pages 253–260, Springer, Cham, 2017.

[18] NGSolve. Software hosted at https://ngsolve.org.

[19] O. Steinbach. Space-time finite element methods for parabolic problems. *Comput. Meth. Appl. Math.*, 15: 551–566, 2015.

[20] K. Voronin, C. S. Lee, M. Neumüller, P. Sepúlveda, and P. Vassilevski. Space-time discretizations using constrained first–order system least squares (CFOSLS). *J. Comput. Phys.*, 373: 863–876, 2018.

Ulrich Langer, Svetlana Matculevich, and Sergey Repin

5 Adaptive space-time isogeometric analysis for parabolic evolution problems

Abstract: The paper proposes new locally stabilized space-time isogeometric analysis approximations to initial boundary value problems of the parabolic type. Previously, similar schemes (but weighted with a global mesh parameter) have been presented and studied by U. Langer, M. Neumüller, and S. Moore (2016). The current work devises a localized version of this scheme, which is suited for adaptive mesh refinement. We establish coercivity, boundedness, and consistency of the corresponding bilinear form. Using these fundamental properties together with standard approximation error estimates for B-splines and NURBS, we show that the space-time isogeometric analysis solutions generated by the new scheme satisfy asymptotically optimal a priori discretization error estimates. Error indicators used for mesh refinement are based on a posteriori error estimates of the functional type that have been introduced by S. Repin (2002), and later rigorously studied in the context of isogeometric analysis by U. Langer, S. Matculevich, and S. Repin (2017). Numerical results discussed in the paper illustrate an improved convergence of global approximation errors and respective error majorants. They also confirm the local efficiency of the error indicators produced by the error majorants.

Keywords: parabolic initial-boundary value problems, locally stabilized space-time isogeometric analysis (IgA), a priori and a posteriori error estimates, adaptive IgA

MSC 2010: 35K20, 65M15, 65M60, 65M55

1 Introduction

Time-dependent problems governed by parabolic partial differential equations (PDEs) are typical models in many scientific and engineering applications. This fact trig-

Acknowledgement: This research is supported by the Austrian Science Fund (FWF) through the NFN S117–3 project. We would like to thank A. Mantzaflaris for his permanent support of the open-source C++ library G+Smo [25] that was used to implement the adaptive space-time IgA schemes and to perform all the numeral tests presented in this work. The authors thank the anonymous reviewer for careful reading and many helpful comments and valuable suggestions.

Ulrich Langer, Svetlana Matculevich, Johann Radon Institute for Computational and Applied Mathematics, Altenberger Str. 69, 4040 Linz, Austria, e-mails: ulrich.langer@ricam.oeaw.ac.at, smatculevich@ricam.oeaw.ac.at
Sergey Repin, University of Jyvaskyla, Jyvaskyla, Finland; and Peter the Great St. Petersburg Polytechnic University, Polytechnicheskaya, 29, St. Petersburg, Russia, e-mail: serepin@jyu.fi

https://doi.org/10.1515/9783110548488-005

gers their active investigation. This paper is focused on the numerical treatment of parabolic problems by means of *isogeometric analysis (IgA)* [29] combined with a full *space-time approach* that considers time as just another variable; see [20] and [60] for a survey of time-parallel and space-time methods, respectively. Due to the fast development of memory-distributed computations, this approach to the quantitative analysis of evolutionary problems became rather popular. Moreover, this way of treating evolutionary systems is not affected by the curse of sequentiality, which is typical for time-marching schemes, and allows simultaneous space-time adaptivity. Various versions of space-time methods can efficiently be used in combination with parallelization techniques; see, for example, [20, 21, 27, 43].

This paper uses an idea similar to that applied for the derivation of the globally stabilized space-time scheme in [43]. It is based on the testing of the underlying parabolic partial differential equation by means of time-upwind test functions that are scaled by the global discretization parameter h. This approach is motivated by the space-time streamline diffusion method studied in [26, 31, 32]. In contrast to the scheme presented in [43], the current work focuses on the element-wise analysis using "time-upwind" test functions that are scaled by the local discretization parameter h_K. This locally stabilized space-time IgA scheme is well suited for adaptivity.

One of the most attractive features of the IgA method is certainly connected with superior approximation properties of the splines of the highest smoothness leading to the so-called k-refinement in IgA. Indeed, C^{p-1}-smooth splines, or NURBS of the underlying polynomial degree p, deliver the approximation order p in the energy norm with the smallest number of degrees of freedoms (d. o. f.) for sufficiently smooth solutions. Since in many technical applications, solutions are patch-wise smooth or even analytic, it makes sense to use such approximations in the interior of the patches instead of C^0-smooth approximations, whereas the latter ones are more appropriate across interfaces and near singularities. In particular, we exploit the advantages of k-refinement in the functional a posteriori estimates and in the corresponding adaptive IgA strategies. Another attractive inherent feature of IgA is an improved spectral property of the stiffness matrix in the case of its generation by splines of the highest smoothness. The exact representation of technical geometries by splines and NURBS in CAD, and the use of the same basis functions for the approximation of the solution of PDEs, or PDE systems, make the use of IgA very interesting for industrial applications. This is especially useful for adaptive space-time IgA techniques in the case of moving spatial domains and interfaces. In contrast to the sequential approach, moving domains and interfaces in space are fixed in the space-time cylinder. This makes an essential difference in adaptivity; that is, it is a challenge to handle mesh localization in time-stepping methods for moving spatial domains and interfaces, but it is rather straightforward for the space-time approach. We refer the reader to [61, 62] for some interesting industrial applications.

The construction of effective adaptive refinement techniques is highly important for the design of fast and efficient numerical methods for solving PDEs. Adaptivity strongly relies on a reliable and locally efficient a posteriori error estimation. We refer to [1, 3, 45, 53] for an overview of different error estimators. An efficient error indicator is aimed at identifying the areas, where discretization errors are excessively high, in order to refine the mesh and minimize local errors. A smart combination of solvers and error indicators could potentially provide a fully automated refinement algorithm, taking into account special features of the problem and generating a discretization that produces approximate solutions with the desired accuracy.

Due to a tensor-product setting of splines in IgA, the corresponding mesh refinement procedures inherit specific effects and may include a large percentage of superfluous control points. Challenges, arising along with these disadvantages, have triggered the development of local refinement techniques for IgA, such as *T-splines* introduced in [58, 59] and analyzed in [5, 6, 56, 57], *hierarchical* (HB-splines) [18, 35, 64], and *truncated hierarchical B-splines* (THB-splines) [23], *patchwork splines* (PB-splines) [15], *locally refined splines* (LR-splines) [8, 13], *polynomial splines over hierarchical T-meshes* (PHT-splines) [48, 65], et cetera. In the case of elliptic boundary value problems, local refinement IgA techniques were combined with some a posteriori error estimation approaches in several publications, for example, a posteriori error estimates using hierarchical bases in [14, 64], residual-based a posteriori error estimators and their modifications in [9, 30, 36, 65], and goal-oriented error estimators in [11, 37, 38, 63]. In the context of evolutionary problems, there is a huge number of papers on time-stepping methods and the closely related time dG or dPG methods on time-slices, but only a few papers on fully unstructured finite element meshes; see [60] for the relevant references. To our knowledge, there are no papers on fully unstructured and adaptive space-time IgA schemes.

In this work, we utilize fully guaranteed error estimates in terms of several global norms equivalent to the norm of the functional space containing the corresponding generalized solution. These estimates do not use mesh-dependent constants (which must be recalculated in the process of mesh adaptation), and include only global constants characterizing the geometry. Henceforth, we simply call them error majorants. A posteriori error estimates of this functional type were originally introduced in [50, 51] and later applied to various problems; see [45, 53] and references therein. These estimates are valid for any approximation from the admissible functional space. They do not require special properties from approximations (for example, Galerkin orthogonality) or/and from the exact solution (for example, extra regularity beyond the minimal energy class that guarantees the existence of a unique generalized solution) and are valid for any approximation from the admissible function space. Moreover, the majorant also generates efficient indicators of local (element-wise) error distributions over the domain.

The paper presents a new locally stabilized space-time IgA scheme. We first show coercivity (ellipticity), boundedness, and consistency of the locally stabilized bilinear form corresponding to the space-time IgA scheme. These properties imply an a priori estimate of the discretization error by the best approximation error in suitable mesh-dependent norms. The combination of such an estimate with known approximation results for spline spaces yields a priori discretization error estimates in terms of convergence rates under additional regularity assumptions for the solution. It is clear that the locally stabilized space-time IgA schemes are especially suited for adaptivity in space and time simultaneously, which, in this paper, is driven by the functional-type a posteriori error estimates discussed above. By exploiting the universality and efficiency of these error estimates, and by taking advantage of smoothness of the IgA approximations, we aim at constructing fast fully adaptive space-time IgA methods. These two techniques have been already combined in the case of elliptic boundary value problems in [33] and [46] using tensor-based splines and THB-splines [22, 23, 24], respectively. Both works confirmed that the majorants provide not only reliable upper bounds of the total energy error, but also quantitatively sharp indicators of local errors. For the time-dependent problems, the simplest form of such error bounds was derived for the heat equation in [52], and tested for the generalized diffusion equation in [19]. Majorants for approximations to the evolutionary convection-diffusion problem, which may have jumps in time, were considered in [54]. In [47], the authors study the robustness of the majorant to a drastic change in values of the reaction parameter in evolutionary reaction-diffusion problems and provide a comparison of the upper bound of the error (majorant) to the newly introduced minorant. Another extensive discussion on the numerical properties of the above-mentioned error estimates with respect to both time-marching and space-time methods can be found in [28]. Paper [42], that precedes the current study, presents new functional-type a posteriori error estimates in the context of globally weighted space-time IgA schemes introduced in [43]. Numerical examples presented there show the efficiency of functional a posteriori error estimates for IgA approximations with respect to several examples exhibiting different features, and report on the computing costs for these bounds. Moreover, the numerical examples discussed in [42] demonstrate the efficiency of the space-time THB-spline-based adaptive procedure. Therefore, the importance of locally stabilized space-time IgA schemes and the investigation of their numerical properties are rather natural in the context of the construction of fully adaptive schemes for initial-boundary value problems (I-BVPs). Both latter studies are the pioneering works in the direction of full space-time discretizations in the context of IgA.

This work is organized as follows: Section 2 defines the model evolutionary problem, and recapitulates notation and function spaces, which are used throughout the paper. Section 3 presents a concise overview of the IgA framework and respective notions and definitions, whereas Section 4 gives an overview of the preliminary knowledge used later in the paper. Furthermore, in Section 5, we revisit the globally stabilized space-time IgA scheme from [43] and discuss its main properties. Section 6 introduces the new locally stabilized version of the space-time IgA scheme together with the proofs of coercivity, boundedness, and consistency of the bilinear form generated by the IgA scheme. We present a priori discretization error estimates that are a consequence of coercivity, boundedness, consistency, and the known approximation properties of the corresponding IgA spaces. Section 7 recapitulates the functional a posteriori error estimates that are the basis for the adaptive procedures used in our numerical studies. Furthermore, we discuss some implementation details. In Section 8, we present a series of numerical examples possessing different features that emphasize the most distinguished properties of the adaptive scheme. The last part, Section 9, draws final conclusions.

2 Space-time variational formulation

Let $\Omega \subset \mathbb{R}^d$, $d \in \{1, 2, 3\}$ be a bounded domain with Lipschitz continuous boundary $\partial\Omega$ and $(0, T)$, $0 < T < +\infty$, be a given time interval. By $Q := \Omega \times (0, T)$ and $\overline{Q} := Q \cup \partial Q$, we denote the space-time cylinder and its closure. The boundary ∂Q of Q is composed of three parts: $\partial Q := \Sigma \cup \overline{\Sigma}_0 \cup \overline{\Sigma}_T$, where $\Sigma = \partial\Omega \times (0, T)$, $\Sigma_0 = \Omega \times \{0\}$, and $\Sigma_T = \Omega \times \{T\}$.

In this paper, we discuss adaptive space-time IgA approximations of evolutionary problems using the classical *linear parabolic initial-boundary value problem*: find u satisfying the equations

$$\partial_t u - \Delta_x u = f \quad \text{in } Q, \qquad u = 0 \quad \text{on } \Sigma, \qquad u = u_0 \quad \text{on } \overline{\Sigma}_0, \tag{2.1}$$

where ∂_t denotes the time derivative, Δ_x is the spatial Laplace operator, f is a given source function, and u_0 is a given initial state, as a model problem.

Throughout the paper, we use the following spaces of functions and related notation: By

$$\|v\|_Q := \|v\|_{L_2(Q)} = (v, v)_Q^{1/2} \quad \text{and} \quad (v, w)_Q := \int_Q v(x, t) w(x, t)\, dx\, dt,$$

we denote the norm and the scalar product in $L_2(Q)$, respectively, with similar notation used for the product of vector valued functions. By $H^s(Q)$, $1 \le s \in \mathbb{N}$, we denote the standard Sobolev spaces equipped with the norm

$$\|v\|_{H^s(Q)} := \Big(\int_Q \sum_{|\alpha| \le s} |\partial^\alpha v|^2\, dx\, dt \Big)^{1/2},$$

where

$$\partial^\alpha v := \frac{\partial^{|\alpha|} v}{\partial x_1^{\alpha_1} \cdots \partial x_d^{\alpha_d} \partial x_{d+1}^{\alpha_{d+1}}},$$

with $x_{d+1} = t$, $\alpha := (\alpha_1, \dots, \alpha_d, \alpha_{d+1})$ is a multi-index, and $|\alpha| = \sum_{i=1}^{d+1} \alpha_i$.

The H^s seminorm is denoted by

$$|v|_{H^s(Q)} := \Big(\int_Q \sum_{|\alpha|=s} |\partial^\alpha v|^2 \, dx \, dt \Big)^{1/2}.$$

Next, we introduce the spaces

$$\begin{aligned}
H_0^{1,0}(Q) &:= \{u \in L_2(Q) &&: \nabla_x u \in [L_2(Q)]^d, u = 0 \text{ on } \Sigma\},\\
H_{0,\overline{0}}^{1}(Q) = H_{0,\overline{0}}^{1,1}(Q) &:= \{u \in H_0^{1,0}(Q) &&: \partial_t u \in L_2(Q), u = 0 \text{ on } \Sigma_T\},\\
H_{0,\underline{0}}^{1}(Q) &:= \{u \in H_0^{1,0}(Q) &&: \partial_t u \in L_2(Q), u = 0 \text{ on } \Sigma_0\}, \quad \text{and}\\
H_0^{\Delta_x,1}(Q) &:= \{u \in H_0^{1,0}(Q) &&: \Delta_x u \in L_2(Q), \partial_t u \in L_2(Q)\}.
\end{aligned}$$

Finally,

$$H_{0,\underline{0}}^{s}(Q) := H^s(Q) \cap H_{0,\underline{0}}^{1}(Q),$$

and

$$H^{\mathrm{div}_x,0}(Q) := \{\mathbf{y} \in [L_2(Q)]^d : \mathrm{div}_x \mathbf{y} \in L_2(Q)\}$$

equipped with a scalar product

$$(\mathbf{v}, \mathbf{w})_{\mathrm{div}_x,0} := (\mathbf{v}, \mathbf{w})_Q + (\mathrm{div}_x \mathbf{v}, \mathrm{div}_x \mathbf{w})_Q.$$

Since Ω is bounded, we have the Friedrichs inequality $\|w\|_\Omega \le C_F \|\nabla_x w\|_\Omega$ for all $w \in H_0^1(\Omega)$, which also implies $\|w\|_Q \le C_F \|\nabla_x w\|_Q$ for all $w \in H_0^{1,0}(Q)$.

The standard space-time variational formulation of the initial-boundary value problem (2.1); that is, find $u \in H_0^{1,0}(Q)$ such that

$$a(u, w) = \ell(w), \quad \forall w \in H_{0,\overline{0}}^{1}(Q), \tag{2.2}$$

with the bilinear form

$$a(u, w) := (\nabla_x u, \nabla_x w)_Q - (u, \partial_t w)_Q,$$

and the linear functional

$$\ell(w) := (f, w)_Q + (u_0, w)_{\Sigma_0}$$

was considered in [40]. Here and later on,

$$(u_0, w)_{\Sigma_0} := \int_{\Sigma_0} u_0(x) w(x, 0) \, dx.$$

It was shown that problem (2.2) has a unique solution with an additional regularity, provided that $u_0 \in L_2(\Omega)$ and $f \in L_{2,1}(Q)$, where

$$L_{2,1}(Q) := \left\{ v \in L_1(Q) : \int_0^T \|v(t)\|_\Omega \, dt < \infty \right\}.$$

Moreover, if $f \in L_2(Q)$ and $u_0 \in H_0^1(\Omega)$, then (2.2) is uniquely solvable in $H_0^{\Delta_x,1}(Q)$, and the solution u continuously depends on t in the norm of the space $H_0^1(\Omega)$; see, for example, [39] and [40, Theorem 2.1]. Furthermore, according to [40, Remark 2.2], $\|\nabla_x u(\cdot, t)\|_\Omega^2$ is an absolutely continuous function of $t \in [0, T]$ for any $u \in H_0^{\Delta_x,1}(Q)$.

Throughout the paper, we assume that $f \in L_2(Q)$ and $u_0 \in H_0^1(\Sigma_0)$. That is, we know that the solution u of the space-time variational problem (2.2) belongs to $H_0^{\Delta_x,1}(Q)$. In other words, we consider the case of maximal parabolic regularity. Then, without a loss of generality, we can assume homogeneous initial conditions $u_0 = 0$; cf. [27]. We refer the reader to [17] for an overview of recent results on maximal parabolic regularity.

3 IgA framework

For the convenience of the reader, we briefly recall the general concept of the IgA technology, the definition of B-splines, NURBS, and THB-splines, and their use in the geometrical representation of the space-time cylinder Q. Moreover, we recall the construction of the IgA trial and test spaces, which are used to approximate solutions, satisfying (2.2) under the condition of maximal parabolic regularity. For a detailed discussion of the IgA basics, we refer the reader to [7, 10, 29].

Let $2 \le p \in \mathbb{N}$ be the polynomial degree, and $n \in \mathbb{N}^+$ denote the number of basis functions used to construct a piecewise polynomial *B-spline curve* $C(\xi)$ in $\mathbb{R}^d$; that is, $C(\xi) := \sum_{i=1}^n \mathbf{C}_i \hat{B}_{i,p}$, where $\hat{B}_{i,p}$ are B-spline basis functions and $\mathbf{C}_i$ are the so-called *control points*. B-spline functions are derived from a *knot-vector*, a nondecreasing set of coordinates in the parameter domain, written as $\Xi = \{\xi_1, \dots, \xi_{n+p+1}\}$, $\xi_i \in \mathbb{R}$, where $\xi_1 = 0$ and $\xi_{n+p+1} = 1$. The knots can be repeated, and the *multiplicity* of the ith knot is indicated by m_i. Throughout the paper, we consider only *open* knot vectors; that is, multiplicities m_1 and m_{n+p+1} of the first and the last knots, respectively, are equal to $p + 1$. The *univariate* B-spline basis functions $\hat{B}_{i,p} : \hat{Q} := (0, 1)^{d+1} \to \mathbb{R}$ are defined by means of the *Cox–de Boor formula*,

$$\hat{B}_{i,p}(\xi) := \frac{\xi - \xi_i}{\xi_{i+p} - \xi_i} \hat{B}_{i,p-1}(\xi) + \frac{\xi_{i+p+1} - \xi}{\xi_{i+p+1} - \xi_{i+1}} \hat{B}_{i+1,p-1}(\xi),$$

with $\hat{B}_{i,0}(\xi) := \{1 \text{ if } \xi_i \le \xi \le \xi_{i+1}, \text{ and } 0 \text{ otherwise}\}$, where a division by zero is defined to be zero. These basis functions are $(p - m_i)$-times continuously differentiable across the

ith knot with multiplicity m_i. Hence, $\hat{B}_{i,p}$ is C^{p-1} continuous for every inner knot with $m_i = 1$. For the knots lying on the boundary of the parametric domain, the multiplicity is $p+1$, which makes the B-spline discontinuous on the patch interfaces; a detailed illustration of B-splines is available in the original works [29, Figures 5 and 6] or [4, Figure 1]. We note that the analysis provided in this paper is valid only for domains represented by a single patch.

The *multivariate* B-splines on the space-time parameter domain $\hat{Q} := (0,1)^{d+1}$, $d = \{1,2,3\}$, are defined as a tensor-product of the corresponding univariate B-splines; that is, $\hat{B}_{i,p}(\boldsymbol{\xi}) := \prod_{\alpha=1}^{d+1} \hat{B}_{i_\alpha,p_\alpha}(\xi^\alpha)$, where $i \in \mathcal{I} := \{i = (i_1,\dots,i_{d+1}) : i_\alpha = 1,\dots,n_\alpha\}$, and $\boldsymbol{\xi} \in \hat{Q}$. The univariate and multivariate NURBS basis functions $\hat{R}_{i,p} : \hat{Q} \to \mathbb{R}$ are defined via B-splines by means of the weighting function $W : \hat{Q} \to \mathbb{R}$, $W(\boldsymbol{\xi}) := \sum_{i\in\mathcal{I}} w_i \hat{B}_{i,p}(\boldsymbol{\xi})$, where $w_i \in \mathbb{R}^+$, satisfying $\sum_{i\in\mathcal{I}} w_i = 1$; that is, $\hat{R}_{i,p}(\boldsymbol{\xi}) := w_i \hat{B}_{i,p}(\boldsymbol{\xi})/W(\boldsymbol{\xi})$.

To recall basic definitions related to THB-splines, we follow the structure outlined in [22], and consider a finite sequence of nested d-variate tensor-product spline spaces $\widehat{V}^0 \subset \cdots \subset \widehat{V}^N$, defined on the axis-aligned box-domain $\widehat{\Omega}^0 \subset \mathbb{R}^d$. To each space V^ℓ, we assign a tensor-product B-spline basis of degree $\boldsymbol{p}$, $\{\hat{B}^\ell_{\boldsymbol{i},\boldsymbol{p}}\}_{\boldsymbol{i}\in\mathcal{I}^\ell}$, $\mathcal{I}^\ell := \{\boldsymbol{i} = (i_1,\dots,i_d), i_k = 1,\dots,n^\ell_k \text{ for } k = 1,\dots,d\}$, where $\mathcal{I}^\ell$ is a set of multi-indices for each level, and n^ℓ_k denotes the number of univariate B-spline basis functions in the k-th coordinate direction. After assuming that $\mathcal{I}^\ell$ has a fixed ordering and rewriting the basis as $\hat{\mathbf{B}}^\ell(\boldsymbol{\xi}) = (\hat{B}^\ell_{\boldsymbol{i},\boldsymbol{p}}(\boldsymbol{\xi}))_{\boldsymbol{i}\in\mathcal{I}^\ell}$, it can be considered as a column-vector of basis functions. Then a spline function $s : \widehat{\Omega}^0 \to \mathbb{R}^m$ is defined by $\hat{\mathbf{B}}^\ell(\boldsymbol{\xi})$ and a coefficient matrix C^ℓ, i. e., $s(\boldsymbol{\xi}) = \sum_{\boldsymbol{i}\in\mathcal{I}^\ell} \hat{B}^\ell_{\boldsymbol{i},\boldsymbol{p}}(\boldsymbol{\xi}) c^\ell_i = \hat{\mathbf{B}}^\ell(\boldsymbol{\xi})^{\mathrm{T}} C^\ell$, where $c^\ell_i \in \mathbb{R}^m$ are row-coefficients of C^ℓ.

Since $\widehat{V}^\ell \subset \widehat{V}^{\ell+1}$, the basis $\hat{\mathbf{B}}^\ell$ can be represented by a linear combination of $\hat{\mathbf{B}}^{\ell+1}$, namely, $s(\boldsymbol{\xi}) = \hat{\mathbf{B}}^\ell(\boldsymbol{\xi})^{\mathrm{T}} C^\ell = \hat{\mathbf{B}}^{\ell+1}(\boldsymbol{\xi})^{\mathrm{T}} R^{\ell+1} C^\ell$, where $R^{\ell+1}$ is a refinement matrix. Its entries can be obtained from B-splines refinement rules; see [49]. Along with nested spaces, a corresponding sequence of nested domains is considered $\widehat{\Omega}^0 \supseteq \cdots \supseteq \widehat{\Omega}^N$, where each $\widehat{\Omega}^\ell \subset \mathbb{R}^d$ is covered by a collection of cells with respect to the tensor-product grid of level ℓ. In this work, we focus on a dyadic cell refinement with uniform degrees $p_\alpha = p$ for all levels and coordinate directions, therefore, $\boldsymbol{p} = p$ in further exposition.

Let the *characteristic matrix* X^ℓ of $\hat{\boldsymbol{B}}^\ell(\boldsymbol{\xi})$ with respect to Ω^ℓ and $\Omega^{\ell+1}$ be defined as

$$X^\ell := \operatorname{diag}(x^\ell_{\boldsymbol{i}})_{\boldsymbol{i}\in\mathcal{I}^\ell}, \quad x^\ell_{\boldsymbol{i}} := \begin{cases} 1, & \text{if } \operatorname{supp}\hat{B}^\ell_{\boldsymbol{i},\boldsymbol{p}} \subseteq \Omega^\ell \wedge \operatorname{supp}\hat{B}^\ell_{\boldsymbol{i},\boldsymbol{p}} \nsubseteq \Omega^{\ell+1}, \\ 0, & \text{otherwise.} \end{cases}$$

Next, for each level ℓ, the set of the indices of *active functions* can be defined by $\mathcal{I}^\ell_* := \{\mathcal{I}^\ell : x^\ell_{\boldsymbol{i}} = 1\}$.

To store the indices of all active functions at all hierarchical levels, we define an *index set* $\boldsymbol{\mathcal{I}} := \{(\ell, \boldsymbol{i}) : \ell \in \{0,\dots,N\}, \boldsymbol{i} \in \mathcal{I}^\ell_*\}$. The initial hierarchical data structure is defined by the tensor-product mesh $\widehat{\Omega}^0$; see Figure 5.2a. In particular, we illustrate the knot lines of the spaces $V^0 \subset \cdots \subset V^3$, where the levels increase from the left to the right. By means of the insertion operation, new subdomains can be added to obtain new representations $\widehat{\Omega}^0 \supseteq \cdots \supseteq \widehat{\Omega}^3$. The sets of active basis functions $\mathcal{I}_*^\ell$ and the characteristic matrices X^ℓ for all levels ℓ are extracted simultaneously with the new box insertion and initialization of the basis. Figure 5.2b illustrates meshes at the refinement levels $\ell = 0, 1, 2, 3$.

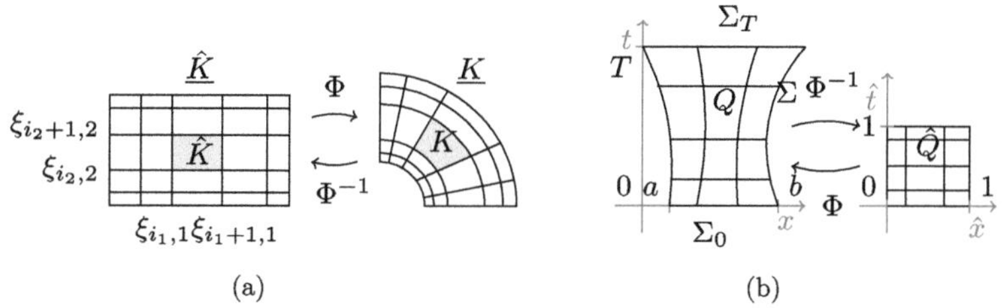

Figure 5.1: (a) Illustration of the element K, its support extensions $\underline{K}$, and their counterparts on the parametric domain $\hat{K}$ and $\underline{\hat{K}}$. (b) Mapping of the reference domain $\hat{Q}$ to the space-time cylinder Q.

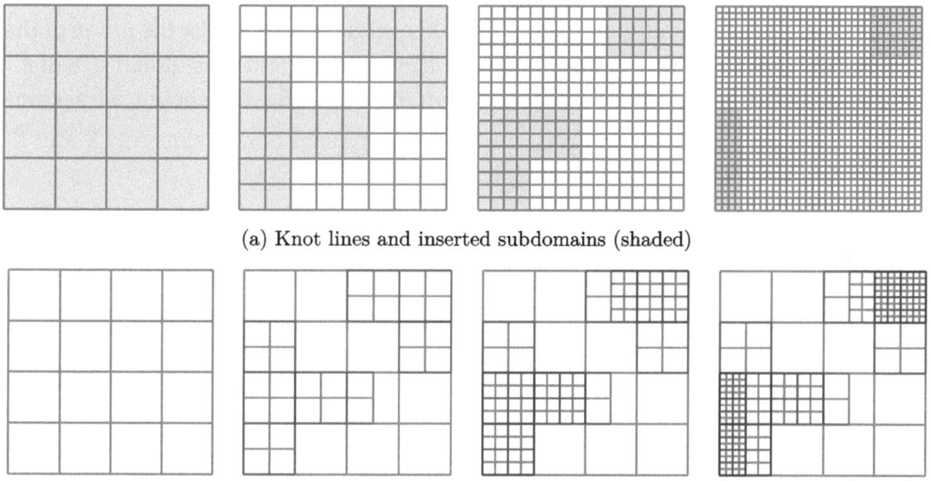

(a) Knot lines and inserted subdomains (shaded)

(b) Hierarchical meshes for refinement levels

Figure 5.2: Hierarchical levels $\ell = 0, 1, 2, 3$.

The THB-spline basis related to hierarchical domains is defined as

$$\hat{\mathbf{T}}(\boldsymbol{\xi}) = (\mathcal{K}_{\boldsymbol{i}}^\ell(\boldsymbol{\xi}))_{(l,\boldsymbol{i})\in\mathcal{I}}, \quad \mathcal{K}_{\boldsymbol{i}}^\ell(\boldsymbol{\xi}) = \operatorname{trunc}^N(\operatorname{trunc}^{N-1}(\ldots\operatorname{trunc}^{\ell+1}(\hat{B}_{\boldsymbol{i},\boldsymbol{p}}^\ell(\boldsymbol{\xi})))),$$

where the *truncation* of any function $s(\boldsymbol{\xi}) \in \widehat{V}^\ell$ with respect to the level $\ell+1$ is defined by $\operatorname{trunc}^{\ell+1}(s(\boldsymbol{\xi})) = \hat{\mathbf{B}}^{\ell+1}(\boldsymbol{\xi})^{\mathrm{T}}(I^{\ell+1} - X^{\ell+1})R^{\ell+1}C^\ell$. Here, $I^{\ell+1}$ denotes an identity matrix $I^{\ell+1}$ of size $|I^{\ell+1}| \times |I^{\ell+1}|$, the multiplication of $R^{\ell+1}$ by C^ℓ represents $s(\boldsymbol{\xi})$ with respect to the level $\ell+1$, and an additional multiplication by $(I^{\ell+1} - X^{\ell+1})$ performs the truncation operation. For a detailed discussion of the truncation operation, we refer the reader to [22, 23, 24]. For the case of univariate quadratic spline basis functions, effects of truncation are described in [22, Figure 2].

In association with the knot-vectors

$$\Xi^\alpha := \{\xi^\alpha := (\xi_1^\alpha, \dots, \xi_{n^\alpha+p^\alpha+1}^\alpha) \in \mathbb{R}^{n^\alpha+p^\alpha+1}\}$$

with the index $\alpha = 1, \dots, d+1$, we define a mesh partitioning $\hat{\mathcal{K}}_h$ of $\hat{Q}$ into $(d+1)$-dimensional open knot-spans (elements)

$$\begin{aligned}\hat{\mathcal{K}}_h &= \hat{\mathcal{K}}_h(\Xi^1, \dots, \Xi^{d+1}) \\ &:= \{\hat{K} = \otimes_{\alpha=1}^{d+1}(\xi_{i_\alpha}^\alpha, \xi_{i_{\alpha+1}}^\alpha) : \hat{K} \neq \emptyset, i_\alpha = p_\alpha + 1, \dots, n_\alpha - 1\}.\end{aligned}$$

A nonempty element $\hat{K} \in \hat{\mathcal{K}}_h$ is characterized by its diameter $\hat{h}_{\hat{K}}$. To $\hat{K}$, we associate $\underline{\hat{K}} \subset \hat{Q}$, defined as

$$\underline{\hat{K}} = \otimes_{\alpha=1}^{d+1}(\xi_{i_\alpha-p_\alpha+}^\alpha, \xi_{i_\alpha+p_\alpha+1}^\alpha) \cap \hat{Q}.$$

The set $\underline{\hat{K}}$ represents the support extension of $\hat{K}$, and is constructed by the union of the supports of basis functions intersecting with $\hat{K}$; see Figure 5.1a. The global size of $\hat{\mathcal{K}}_h$ is denoted by $\hat{h} := \max_{\hat{K}\in\hat{\mathcal{K}}_h}\{\hat{h}_{\hat{K}}\}$, where $\hat{h}_{\hat{K}} := \operatorname{diam}(\hat{K})$. For the time being, we assume locally quasiuniform meshes; that is, the ratio of two neighboring elements $\hat{K}$ and $\hat{K}'$ satisfies the inequality $c_1 \leq \hat{h}_{\hat{K}}/\hat{h}_{\hat{K}'} \leq c_2$ with constants $c_1, c_2 > 0$.

The physical space-time domain $Q \subset \mathbb{R}^{d+1}$ is defined from the parametric domain $\hat{Q} = (0,1)^{d+1}$ by the geometrical mapping $\Phi : \hat{Q} \to Q := \Phi(\hat{Q}) \subset \mathbb{R}^{d+1}$, $\Phi(\boldsymbol{\xi}) := \sum_{i\in\mathcal{I}} \hat{R}_{i,p}(\boldsymbol{\xi})\mathbf{P}_i$, where $\{\mathbf{P}_i\}_{i\in\mathcal{I}} \in \mathbb{R}^{d+1}$ are the control points; see Figure 5.1b. For each $\hat{K} \in \hat{\mathcal{K}}_h$ and $\underline{\hat{K}}$, we obtain an element and a support extension on the physical domain $K = \Phi(\hat{K}) := \{\Phi(\xi) : \xi \in \hat{K}\}$, and $\underline{K} := \Phi(\underline{\hat{K}})$, respectively. The physical mesh $\mathcal{K}_h := \{K = \Phi(\hat{K}) : \hat{K} \in \hat{\mathcal{K}}_h\}$ is defined on the space-time cylinder Q. The global mesh-size is denoted by

$$h := \max_{K\in\mathcal{K}_h}\{h_K\}, \quad h_K := \|\nabla_x\Phi\|_{L_\infty(K)}\hat{h}_{\hat{K}}. \tag{3.1}$$

For a priori error estimates derivation, we assume that

$$h_K \leq h \leq C_u h_K. \tag{3.2}$$

The set of facets corresponding to $\mathcal{K}_h$ is denoted by $\mathcal{E}_h$, and can be grouped into the inner facets

$$\mathcal{E}_h^I = \{E \in \mathcal{E}_h^K : \exists K, K' \in \mathcal{K}_h : E = \partial K \cap \partial K' \wedge E \not\subset \partial Q\},$$

and the facets intersecting with the boundary,

$$\mathcal{E}_h^{\partial Q} = \{E \in \mathcal{E}_h^K : \exists K, K' \in \mathcal{K}_h : E = \partial K \cap \partial K' \wedge E \cap \partial Q \neq \emptyset\}.$$

The latter one contains

$$\mathcal{E}_h^{\Sigma} = \{E \in \mathcal{E}_h^K : \exists K, K' \in \mathcal{K}_h : E = \partial K \cap \partial K' \wedge E \cap \Sigma \neq \emptyset\} \quad \text{and}$$
$$\mathcal{E}_h^{\Sigma_T} = \{E \in \mathcal{E}_h^K : \exists K, K' \in \mathcal{K}_h : E = \partial K \cap \partial K' \wedge E \cap \Sigma_T \neq \emptyset\}.$$

Let $\mathcal{E}_h^K$ denote the set of facets of the local element $K \in \mathcal{K}_h$; that is,

$$\mathcal{E}_h^K := \{E \in \mathcal{E}_h : E \cap \partial K \neq \emptyset, K \in \mathcal{K}_h\}.$$

The space-time IgA spaces are constructed by a push-forward of the basis functions defined on the parametric domain

$$V_h := \operatorname{span}\{\phi_{h,i} := \hat{\phi}_{h,i} \circ \Phi^{-1}\}_{i \in \mathcal{I}}. \tag{3.3}$$

Here, $\hat{\phi}_{h,i}$ is a spline basis function of a degree p of the space $\hat{\mathcal{S}}_h^p$ on the parametric domain, and Φ is assumed to be invertible in Q, with smooth inverse on each element $K \in \mathcal{K}_h$; see [4, 7] and references therein. Moreover, we introduce the subspace $V_{0h} := V_h \cap H^1_{0,\underline{0}}(Q)$ of functions satisfying homogeneous initial and boundary conditions.

4 Preliminary results

First, we recall some fundamental relations, such as scaled trace and inverse inequalities, which are important for the derivation of a priori discretization error estimates for the space-time IgA scheme presented in Section 6.

Lemma 4.1. [16, *Theorem 3.2*] *Let $K \in \mathcal{K}_h$. Then the* ***scaled trace inequality***

$$\|v\|_{\partial K} \leq C_{\mathrm{tr}} h_K^{-1/2} (\|v\|_K + h_K \|\nabla v\|_K) \tag{4.1}$$

holds for all $v \in H^1(K)$, where $\nabla = (\nabla_x, \partial_t)$, h_K is a local mesh size (cf. (3.1)*), and C_{tr} is a positive constant independent of $K \in \mathcal{K}_h$.*

Lemma 4.2. [4, *Theorem 4.1*] *Let $K \in \mathcal{K}_h$. Then the* ***inverse inequalities***

$$\|\nabla_x v_h\|_K \leq C_{\mathrm{inv},1} h_K^{-1} \|v_h\|_K, \quad \text{and} \quad \|v_h\|_{\partial K} \leq C_{\mathrm{inv},0} h_K^{-1/2} \|v_h\|_K \tag{4.2}$$

hold for all $v_h \in V_h$, where $C_{\mathrm{inv},0}$ and $C_{\mathrm{inv},1}$ are positive constants independent of $K \in \mathcal{K}_h$, and $h_K := \operatorname{diam}_{K \in \mathcal{K}_h}$ is a local mesh size.

For completeness, we recall fundamental results on the approximation properties of spaces generated by NURBS; see [4, Section 3]. It states the existence of a projection operator that provides asymptotically optimal approximation results.

Lemma 4.3. [4, *Theorem 3.1*] *Let* ℓ *and* s *be integers such that* $0 \le \ell \le s \le p+1$, *and* $u \in H^s_{0,\underline{0}}(Q)$. *Then there exists a projection operator* $\Pi_h : H^s_{0,\underline{0}}(Q) \to V_{0h}$ *such that*

$$|v - \Pi_h v|^2_{H^\ell(K)} \le C^2_{\ell,s} h_K^{2(s-\ell)} \sum_{i=0}^{s} c_K^{2(i-\ell)} |v|^2_{H^i(\underline{K})}, \quad \forall v \in L_2(Q) \cap H^\ell(\underline{K}), \tag{4.3}$$

where K *and* $\underline{K}$ *are an element and its support extension,* $C_{\ell,s}$ *is a dimensionless constant that depends on* s, ℓ, p, *and the shape regularity of* K, *described by* Φ *and its gradient,* h_K *is a local mesh size (see* (3.1)*), and* $c_K := \|\nabla_x \Phi\|_{L_\infty(\Phi^{-1}(\hat{\underline{K}}))}$.

Unlike the classical finite element spaces of degree p, Lemma 4.3 provides error bounds, where the ℓth-order seminorm of the error $v - \Pi_h v$ is controlled by the full s^{th}-order norm of v. In particular, the following formulations of (4.3) will be used:

$$\|v - \Pi_h v\|^2_{L_2(K)} \le C^2_{0,s} h_K^{2s} \sum_{i=0}^{s} c_K^{2i} |v|^2_{H^i(\underline{K})}, \tag{4.4}$$

$$|v - \Pi_h v|^2_{H^1(K)} \le C^2_{1,s} h_K^{2(s-1)} c_K^{-2} \sum_{i=0}^{s} c_K^{2i} |v|^2_{H^i(\underline{K})}, \tag{4.5}$$

$$|v - \Pi_h v|^2_{H^2(K)} \le C^2_{2,s} h_K^{2(s-2)} c_K^{-4} \sum_{i=0}^{s} c_K^{2i} |v|^2_{H^i(\underline{K})}, \tag{4.6}$$

for any $v \in L_2(Q) \cap H^\ell(\underline{K})$.

5 Globally stabilized schemes

Globally stabilized space-time IgA schemes for parabolic equations have been presented and analyzed in [43]. In particular, it was shown that the corresponding discrete bilinear form is coercive (elliptic) with respect to a discrete energy norm, bounded, consistent, and that the computed IgA approximations satisfy optimal a priori discretization error estimates. To derive a *globally stabilized discrete IgA space-time scheme*, the authors considered time-upwind test function $v_h + \delta_h \partial_t v_h$, $\delta_h = \theta h$, $v_h \in V_{0h}$, such that $\theta > 0$ is an auxiliary constant, and h is the global mesh-size; see (3.1). This implies the stabilized space-time IgA scheme: find $u_h \in V_{0h}$, satisfying

$$a_h(u_h, v_h) = l_h(v_h), \quad \forall v_h \in V_{0h}, \tag{5.1}$$

where

$$a_h(u_h, v_h) := (\partial_t u_h, v_h + \delta_h \partial_t v_h)_Q + (\nabla_x u_h, \nabla_x (v_h + \delta_h \partial_t v_h))_Q,$$

and

$$l_h(v_h) := (f, v_h + \delta_h \partial_w v_h)_Q.$$

The name of the scheme is inspired by the performed stabilization approach. That is, the upwind function is dependent on the global mesh-size and, therefore, is proportional to the same h on each element K, even when adaptive refinement is performed; see the corresponding study in [42]. Stabilization of (2.2) is crucial for proving the coercivity of the bilinear form, which yields uniqueness and existence of the discrete solution $u_h \in V_{0h}$. Combining coercivity and boundedness properties of $a_h(\cdot,\cdot)$ with the consistency of the scheme and approximation results for IgA spaces, we obtain the corresponding a priori error estimate with respect to the norm

$$\|v_h\|_h^2 := \|\nabla_x v_h\|_Q^2 + \delta_h \|\partial_t v_h\|_Q^2 + \|v_h\|_{\Sigma_T}^2 + \delta_h \|\nabla_x v_h\|_{\Sigma_T}^2,$$

which is presented in Theorem 5.1 below.

Theorem 5.1. [41, 43] *Let $u \in H_0^s(Q) := H^s(Q) \cap H_0^{1,0}(Q)$, $s \in \mathbb{N}$, $s \geq 2$, be the exact solution to (2.2), and let $u_h \in V_{0h}$ be the solution to (5.1) with some fixed parameter $\theta > 0$. Then, the a priori discretization error estimate,*

$$\|u - u_h\|_h \leq C h^{r-1} \|u\|_{H^r(Q)},$$

holds with $r = \min\{s, p+1\}$ and some generic constant $C > 0$ independent of h.

6 Locally stabilized IgA scheme and a priori error analysis

In this section, we assume that $p \geq 2$ and $m \leq p-1$, which yields that $V_{0h} \subset C^1(\overline{Q})$, providing the inclusion $V_{0h} \subset H_{0,\underline{0}}^{\Delta_x,1}(Q) := H_0^{\Delta_x,1}(Q) \cap H_{0,\underline{0}}^{1,1}(Q)$. Since $f \in L^2(Q)$, the solution u of (2.2) belongs to $H_{0,\underline{0}}^{\Delta_x,1}(Q)$. Hence, for all $K \in \mathcal{K}_h$, we can use the equation in the strong form; that is, $\partial_t u - \Delta_x u = f$ in $L^2(K)$, and multiply it by the localized test functions

$$v_h + \delta_K \partial_t v_h, \quad \delta_K = \theta_K h_K, \quad \theta_K > 0, \quad h_K := \operatorname{diam}(K),$$

such that

$$(\partial_t u - \Delta_x u, v_h + \delta_K \partial_t v_h)_K = (f, v_h + \delta_K \partial_t v_h)_K, \quad \forall v_h \in V_{0h}.$$

By summing up all the elements in $\mathcal{K}_h$, we obtain the relation

$$\begin{aligned}(\partial_t u - \Delta_x u, v_h)_Q &+ \sum_{K \in \mathcal{K}_h} \delta_K (\partial_t u - \Delta_x u, \partial_t v_h)_K \\ &= (f, v_h)_Q + \sum_{K \in \mathcal{K}_h} \delta_K (f, \partial_t v_h)_K.\end{aligned}$$

Integration by parts with respect to the space variables yield

$$\begin{aligned}\ell_{\mathrm{loc},h}(v_h) &:= (f, v_h)_Q + \sum_{K\in\mathcal{K}_h} \delta_K(f, \partial_t v_h)_K \\ &= (\partial_t u, v_h)_Q + (\nabla_x u, \nabla_x v_h)_Q \\ &\quad + \sum_{K\in\mathcal{K}_h} \delta_K\big((\partial_t u, \partial_t v_h)_K + (\nabla_x u, \nabla_x \partial_t v_h)_K - \langle \boldsymbol{n}_x^{\partial K}\cdot \nabla_x u, \partial_t v_h\rangle_{\partial K}\big) \\ &=: a_{\mathrm{loc},h}(u, v_h),\end{aligned}$$

where $\boldsymbol{n}_x^{\partial K}$ is the external normal vector to ∂K. Here, the last term is nothing else but the duality product

$$\langle \cdot,\cdot\rangle_{\partial K} = \langle \cdot,\cdot\rangle_{H^{-1/2}(\partial K)\times H^{1/2}(\partial K)} : H^{-1/2}(\partial K)\times H^{1/2}(\partial K) \to \mathbb{R},$$

and $H^{-1/2}$ is the dual space to $H^{1/2}$. Thus, we arrive at the finite dimensional problem: find $u_h \in V_{0h}$ satisfying the identity

$$a_{\mathrm{loc},h}(u_h, v_h) = \ell_{\mathrm{oc},h}(v_h), \quad \forall v_h \in V_{0h}, \tag{6.1}$$

where

$$\begin{aligned}a_{\mathrm{loc},h}(u_h, v_h) &:= (\partial_t u_h, v_h)_Q + (\nabla_x u_h, \nabla_x v_h)_Q \\ &\quad + \sum_{K\in\mathcal{K}_h} \delta_K\big((\partial_t u_h, \partial_t v_h)_K + (\nabla_x u_h, \nabla_x \partial_t v_h)_K\big) \\ &\quad - \sum_{K\in\mathcal{K}_h} \delta_K \sum_{E\in\mathcal{E}_h^K\cap\mathcal{E}_h^I} (\boldsymbol{n}_x^E\cdot\nabla_x u_h, \partial_t v_h)_E.\end{aligned}$$

Due to the assumptions $v_{h|\Sigma} = 0$ and $\boldsymbol{n}^E_{x|\Sigma_0\cup\Sigma_T} = \boldsymbol{0}$, contributions of the form $(\boldsymbol{n}_x^E\cdot\nabla_x u_h, \partial_t v_h)_{E\in\mathcal{E}_h^K\cap\mathcal{E}_h^{\partial Q}}$ vanish. The scheme is called locally stabilized, because the stabilization is only performed element-wise over each $K \in \mathcal{K}_h$.

Remark 6.1. A similar approach was applied to conventional discretization techniques, that is, to finite element discretizations. The corresponding investigation can be found in [44]. The authors study the analogous bilinear form of the parabolic problem, where the given data satisfy the general assumptions dictated by the maximal parabolic regularity. The difference of the IgA approach to the conventional one lies in the regularity of numerical fluxes. IgA schemes produce approximations, which satisfy the condition $u_h \in C^{p-1}$ on the physical patch, where $p \geq 2$. Therefore, numerical fluxes have square summable divergence, whereas the finite element approach used in [44] operates with C^0 discretizations; that is, the finite element fluxes do not belong to $H^{\mathrm{div}_x,0}(Q)$.

6.1 Coercivity

Lemma 6.1. *Let the parameters θ_K be sufficiently small; that is, $\theta_K \in (0, \frac{h_K}{dC^2_{\mathrm{inv},1}}]$, where $C_{\mathrm{inv},1}$ is the constant in the inverse inequality (4.2) associated with $K \in \mathcal{K}_h$. Then the bilinear form $a_{\mathrm{loc},h}(\cdot,\cdot) : V_{0h} \times V_{0h} \to \mathbb{R}$ is V_{0h}-coercive with respect to the norm*

$$\|v_h\|^2_{\mathrm{loc},h} := \|\nabla_x v_h\|^2_Q + \frac{1}{2}\|v_h\|^2_{\Sigma_T} + \sum_{K\in\mathcal{K}_h} \delta_K \|\partial_t v_h\|^2_K. \tag{6.2}$$

That is, there exists a constant $\mu_{\mathrm{loc},c} > 0$, such that

$$a_{\mathrm{loc},h}(v_h, v_h) \geq \mu_{\mathrm{loc},c}\|v_h\|^2_{\mathrm{loc},h}, \quad \forall v_h \in V_{0h}. \tag{6.3}$$

Proof. Integration by parts of $a_{\mathrm{loc},h}(v_h, v_h)$ yields

$$\begin{aligned}
a_{\mathrm{loc},h}(v_h, v_h) &:= (\partial_t v_h, v_h)_Q + (\nabla_x v_h, \nabla_x v_h)_Q \\
&\quad + \sum_{K\in\mathcal{K}_h} \delta_K \Big\{ (\partial_t v_h, \partial_t v_h)_K + (\nabla_x v_h, \nabla_x \partial_t v_h)_K - \sum_{E\in\mathcal{E}^I_h} (\boldsymbol{n}^E_x \cdot \nabla_x u_h, \partial_t v_h)_E \Big\} \\
&= \frac{1}{2}\|v_h\|^2_{\Sigma_T} + \|\nabla_x v_h\|^2_Q + \sum_{K\in\mathcal{K}_h} \delta_K \{ \|\partial_t v_h\|^2_K - (\Delta_x v_h, \partial_t v_h)_K \}.
\end{aligned} \tag{6.4}$$

To prove coercivity, we need to estimate the last term in (6.4). By using (4.2) and Young's inequality, we arrive at

$$\begin{aligned}
\sum_{K\in\mathcal{K}_h} \delta_K (\Delta_x v_h, \partial_t v_h)_K &\leq \Big(\sum_{K\in\mathcal{K}_h} \delta_K \|\Delta_x v_h\|^2_K \Big)^{1/2} \Big(\sum_{K\in\mathcal{K}_h} \delta_K \|\partial_t v_h\|^2_K \Big)^{1/2} \\
&\leq \Big(\sum_{K\in\mathcal{K}_h} \delta_K d \sum_{l=1}^{d} \|\partial^2_{x_l} v_h\|^2_K \Big)^{1/2} \Big(\sum_{K\in\mathcal{K}_h} \delta_K \|\partial_t v_h\|^2_K \Big)^{1/2} \\
&\leq \Big(\sum_{K\in\mathcal{K}_h} \theta_K h_K d \sum_{l=1}^{d} C^2_{\mathrm{inv},1} h_K^{-2} \|\partial_{x_l} v_h\|^2_K \Big)^{1/2} \Big(\sum_{K\in\mathcal{K}_h} \delta_K \|\partial_t v_h\|^2_K \Big)^{1/2} \\
&\leq \frac{d}{2} \max_{K\in\mathcal{K}_h} \Big(\frac{\theta_K}{h_K} C^2_{\mathrm{inv},1} \Big) \Big(\|\nabla_x v_h\|^2_Q + \sum_{K\in\mathcal{K}_h} \delta_K \|\partial_t v_h\|^2_K \Big).
\end{aligned}$$

Therefore, $a_{\mathrm{loc},h}(v_h, v_h)$ can be bounded from below as follows:

$$\begin{aligned}
a_{\mathrm{loc},h}(v_h, v_h) &\geq \frac{1}{2}\|v_h\|^2_{\Sigma_T} \\
&\quad + \Big(1 - \frac{d}{2} \max_{K\in\mathcal{K}_h} \frac{\theta_K}{h_K} C^2_{\mathrm{inv},1} \Big) \Big(\sum_{K\in\mathcal{K}_h} \delta_K \|\partial_t v_h\|^2_K + \|\nabla_x v_h\|^2_Q \Big) \\
&\geq \frac{1}{2}\|v_h\|^2_{\mathrm{loc},h},
\end{aligned}$$

provided that $\theta_K \in (0, \frac{h_K}{dC^2_{\mathrm{inv},1}}]$ for all $K \in \mathcal{K}_h$. □

Remark 6.2. The computation of the constants $C_{\mathrm{inv},1}$ in inverse inequalities corresponds to the question of accurate estimation of maximal eigenvalues for generalized eigenvalue problems for the considered differential equations. In [34], the authors applied symbolic computation methods to this problem defined on square elements and were able to improve the previously known upper bounds in [55].

V_{0h}-coercivity of $a_{\mathrm{loc},h}(\cdot,\cdot)$ implies uniqueness, and uniqueness provides the existence of the IgA solution $u_h \in V_{0h}$. From Lemma 6.1, it also immediately follows that the system matrix of the linear system generated by the bilinear form is positive definite.

6.2 Boundedness

To prove a priori discretization error estimates, we need to show the uniform boundedness of the localized bilinear form $a_{\mathrm{loc},h}(\cdot,\cdot)$ on $V_{0h,*} \times V_{0h}$, where $V_{0h,*} := H^1_{0,\underline{0}}(Q) \cap H^{\Delta_x,1}_{0,\underline{0}}(Q) + V_{0h}$ is equipped with the norm

$$\|v\|^2_{\mathrm{loc},h,*} := \|v\|^2_{\mathrm{loc},h} + \sum_{K\in\mathcal{K}_h} (\delta_K^{-1}\|v\|^2_K + \delta_K\|\Delta_x v\|^2_K).$$

Lemma 6.2. *Assume that $\theta_K \in (0, \frac{h_K}{dC^2_{\mathrm{inv},1}}]$, $K \in \mathcal{K}_h$. Then, the bilinear form $a_{\mathrm{loc},h}(\cdot,\cdot)$ is uniformly bounded on $V_{0h,*} \times V_{0h}$; that is, there exists a positive constant $\mu_{\mathrm{loc},b}$ that does not depend on h_K, such that*

$$|a_{\mathrm{loc},h}(u, v_h)| \le \mu_{\mathrm{loc},b}\|u\|_{\mathrm{loc},h,*}\|v_h\|_{\mathrm{loc},h}, \quad \forall u \in V_{0h,*}, \quad \forall v_h \in V_{0h}. \tag{6.5}$$

Proof. For the first term in $a_{\mathrm{loc},h}(u, v_h)$, we apply integration by parts with respect to time and the Cauchy inequality:

$$\begin{aligned}(\partial_t u, v_h)_Q &\le \|u\|_{\Sigma_T}\|v_h\|_{\Sigma_T} + \sum_{K\in\mathcal{K}_h} \delta_K^{-1/2}\|u\|_K \delta_K^{1/2}\|\partial_t v_h\|_K \\ &\le \Big(\|u\|^2_{\Sigma_T} + \sum_{K\in\mathcal{K}_h}\delta_K^{-1}\|u\|^2_K\Big)^{1/2}\Big(\|v_h\|^2_{\Sigma_T} + \sum_{K\in\mathcal{K}_h}\delta_K\|\partial_t v_h\|^2_K\Big)^{1/2}.\end{aligned}$$

Estimating the second, the third, and the last term by Cauchy's inequalities and combining the obtained results, the bilinear form can be bounded as follows:

$$\begin{aligned}&|a_{\mathrm{loc},h}(u, v_h)| \\ &\quad\le \Big(\|u\|^2_{\Sigma_T} + \|\nabla_x u\|^2_Q + \sum_{K\in\mathcal{K}_h}\{\delta_K^{-1}\|u\|^2_K + \delta_K(\|\partial_t u\|^2_K + \|\Delta_x u\|^2_K)\}\Big)^{1/2} \\ &\quad\times\Big(\|v_h\|^2_{\Sigma_T} + \|\nabla_x v_h\|^2_Q + 3\sum_{K\in\mathcal{K}_h}\delta_K\|\partial_t v_h\|^2_K\Big)^{1/2} \\ &\quad\le \mu_{\mathrm{loc},b}\|u\|_{\mathrm{loc},h,*}\|u_h\|_{\mathrm{loc},h},\end{aligned}$$

with $\mu_{\mathrm{loc},b} = 3$. □

6.3 Approximation properties

Estimate (4.3) implies a priori estimates of the interpolation error $u - \Pi_h u$, measured in terms of the L_2 norm and the mesh-dependent norms $\|\cdot\|_{\mathrm{loc},h}$ and $\|\cdot\|_{\mathrm{loc},h,*}$, which we later need to obtain an a priori estimate for $u - u_h$.

Lemma 6.3. *Let ℓ and s be integers such that $0 \le \ell \le s \le p+1$, and assume $u \in H^s_{0,\underline{0}}(Q)$. Then there exist a projection operator $\Pi_h : H^s_{0,\underline{0}}(Q) \to V_{0h}$ (see Lemma 4.3), and positive constants C_1 and C_2, such that the quasi interpolation error estimates*

$$\|u - \Pi_h u\|^2_{\mathrm{loc},h} \le C_1 \sum_{K\in\mathcal{K}_h} h_K^{2(s-1)} \sum_{i=0}^{s} c_K^{2i} |u|^2_{H^i(\underline{K})} \quad \text{and} \tag{6.6}$$

$$\|u - \Pi_h u\|^2_{\mathrm{loc},h,*} \le C_2 \sum_{K\in\mathcal{K}_h} h_K^{2(s-1)} \sum_{i=0}^{s} c_K^{2i} |u|^2_{H^i(\underline{K})} \tag{6.7}$$

hold for all $u \in L_2(Q) \cap H^s(\underline{K})$.

Proof. To prove (6.6) and (6.7), we need to provide estimates for each term in the norm $\|u - \Pi_h u\|_{\mathrm{loc},h}$. To bound the first term, we use (4.5); that is,

$$\|\nabla_x (u - \Pi_h u)\|^2_Q \le C^2_{1,s} \Big(\max_{K\in\mathcal{K}_h} c_K^{-2}\Big) \sum_{K\in\mathcal{K}_h} h_K^{2(s-1)} \sum_{i=0}^{s} c_K^{2i} |u|^2_{H^s(\underline{K})}. \tag{6.8}$$

To show the next one, we introduce the set

$$\mathcal{K}_h^{\Sigma_T} := \{K \in \mathcal{K}_h | \partial K \cap \Sigma_T \ne \emptyset\}.$$

In view of (4.1), (3.2), (4.4), and (4.5), the estimate of the norm on Σ_T reads as follows:

$$\begin{aligned}
\|u - \Pi_h u\|^2_{\Sigma_T} &\le \sum_{K\in\mathcal{K}_h^{\Sigma_T}} C^2_{\mathrm{tr}} \big(h_K^{-1} \|u - \Pi_h u\|^2_K + h_K |u - \Pi_h u|^2_{H^1(K)}\big) \\
&\le \max_{K\in\mathcal{K}_h^{\Sigma_T}} \{C^2_{\mathrm{tr}}\} C_u \Big(h^{-1} \sum_{K\in\mathcal{K}_h^{\Sigma_T}} \|u - \Pi_h u\|^2_K + h \sum_{K\in\mathcal{K}_h^{\Sigma_T}} |u - \Pi_h u|^2_{H^1(K)} \Big) \\
&\le \max_{K\in\mathcal{K}_h^{\Sigma_T}} \{C^2_{\mathrm{tr}}\} C_u \Big(C^2_{0,s} \sum_{K\in\mathcal{K}_h^{\Sigma_T}} h_K^{2s-1} \sum_{i=0}^{s} c_K^{2i} |u|^2_{H^i(\underline{K})} \\
&\quad + C^2_{1,s} \sum_{K\in\mathcal{K}_h^{\Sigma_T}} h_K^{2s-1} \sum_{i=0}^{s} c_K^{2(i-1)} |u|^2_{H^i(\underline{K})} \Big) \\
&\le C_{\Sigma_T} \sum_{K\in\mathcal{K}_h^{\Sigma_T}} h_K^{2s-1} \sum_{i=0}^{s} c_K^{2i} |u|^2_{H^i(\underline{K})},
\end{aligned} \tag{6.9}$$

where

$$C_{\Sigma_T} = C_u \max_{K \in \mathcal{K}_h^{\Sigma_T}} \{C_{\mathrm{tr}}^2 (C_{0,s}^2 + C_{1,s}^2 c_K^{-2})\}. \tag{6.10}$$

Finally, by using (3.2) and (4.5), and applying similar arguments, we derive the bound of the third term in $\|u - \Pi_h u\|_{\mathrm{loc},h}^2$:

$$\begin{aligned} \sum_{K \in \mathcal{K}_h} \delta_K \|\partial_t (u - \Pi_h u)\|_K^2 &\le \max_{K \in \mathcal{K}_h} \delta_K \sum_{K \in \mathcal{K}_h} |u - \Pi_h u|_{H^1(K)}^2 \\ &\le C_{1,s}^2 \max_{K \in \mathcal{K}_h} \{\delta_K c_K^{-2}\} \sum_{K \in \mathcal{K}_h} h_K^{2(s-1)} \sum_{i=0}^{s} c_K^{2i} |u|_{H^s(\underline{K})}^2. \end{aligned} \tag{6.11}$$

Combining (6.8), (6.9), and (6.11), we obtain the bound

$$\|u - \Pi_h u\|_{\mathrm{loc},h}^2 \le C_1 \sum_{K \in \mathcal{K}_h} h_K^{2(s-1)} \sum_{i=0}^{s} c_K^{2i} |u|_{H^i(\underline{K})}^2, \tag{6.12}$$

where $C_1 = \max_{K \in \mathcal{K}_h} \{C_{1,s}^2 (1 + \delta_K) c_K^{-2} + C_{\Sigma_T}\}$, and C_{Σ_T} is a constant defined in (6.9).

To prove (6.7), we need to estimate the quantities $\sum_{K \in \mathcal{K}_h} \delta_K^{-1} \| \cdot \|_K^2$ and $\delta_K \|\Delta_x v\|_K^2$ included into $\| \cdot \|_{\mathrm{loc},h,*}$. In view of (4.4), we obtain

$$\sum_{K \in \mathcal{K}_h} \delta_K^{-1} \|u - \Pi_h u\|_K^2 \le C_{0,s}^2 C_u \max_{K \in \mathcal{K}_h} \frac{h_K}{\theta_K} \sum_{K \in \mathcal{K}_h} h_K^{2(s-1)} \sum_{i=0}^{s} c_K^{2i} |u|_{H^i(\underline{K})}^2.$$

Since $\theta_K \le h_K / (d C_{\mathrm{inv},1}^2)$, we use (4.6) and bound the second term as follows:

$$\begin{aligned} \sum_{K \in \mathcal{K}_h} \delta_K \|\Delta_x (u - \Pi_h u)\|_K^2 &\le \sum_{K \in \mathcal{K}_h} \frac{h_K}{d C_{\mathrm{inv},1}^2} h_K d |u - \Pi_h u|_{H^2(K)}^2 \\ &\le C_{2,s}^2 \max_{K \in \mathcal{K}_h} \{C_{\mathrm{inv},1}^{-2} c_K^{-4}\} \sum_{K \in \mathcal{K}_h} h_K^{2(s-1)} \sum_{i=0}^{s} c_K^{2i} \|u\|_{H^s(\underline{K})}^2. \end{aligned}$$

Thus, we obtain

$$\|u - \Pi_h u\|_{\mathrm{loc},h,*}^2 \le C_2 \max_{K \in \mathcal{K}_h} \frac{h_K}{\theta_K} \sum_{K \in \mathcal{K}_h} h_K^{2(s-1)} \sum_{i=0}^{s} c_K^{2i} |u|_{H^i(\underline{K})}^2,$$

where

$$C_2 = \max_{K \in \mathcal{K}_h} \left\{ C_{\Sigma_T} + C_{2,s}^2 C_{\mathrm{inv},1}^{-2} c_K^{-4} + C_{0,s}^2 C_u \frac{h_K}{\theta_K} \right\}, \tag{6.13}$$

and C_{Σ_T} is defined in (6.9). □

6.4 Consistency

Lemma 6.4. *If the solution $u \in H_{0,\underline{0}}^{\Delta_x,1}(Q)$, then it satisfies the consistency identity*

$$a_{\mathrm{loc},h}(u, v_h) = \ell_{\mathrm{loc},h}(v_h), \quad \forall v_h \in V_{0h}. \tag{6.14}$$

Proof. The consistency identity (6.14) was derived along with the derivation of the space-time IgA scheme (6.1). □

6.5 A priori discretization error estimates

Lemma 6.5. *Let $u \in H_{0,\underline{0}}^{\Delta_x,1}(Q)$ be the exact solution of (2.2), and $u_h \in V_{0h}$ be the approximate solution generated by (6.1). Then the best approximation estimate*

$$\|u - u_h\|_{\mathrm{loc},h} \le \left(1 + \frac{\mu_{\mathrm{loc},b}}{\mu_{\mathrm{loc},c}}\right) \inf_{v_h \in V_{0h}} \|u - v_h\|_{\mathrm{loc},h,*} \tag{6.15}$$

holds, where $\mu_{\mathrm{loc},c}$ and $\mu_{\mathrm{loc},b}$ are the constants from Lemmas 6.1 and 6.2, respectively.

Proof. By the triangle inequality, the estimate of the discretization error $u - u_h$ reads as

$$\|u - u_h\|_{\mathrm{loc},h} \le \|u - \Pi_h u\|_{\mathrm{loc},h} + \|\Pi_h u - u_h\|_{\mathrm{loc},h}. \tag{6.16}$$

The first term on the right hand side of (6.16) can easily be estimated by means of Lemma 6.3. For the estimation of $\|\Pi_h u - u_h\|_{\mathrm{loc},h}$, we first use the V_{0h}–ellipticity of $a_{\mathrm{loc},h}(\cdot,\cdot)$ with respect to the norm $\|\cdot\|_{\mathrm{loc},h}$ (see Lemma 6.1); that is,

$$\mu_{\mathrm{loc},c}\|\Pi_h u - u_h\|_{\mathrm{loc},h}^2 \le a_{\mathrm{loc},h}(\Pi_h u - u_h, \Pi_h u - u_h).$$

Next, by means of the Galerkin orthogonality $a_{\mathrm{loc},h}(u - u_h, v_h) = 0$, $\forall v_h \in V_{0h}$, which directly follows from consistency (6.14),

$$\begin{aligned}\mu_{\mathrm{loc},c}\|\Pi_h u - u_h\|_{\mathrm{loc},h}^2 &\le a_{\mathrm{loc},h}(\Pi_h u - u_h, \Pi_h u - u_h)\\ &= a_{\mathrm{loc},h}(\Pi_h u - u, \Pi_h u - u_h).\end{aligned}$$

Finally, we apply Lemma 6.1, and obtain the estimate

$$\mu_{\mathrm{loc},c}\|\Pi_h u - u_h\|_{\mathrm{loc},h}^2 \le \mu_{\mathrm{loc},b}\|\Pi_h u - u\|_{\mathrm{loc},h,*}\|\Pi_h u - u_h\|_{\mathrm{loc},h},$$

which automatically yields

$$\|\Pi_h u - u_h\|_{\mathrm{loc},h} \le \frac{\mu_{\mathrm{loc},b}}{\mu_{\mathrm{loc},c}}\|\Pi_h u - u\|_{\mathrm{loc},h,*}. \tag{6.17}$$

Combining $\|\Pi_h u - u\|_{\mathrm{loc},h} \le \|\Pi_h u - u\|_{\mathrm{loc},h,*}$, (6.17), and (6.16), we arrive at

$$\|u - u_h\|_{\mathrm{loc},h} \le \left(1 + \frac{\mu_{\mathrm{loc},b}}{\mu_{\mathrm{loc},c}}\right)\|u - \Pi_h u\|_{\mathrm{loc},h,*}.$$

□

Theorem 6.1. *Let $p \geq 2$, $u \in H_0^s(Q)$, $s \geq 2$, be the exact solution of* (2.2), *and $u_h \in V_{0h}$ be the solution of* (6.1) *with $\theta_K \in (0, \frac{h_K}{dC_{\text{inv},1}^2}]$, $K \in \mathcal{K}_h$. Then the a priori discretization error estimate*

$$\|u - u_h\|_{\text{loc},h}^2 \leq C \sum_{K \in \mathcal{K}_h} h_K^{2(s-1)} \sum_{i=0}^{r} c_K^{2i} |u|_{H^i(\underline{K})}^2$$

holds, where

$$C = (1 + \mu_{\text{loc},b}/\mu_{\text{loc},c})^2 C_2$$

is a constant independent of h, $r = \min\{s, p+1\}$, and p denotes the polynomial degree of the THB-splines; $\mu_{\text{loc},b}$ and $\mu_{\text{loc},c}$ are the constants in the boundedness (6.3) *and coercivity* (6.5) *inequalities, respectively.*

Proof. Application of estimate (6.7) yields

$$\|u - u_h\|_{\text{loc},h}^2 \leq \left(1 + \frac{\mu_{\text{loc},b}}{\mu_{\text{loc},c}}\right)^2 C_2 \sum_{K \in \mathcal{K}_h} h_K^{2(s-1)} \sum_{i=0}^{s} c_K^{2i} |u|_{H^i(\underline{K})}^2,$$

where C_2 is defined in (6.13). □

7 A posteriori discretization error estimates

In this section, we discuss implementations of the numerical scheme (discussed above), and estimates used to control the quality of the approximations. We also present the experimental order of convergence (eoc),

$$\text{eoc}(\|e\|_{\text{loc},h}^2) := \frac{\log(\|e^{(k)}\|_{\text{loc},h}^2 / \|e^{(k-1)}\|_{\text{loc},h}^2)}{\log(N_{\text{d. o. f.}}^{-1/(d+1)}(k) / N_{\text{d.o.f}}^{-1/(d+1)}(k-1))},$$

in terms of the error norm (6.2), where $k \in \mathbb{N}^+$ denotes the number of the refinement steps, $N_{\text{d. o. f.}}(k)$ is the number of degrees of freedom of u_h corresponding to the refinement k, and d is the dimension of the considered spatial domain. Also, we discuss computational properties of the majorants $\overline{\mathrm{M}}^{\mathrm{I}}$ and $\overline{\mathrm{M}}^{\mathrm{II}}$ that follow from [52] and of the error identity $\mathbb{E}$ [2]. Moreover, we compare time expenditures that are required for getting approximations of the solution with the time spent for computing efficient error bounds.

As before, we assume that $u_h \in V_{0h} := V_h \cap H_{0,\underline{0}}^{\Delta_x,1}(Q)$, and define

$$u_h(x,t) = u_h(x_1, \ldots, x_{d+1}) := \sum_{i \in \mathcal{I}} \mathrm{u}_{h,i} \phi_{h,i}(x_1, \ldots, x_{d+1}),$$

where $\underline{\mathrm{u}}_h := [\mathrm{u}_{h,i}]_{i \in \mathcal{I}} \in \mathbb{R}^{|\mathcal{I}|}$ contains the control points to be defined by the linear system generated by (6.1). The system is solved by means of a sparse direct LU factorization. This choice of the solver is motivated by our intention to provide a fair comparison of time expenditures used for solving the systems generating u_h and the auxiliary functions $\boldsymbol{y}_h$ (flux reconstruction for the majorant $\overline{\mathrm{M}}^{\mathrm{I}}$) and the function w_h (reconstructed for the improved majorant $\overline{\mathrm{M}}^{\mathrm{II}}$). Such a comparison is conducted to show that higher smoothness of approximations per degrees of freedom (provided by IgA) allows us to make the computation of a posteriori error estimates much faster than the approximation of u_h.

Approximation properties of u_h are analyzed by studying the error $e = u - u_h$ in terms of the norm (6.2) and the norm

$$|||e|||^2 := \|\nabla_x e\|_Q^2 + \|e\|_{\Sigma_T}^2.$$

The choice of functional error estimates among other approaches is motivated by their universality and ability to provide guaranteed, sharp, and fully computable bounds of approximation errors. Moreover, functional type error majorants for the I-BVPs are formed by integrals associated with the whole space-time cylinder and, therefore, they are well adapted to the analysis of fully unstructured (in space and time) IgA schemes.

The norm $|||e|||^2$ is controlled by the first form of the majorant (see, for example, [52])

$$\begin{aligned}\overline{\mathrm{M}}^{\mathrm{I}}(u_h, \boldsymbol{y}_h) &:= (1+\beta)\|\boldsymbol{y}_h - \nabla_x u_h\|_Q^2 + \left(1 + \frac{1}{\beta}\right) C_{\mathrm{F}}^2 \|\mathrm{div}_x \boldsymbol{y}_h + f - \partial_t u_h\|_Q^2 \\ &= (1+\beta)\overline{\mathrm{m}}_{\mathrm{d}}^{\mathrm{I},2} + \left(1 + \frac{1}{\beta}\right) C_{\mathrm{F}}^2 \overline{\mathrm{m}}_{\mathrm{eq}}^{\mathrm{I},2}.\end{aligned}$$

Here, $\beta > 0$ is a free parameter, and $\boldsymbol{y}_h$ is an auxiliary vector-valued function, which is to be defined by a suitable reconstruction of the flux. The space $Y_h \subset H^{\mathrm{div}_x,0}(Q)$ describes free parameters that are needed to be chosen in such a way that $\overline{\mathrm{M}}^{\mathrm{I}}(u_h, \boldsymbol{y}_h)$ is minimized. This estimate has been derived by purely functional analysis of the corresponding integral identity. Hence, it does not contain mesh dependent constants and is valid for a wide class of approximations. The space $Y_h \equiv \oplus^d \mathcal{S}_h^q := \{\boldsymbol{\psi}_{h,i} := \hat{\boldsymbol{\psi}}_{h,i} \circ \Phi^{-1}\}$, where $\hat{\boldsymbol{\psi}}_{h,i}$ is a vector–valued spline basis function of the degree q of the approximation space $\oplus^d \hat{\mathcal{S}}_h^q$ that is defined on the parametric domain. The sharpest estimate is obtained by the minimization of $\overline{\mathrm{M}}^{\mathrm{I}}(u_h, \boldsymbol{y}_h)$ with respect to $\boldsymbol{y}_h(x,t) := \boldsymbol{y}_h(x_1, \dots, x_{d+1}) = \sum_{i \in \mathcal{I} \times d} \mathrm{y}_{h,i} \boldsymbol{\psi}_{h,i}(x_1, \dots, x_{d+1})$, where $\underline{\mathbf{y}}_h := [\mathrm{y}_{h,i}]_{i \in \mathcal{I} \times d} \in \mathbb{R}^{d|\mathcal{I}|}$. Notice that $\underline{\mathbf{y}}_h$ is a vector of the dimension $d|\mathcal{I}|$, defined by the system of linear algebraic equations

$$(C_{\mathrm{F}}^2 \mathrm{Div}_h + \beta \mathrm{M}_h)\underline{\mathbf{y}}_h = -C_{\mathrm{F}}^2 \underline{\mathrm{z}}_h + \beta \underline{\mathrm{g}}_h, \tag{7.1}$$

where

$$\begin{aligned}\mathrm{Div}_h &:= [(\mathrm{div}_x \boldsymbol{\psi}_{h,i}, \mathrm{div}_x \boldsymbol{\psi}_{h,j})_Q]_{i,j=1}^{d|\mathcal{I}|}, & \underline{\mathrm{z}}_h &:= [(f - v_t, \mathrm{div}_x \boldsymbol{\psi}_{h,j})_Q]_{j=1}^{d|\mathcal{I}|}, \\ \mathrm{M}_h &:= [(\boldsymbol{\psi}_{h,i}, \boldsymbol{\psi}_{h,j})_Q]_{i,j=1}^{d|\mathcal{I}|}, & \underline{\mathrm{g}}_h &:= [(\nabla_x v, \boldsymbol{\psi}_{h,j})_Q]_{j=1}^{d|\mathcal{I}|}.\end{aligned}$$

The optimal value for β reads as $\overline{\beta} := C_{\mathrm{F}} \overline{\mathrm{m}}^{\mathrm{I}}_{\mathrm{eq}} / \overline{\mathrm{m}}^{\mathrm{I}}_{\mathrm{d}}$. According to the numerical results obtained in [33, 42, 46], the most efficient majorant reconstruction is obtained if $q \gg p$. At the same time, the approximation u_h is reconstructed on the mesh $\mathcal{K}_h$, whereas a coarser mesh $\mathcal{K}_{Mh}$, $M \in \mathbb{N}^+$ is used to recover $\boldsymbol{y}_h$. This helps us to minimize the number of d. o. f. used in the representation of $\boldsymbol{y}_h$. The initial mesh $\mathcal{K}_h^0$ and the corresponding basis functions are assumed to be given via the geometry representation of the computational domain. Throughout the set of numerical examples, $\mathcal{K}_h^0$ is generated by $N_{\mathrm{ref},0}$ initial uniform refinements before actual testing. In our implementation, the linear system (7.1) is solved by a sparse direct $\mathrm{LDL}^\top$ factorization.

If the solution of the problem rapidly changes in time, the residual $\overline{\mathrm{m}}^{\mathrm{I},2}_{\mathrm{eq}}$ in $\overline{\mathrm{M}}^{\mathrm{I}}$ might be large in comparison to $\overline{\mathrm{m}}^{\mathrm{I},2}_{\mathrm{d}}$, even after minimization of $\overline{\mathrm{M}}^{\mathrm{I}}$. There, we introduce an advanced form of the majorant $\overline{\mathrm{M}}^{\mathrm{II}}(u_h, \boldsymbol{y}_h, w_h)$ for the quantity $\|\nabla_x e\|^2$, which has the form

$$\overline{\mathrm{M}}^{\mathrm{II}}(u_h, \boldsymbol{y}_h, w_h) := \|w_h - u_h\|^2_{\Sigma_T} + 2\mathcal{F}(u_h, w_h) + (1+\beta)\|\mathbf{r}^{\mathrm{II}}_{\mathrm{d}}\|^2_Q + C_{\mathrm{F}}^2\left(1 + \frac{1}{\beta}\right)\|\mathbf{r}^{\mathrm{II}}_{\mathrm{eq}}\|^2_Q,$$

where

$$\mathbf{r}^{\mathrm{II}}_{\mathrm{d}}(u_h, \boldsymbol{y}_h, w_h) := \boldsymbol{y}_h + \nabla_x w_h - 2\nabla_x u_h,$$

$$\mathbf{r}^{\mathrm{II}}_{\mathrm{eq}}(\boldsymbol{y}_h, w_h) := \operatorname{div}_x \boldsymbol{y}_h + f - \partial_t w_h,$$

and

$$\mathcal{F}(u_h, w_h) := (\nabla_x u_h, \nabla_x (w_h - u_h)) + (\partial_t u_h - f, w_h - u_h).$$

The majorant contains an additional function w_h, which can be taken as the solution of (6.1) on an extended approximation space $W_{0h} := W_h \cap H_0^{\Delta_x,1}(Q)$, with $W_h \equiv \mathcal{S}_h^r := \{\chi_{h,i} := \hat{\chi}_{h,i} \circ \Phi^{-1}\}$, where $\hat{\chi}_{h,i}$ is a spline basis function of degree r of the approximation space $\hat{\mathcal{S}}_h^r$, defined on the parametric domain. The function w_h can be represented by $w_h(x,t) = w_h(x_1, \ldots, x_{d+1}) := \sum_{i \in \mathcal{I}} \mathrm{w}_{h,i}\chi_{h,i}$. Here, $\underline{\mathrm{w}}_h := [\mathrm{w}_{h,i}]_{i \in \mathcal{I}} \in \mathbb{R}^{|\mathcal{I}|}$ is the vector of control points of w_h, defined by the linear system $\mathrm{K}_h^{(r)} \underline{\mathrm{w}}_h = \underline{\mathrm{f}}_h^{(r)}$, where $\mathrm{K}_h^{(r)} := [a_{\mathrm{loc},h}(\chi_{h,i}, \chi_{h,j})]_{i,j \in \mathcal{I}}$, $\underline{\mathrm{f}}_h^{(r)} := [l_{\mathrm{loc},h}(\chi_{h,i})]_{i \in \mathcal{I}}$. Since $\partial_t w_h$ is approximated by a richer space, the term $\|\mathbf{r}^{\mathrm{II}}_{\mathrm{eq}}(\boldsymbol{y}_h, w_h)\|^2_Q$ is expected to be smaller than $\|\mathbf{r}_{\mathrm{eq}}(\boldsymbol{y}_h, u_h)\|^2_Q$. Therefore, the advanced error bound $\overline{\mathrm{M}}^{\mathrm{II}}$ improves the quantitative efficiency of the global error estimation in comparison to $\overline{\mathrm{M}}^{\mathrm{I}}$. The optimal parameter β is defined by the relation $\beta := C_{\mathrm{F}}\|\mathbf{r}^{\mathrm{II}}_{\mathrm{eq}}\|_Q / \|\mathbf{r}^{\mathrm{II}}_{\mathrm{d}}\|_Q$.

One more error norm used in our analysis is generated by the partial differential operator $\mathcal{L} := \partial_t - \Delta_x$. That is,

$$\|e\|^2_{\mathcal{L}} := \|\Delta_x e\|^2_Q + \|\partial_t e\|^2_Q + \|\nabla_x e\|^2_{\Sigma_T}$$

meets the identity [2]:

$$\|e\|_{\mathcal{L}}^2 = \mathbb{E}(u_h) := \left\|\nabla_x(u_0 - u_h)\right\|_{\Sigma_0}^2 + \|\Delta_x u_h + f - \partial_t u_h\|_Q^2.$$

It should be noted that the error identity and the majorants $\overline{\mathrm{M}}^{\mathrm{I}}$ and $\overline{\mathrm{M}}^{\mathrm{II}}$ control different error norms. The identity uses the strongest norm, but it is valid only under the conditions of maximal parabolic regularity. Quantitatively (in terms of the efficiency index), the majorant $\overline{\mathrm{M}}^{\mathrm{II}}$ is better than $\overline{\mathrm{M}}^{\mathrm{I}}$, but it is more computationally expensive. We again emphasize that the error identity $\mathbb{E}$ controls the error in a stronger norm than the error majorants $\overline{\mathrm{M}}^{\mathrm{I}}$ and $\overline{\mathrm{M}}^{\mathrm{II}}$. Both majorants control the same error. Both the error identity and the majorants can be localized in terms of contributions from the mesh element. Thus, they can easily be used for designing adaptive procedures.

8 Numerical experiments

Below we study the behavior of the above-discussed error control tools within a series of benchmark examples. The marking of the elements from $\mathcal{K}_h$ in the adaptive procedure is driven by the bulk marking criterion (also known as Dörfler's marking [12]) denoted by $\mathbb{M}_{\mathrm{BULK}}(\sigma)$, $\sigma \in [0,1]$. Finally, the effectiveness of the error estimators is evaluated by efficiency indices. That is, $I_{\mathrm{eff}}(\mathrm{Est}) := \frac{\mathrm{Est}}{[e]_Q}$, where Est denotes the chosen error bound/identity, and $[\cdot]_Q$ corresponds to the error measure that is controlled by Est. First, we present a rather simple example, which is intended to demonstrate important properties of the numerical scheme. More complicated problems with nontrivial geometries, and singular solutions are considered at the end of the section. The implementation was carried out using the open-source C++ library G+Smo [25].

8.1 Example 1: polynomial solution

We consider problem (2.1) with

$$f(x,t) = -(1-x)x^2(1-2t) - (2-6x)(1-t)t, \quad (x,t) \in Q := (0,1)^2.$$

In this case, the exact solution is the polynomial

$$u(x,t) = (1-x)x^2(1-t)t, \quad (x,t) \in \overline{Q} := [0,1]^2.$$

It is easy to see that u satisfies homogeneous initial and boundary conditions.

The initial mesh is obtained by one global refinement ($N_{\mathrm{ref},0} = 1$). Furthermore, refinements are done with eight steps (hence $N_{\mathrm{ref}} = 8$). The approximation space for u_h is S_h^2. For the auxiliary functions, we assume that both y_h and w_h belong to S_{5h}^3.

The choice of the parameters in the discretization spaces is motivated by the intention to keep computational efforts in the computation of the error estimates considerably low. At the same, the degree of B-splines reconstructing the auxiliary flux must be at least one time higher than the degree of basis functions chosen to approximate u_h. The coarseness of the mesh chosen for the flux discretization space (in this case S^3_{5h}) depends on the complexity of the problem, exact solution, geometry, et cetera. In the numerical examples below, we keep this mesh as coarse as possible until the efficiency indices of the error estimates remain relatively close to one. Table 5.2 illustrates the ratio between the time spent on approximating u_h to the time spent for its error estimation; that is, $t_{\text{appr.}}/t_{\text{er.est.}}$ along with the total time needed for assembling and solving systems determining the d. o. f. of u_h, y_h, and w_h. Table 5.1 illustrates the convergences of the different error measures. That means, that $\|e\|_Q$ is bounded by the majorants $\overline{\mathrm{M}}^{\mathrm{I}}$ and $\overline{\mathrm{M}}^{\mathrm{II}}$, $\|e\|_{\mathrm{loc},h}$, and $\|e\|_{\mathcal{L}}$ controlled by $\mathbb{E}$, with respect to the uniform refinement procedure and adaptive refinement with different bulk parameters. Obviously, the number of degrees of freedom in the case of uniform refinement is much higher than in adaptive refinement. In particular, we consider $\sigma = 0.4$ and $\sigma = 0.6$. Here, the choice of the first value is justified by the previous studies [46], where such bulk parameter provided the most optimal numerical results. The second value is chosen for the comparison of the adaptive meshes. Both cases provide slightly improved convergences in comparison to the expected $O(h^2)$ for $\|e\|_{\mathrm{loc},h}$ and $O(h)$ for $\|e\|_{\mathcal{L}}$. The time expenses for the u_h naturally get lower in the case of $\sigma = 0.6$, since the d. o. f. (u_h) does not grow as fast as in the case with $\sigma = 0.6$.

Table 5.1: *Example 1.* Efficiency of $\overline{\mathrm{M}}^{\mathrm{I}}$, $\overline{\mathrm{M}}^{\mathrm{II}}$, and $\mathbb{E}$ and for $u_h \in S^2_h$ and $y_h, w_h \in S^3_{5h}$, and order of convergence for $\|e\|_{\mathrm{loc},h}$ and $\|e\|_{\mathcal{L}}$ ($N_{\mathrm{ref},0} = 1$).

ref.	$\|\nabla_x e\|_Q$	$I_{\mathrm{eff}}(\overline{\mathrm{M}}^{\mathrm{I}})$	$I_{\mathrm{eff}}(\overline{\mathrm{M}}^{\mathrm{II}})$	$\|e\|_{\mathrm{loc},h}$	eoc	$\|e\|_{\mathcal{L}}$	eoc	$I_{\mathrm{eff}}(\mathbb{E})$
(a) uniform refinement								
2	2.5516×10^{-3}	1.07	1.03	2.5524×10^{-3}	3.43	7.9057×10^{-3}	1.71	1.00
4	1.5947×10^{-4}	1.39	1.20	1.5947×10^{-4}	2.36	1.9764×10^{-4}	1.18	1.00
6	9.9670×10^{-4}	1.00	1.00	9.9670×10^{-6}	2.09	4.9411×10^{-3}	1.05	1.00
8	6.2294×10^{-7}	1.00	1.00	6.2294×10^{-7}	2.02	1.2353×10^{-3}	1.01	1.00
(b) adaptive refinement, $\sigma = 0.4$								
2	2.5516×10^{-3}	1.07	1.03	2.5520×10^{-3}	3.43	7.9057×10^{-2}	1.71	1.00
4	2.2743×10^{-4}	1.41	1.19	2.2745×10^{-4}	2.36	2.1712×10^{-2}	1.37	1.00
6	2.9936×10^{-5}	1.09	1.02	2.9936×10^{-5}	2.71	7.9512×10^{-3}	1.28	1.00
8	4.9501×10^{-6}	1.12	1.05	4.9501×10^{-6}	1.51	3.1138×10^{-3}	0.93	1.00
(c) adaptive refinement, $\sigma = 0.6$								
2	2.5516×10^{-3}	1.07	1.03	2.5520×10^{-3}	3.43	7.9057×10^{-2}	1.71	1.00
4	3.3298×10^{-4}	1.30	1.11	3.3305×10^{-4}	1.80	2.5410×10^{-2}	1.22	1.00
6	5.9048×10^{-5}	1.34	1.14	5.9050×10^{-5}	3.18	1.0976×10^{-2}	1.60	1.00
8	2.3071×10^{-5}	1.25	1.10	2.3072×10^{-5}	2.06	6.7335×10^{-3}	1.41	1.00

Table 5.2: *Example 1*. Assembling and solving time (in seconds) spent for the systems defining d. o. f. of $u_h \in S_h^2$ and $y_h, w_h \in S_{5h}^3$ ($N_{\mathrm{ref},0} = 1$).

ref.	d. o. f.			t_{as}			t_{sol}			$\frac{t_{\mathrm{appr.}}}{t_{\mathrm{er.est.}}}$
	u_h	y_h	w_h	u_h	y_h	w_h	u_h	y_h	w_h	
(a) $\sigma = 0.4$										
4	240	25	25	2.64×10^{-1}	1.44×10^{-2}	1.08×10^{-2}	4.01×10^{-3}	2.36×10^{-4}	1.15×10^{-4}	18.3
6	2,027	25	25	2.42	1.81×10^{-2}	1.61×10^{-2}	1.94×10^{-1}	2.23×10^{-4}	1.34×10^{-4}	142.66
8	11,512	76	76	$1.35 \times 10^{+1}$	1.97×10^{-1}	1.68×10^{-1}	3.39	7.13×10^{-4}	3.76×10^{-4}	85.42
				$t_{\mathrm{as}}(u_h): t_{\mathrm{as}}(y_h): t_{\mathrm{as}}(w_h)$			$t_{\mathrm{sol}}(u_h): t_{\mathrm{sol}}(y_h): t_{\mathrm{sol}}(w_h)$			
				80.13	1.17	1.00	9,021.85	1.90	1.00	
(b) $\sigma = 0.6$										
4	206	25	25	2.27×10^{-1}	1.94×10^{-2}	1.54×10^{-2}	3.35×10^{-3}	1.66×10^{-4}	8.60×10^{-5}	11.77
6	896	25	25	1.07	1.20×10^{-2}	1.96×10^{-2}	4.17×10^{-2}	2.30×10^{-4}	1.25×10^{-4}	90.89
8	2,706	79	79	3.44	1.65×10^{-1}	1.37×10^{-1}	2.69×10^{-1}	6.72×10^{-4}	7.31×10^{-4}	22.3
				$t_{\mathrm{as}}(u_h): t_{\mathrm{as}}(y_h): t_{\mathrm{as}}(w_h)$			$t_{\mathrm{sol}}(u_h): t_{\mathrm{sol}}(y_h): t_{\mathrm{sol}}(w_h)$			
				25.20	1.21	1.00	367.33	0.92	1.00	12.25

Moreover, we compare the estimated order of convergence in Figure 5.3. Here, the majorant is reconstructed with auxiliary functions $y_h \in S_h^3$ ($M = 1$) and $y_h \in S_{7h}^3$ ($M = 7$). The numerical test demonstrates that an efficient error estimation and its local indication can be achieved even using auxiliary fluxes on a very coarse mesh (in this particular case, 7 times coarser than the mesh $\mathcal{K}_h$ for u_h).

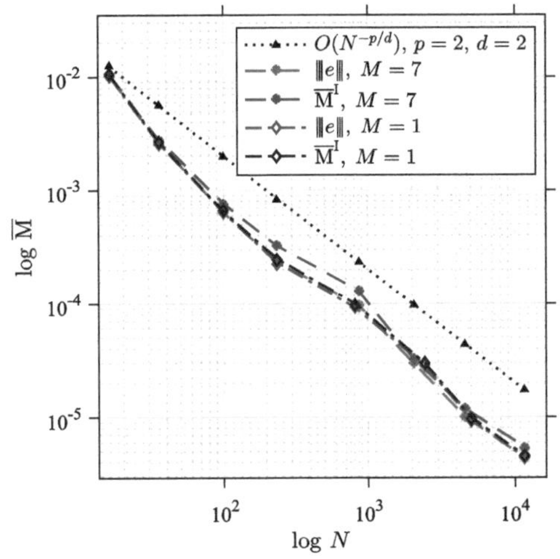

Figure 5.3: *Example 1*. Comparison of the error and majorant $\overline{M}^I$ order of convergence for $y_h \in S_{7h}^3$ and for $y_h \in S_h^3$.

8.2 Example 2: parameterized solution

Now we consider an example with the exact solution depending on the parameters k_1 and k_2. Let $Q = (0,1)^2$ be the unit square, and let the exact solution, the right hand side, and the Dirichlet boundary condition be chosen as follows:

$$
\begin{aligned}
u(x,t) &= \sin k_1\pi x \sin k_2\pi t \quad \text{for } (x,t) \in \overline{Q} = [0,1]^2, \\
f(x,t) &= \sin k_1\pi x(k_2\pi \cos k_2\pi t + k_1^2\pi^2 \sin k_2\pi t) \quad \text{for } (x,t) \in Q = (0,1)^2, \\
u_0(x,t) &= 0 \quad \text{for } (x,t) \in \overline{\Sigma}_0, \quad \text{and} \\
u_D(x,t) &= 0 \quad \text{for } (x,t) \in \Sigma := \partial\Omega \times (0,1).
\end{aligned}
$$

First, we set $k_1 = k_2 = 1$ (*Example 2–1*). We consider eight steps of adaptive refinement ($N_{\text{ref}} = 8$) preceded by three global refinements ($N_{\text{ref},0} = 3$) made to generate the initial mesh $\mathcal{K}_h^0$. For the marking criterion, we chose the bulk parameter $\sigma = 0.4$. The function u_h is approximated both by S_h^2 (case (a)) and S_h^3 (case (b)) spaces, whereas corresponding auxiliary functions by $y_h, w_h \in S_{7h}^4$ and $y_h, w_h \in S_{5h}^6$, respectively; see

Tables 5.3–5.4. Figure 5.4a illustrates different estimated orders of convergence for different approximations $u_h \in S_h^2$ and $u_h \in S_h^3$, which perform slightly better than the expected rates $O(h^2)$ and $O(h^3)$, respectively.

Table 5.3: *Example 2–1.* Efficiency of $\overline{\mathrm{M}}^{\mathrm{I}}$, $\overline{\mathrm{M}}^{\mathrm{II}}$, $\overline{\mathrm{M}}_h^{\mathrm{I}}$, $\mathbb{E}$, and order of convergence of $\|e\|_{\mathrm{loc},h}$ and $\|e\|_{\mathcal{L}}$ for $\sigma = 0.4$ ($N_{\mathrm{ref},0} = 3$).

ref.	$\|\nabla_x e\|_Q$	$I_{\mathrm{eff}}(\overline{\mathrm{M}}^{\mathrm{I}})$	$I_{\mathrm{eff}}(\overline{\mathrm{M}}^{\mathrm{II}})$	$\|e\|_{\mathrm{loc},h}$	eoc	$\|e\|_{\mathcal{L}}$	eoc	$I_{\mathrm{eff}}(\mathbb{E})$
(a) $u_h \in S_h^2$ and $y_h, w_h \in S_{7h}^4$								
4	3.3878×10^{-4}	3.14	1.33	3.5057×10^{-4}	1.96	9.3154×10^{-2}	1.07	1.00
6	4.8136×10^{-5}	4.13	1.70	4.8588×10^{-5}	2.36	3.7361×10^{-2}	1.31	1.00
8	9.2649×10^{-6}	5.78	3.23	9.2835×10^{-6}	3.79	1.7351×10^{-2}	1.79	1.00
(b) $u_h \in S_h^3$ and $y_h, w_h \in S_{5h}^6$								
4	1.3562×10^{-4}	1.64	1.30	1.3591×10^{-4}	3.56	8.9725×10^{-3}	2.89	1.00
6	6.9962×10^{-6}	10.61	10.26	6.9982×10^{-6}	4.17	1.4163×10^{-3}	2.55	1.00
8	3.5507×10^{-7}	3.44	1.24	3.5535×10^{-7}	3.11	1.6376×10^{-4}	2.13	1.00

We also demonstrate the quantitative effectiveness of the error indication provided by $\overline{\mathrm{M}}^{\mathrm{I}}$. In Figure 5.5, a comparison of the meshes illustrates that the refinement based on local values of $\|\nabla_x e\|_K$ (first row) and the indicator $\overline{\mathrm{m}}_{\mathrm{d},K}^{\mathrm{I}}$ (second row) provide similar adaptive patterns.

Next, we set parameters $k_1 = 3$ and $k_2 = 6$ (*Example 2–2*). In this case, the auxiliary variables are approximated by $y_h, w_h \in S_{5h}^7$. Figure 5.4b illustrates the order of convergence of errors and corresponding majorants (for two different marking strategy $\mathbb{M}_{\mathrm{BULK}}(0.4)$ and $\mathbb{M}_{\mathrm{BULK}}(0.6)$), and compares these results to the theoretical one $O(h^2)$. It is easy to see from the plot that the efficiency of the majorant temporarily deteriorates on the first refinement steps (while keeping the same mesh for the flux approximation), but it improves drastically on the last refinements (after starting the refinement of the mesh used for $y_h \in S_{5h}^7$). Such stagnation can be regarded as a guidance on when to start "enriching" the mesh for the dual variable. In general, the choice of the mesh for $\boldsymbol{y}_h$ must be thoughtfully selected, and in some cases coordinated with the mesh of the primal solution. Tables 5.5 and 5.6 compare numerical results obtained for different marking parameters $\sigma = 0.4$ (part (a)) and $\sigma = 0.6$ (part (b)). Finally, Figure 5.6 demonstrates the evolution of meshes associated with the refinement steps 4–6.

8.3 Example 3: Gaussian distribution

As the next test case, we consider the exact solution defined by a sharp local Gaussian distribution:

$$u(x,t) = (x^2 - x)(t^2 - t)e^{-100|(x,t)-(0.8,0.05)|}, \quad (x,t) \in \overline{Q} := [0,1]^2,$$

Table 5.4: *Example 2–1.* Assembling and solving time (in seconds) spent for the systems generating d. o. f. of u_h, y_h, and w_h for $\sigma = 0.4$ ($N_{\mathrm{ref},0} = 3$).

ref.	d. o. f.			t_{as}			t_{sol}			$\frac{t_{\mathrm{appr.}}}{t_{\mathrm{er.est.}}}$
	u_h	y_h	w_h	u_h	y_h	w_h	u_h	y_h	w_h	
(a) $u_h \in S_h^2$ and $y_h, w_h \in S_{7h}^4$										
6	12,935	144	144	$1.55 \times 10^{+1}$	3.97×10^{-1}	3.83×10^{-1}	2.17	2.30×10^{-3}	1.37×10^{-3}	44.25
7	34,037	144	144	$4.90 \times 10^{+1}$	3.98×10^{-1}	3.73×10^{-1}	9.58	3.36×10^{-3}	1.42×10^{-3}	145.95
8	61,258	144	144	$9.37 \times 10^{+1}$	3.80×10^{-1}	3.62×10^{-1}	$2.42 \times 10^{+1}$	2.10×10^{-3}	1.83×10^{-3}	308.55
				$t_{\mathrm{as}}(u_h): t_{\mathrm{as}}(y_h): t_{\mathrm{as}}(w_h)$			$t_{\mathrm{sol}}(u_h): t_{\mathrm{sol}}(y_h): t_{\mathrm{sol}}(w_h)$			
				258.63	1.05	1.00	13,252.51	1.15	1.00	
(b) $u_h \in S_h^3$ and $y_h, w_h \in S_{5h}^6$										
6	13,742	169	169	$1.62 \times 10^{+1}$	7.03×10^{-1}	7.03×10^{-1}	2.11	2.53×10^{-3}	1.43×10^{-3}	25.95
7	35,091	322	322	$5.36 \times 10^{+1}$	5.65	5.52	$1.10 \times 10^{+1}$	9.31×10^{-3}	5.29×10^{-3}	11.41
8	78,561	372	372	$1.91 \times 10^{+2}$	5.61	5.03	$2.40 \times 10^{+1}$	2.51×10^{-2}	7.56×10^{-3}	38.15
				$t_{\mathrm{as}}(u_h): t_{\mathrm{as}}(y_h): t_{\mathrm{as}}(w_h)$			$t_{\mathrm{sol}}(u_h): t_{\mathrm{sol}}(y_h): t_{\mathrm{sol}}(w_h)$			
				37.97	1.11	1.00	3,168.34	3.31	1.00	

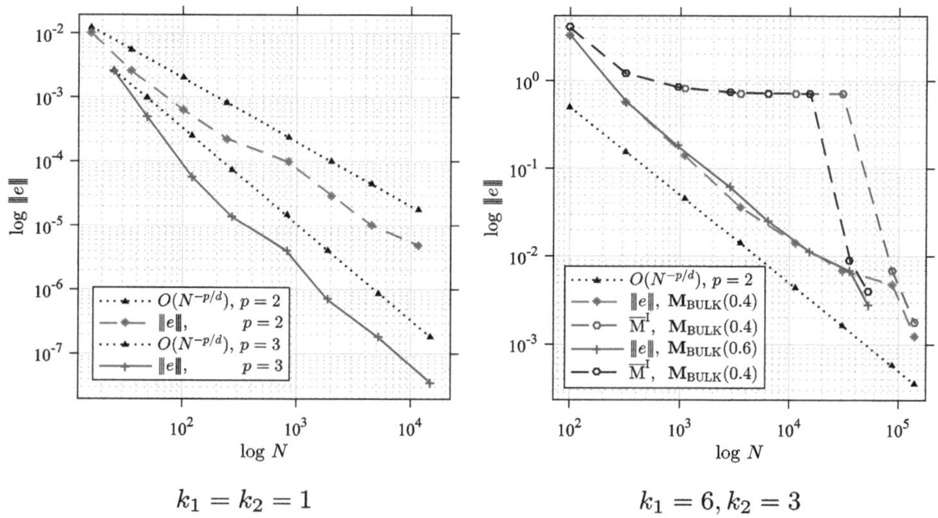

Figure 5.4: *Example 2.* The estimated order of convergence for $u(x,t) = \sin k_1\pi x \sin k_2\pi t$.

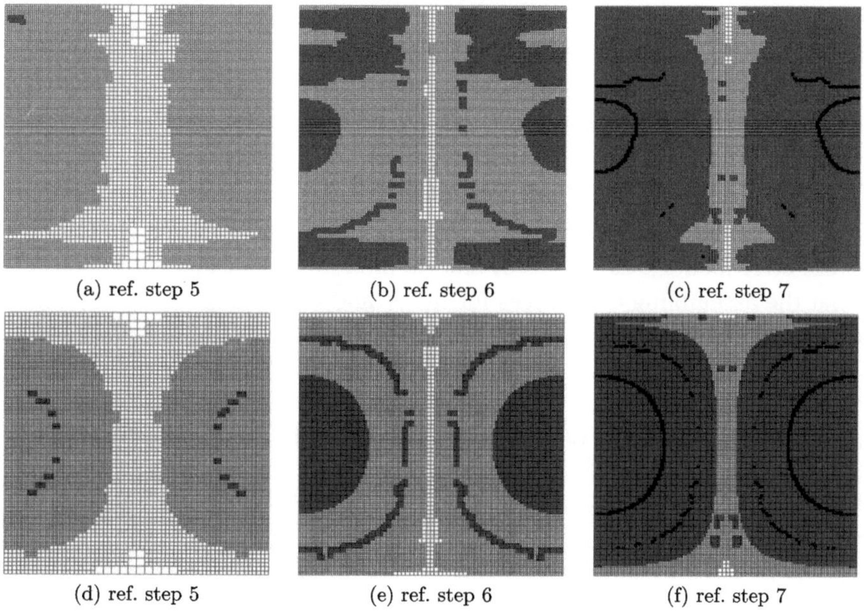

Figure 5.5: *Example 2–1.* Adaptive meshes obtained based on the indicator $\overline{\mathrm{m}}^{\mathrm{I}}_{\mathrm{d},K}$ (top) and on the exact error $\|e\|_{\mathrm{loc},h,K}$ (bottom) with respect to the refinement steps 5–7.

where the peak is located in the point $(x,t) = (0.8, 0.05)$. Then f is computed by substituting u into (2.1). The function u obviously fulfils homogeneous Dirichlet boundary conditions.

Table 5.5: *Example 2–2*. Efficiency of $\overline{\mathrm{M}}^{\mathrm{I}}$, $\overline{\mathrm{M}}^{\mathrm{II}}$, $\mathbb{E}$, and the order of convergence of $\|e\|_{\mathrm{loc},h}$ and $\|e\|_{\mathcal{L}}$ for $u_h \in S_h^2$ and $y_h, w_h \in S_{5h}^7$ ($N_{\mathrm{ref},0} = 3$).

ref.	$\|\nabla_x e\|_Q$	$I_{\mathrm{eff}}(\overline{\mathrm{M}}^{\mathrm{I}})$	$I_{\mathrm{eff}}(\overline{\mathrm{M}}^{\mathrm{II}})$	$\|e\|_{\mathrm{loc},h}$	eoc	$\|e\|_{\mathcal{L}}$	eoc	$I_{\mathrm{eff}}(\mathbb{E})$
(a) $\sigma = 0.4$								
2	5.7161×10^{-1}	2.11	1.38	5.7163×10^{-1}	2.99	$6.2371 \times 10^{+1}$	1.19	1.00
3	1.3927×10^{-1}	5.77	2.20	1.3928×10^{-1}	2.30	$3.1026 \times 10^{+1}$	1.14	1.00
8	1.2298×10^{-3}	1.44	1.16	1.2298×10^{-3}	5.60	2.6917	2.30	1.00
(b) $\sigma = 0.6$								
2	5.7161×10^{-1}	2.11	1.38	5.7163×10^{-1}	2.99	$6.2371 \times 10^{+1}$	1.19	1.00
3	1.7942×10^{-1}	4.69	1.96	1.7945×10^{-1}	2.18	$3.2971 \times 10^{+1}$	1.20	1.00
8	2.7492×10^{-3}	1.44	1.15	2.7492×10^{-3}	4.75	4.0721	1.91	1.00

For the discretization spaces, we use the standard configuration; that is, $u_h \in S_h^2$ for the approximate solution, and $y_h, w_h \in S_h^3$ for the auxiliary functions. We start with four initial global refinements ($N_{\mathrm{ref},0} = 4$), and continue with seven adaptive steps ($N_{\mathrm{ref}} = 7$); see Figure 5.8. As marking criteria, we choose $\mathbb{M}_{\mathrm{BULK}}(0.6)$. The estimated order of convergence is illustrated in Figure 5.7. It confirms that majorants reconstructed with $y_h, w_h \in S_{2h}^6$ are as efficient as the one reconstructed with $y_h, w_h \in S_h^3$. They also drastically improve the convergence order on the first refinement steps.

Numbers exposed in Table 5.7 demonstrate the efficiency of the majorants and the error identity in terms of error estimation and show that $\overline{\mathrm{M}}^{\mathrm{II}}$ is twice sharper than $\overline{\mathrm{M}}^{\mathrm{I}}$, whereas the error identity, as expected, reflects the error $\|e\|_{\mathcal{L}}$ exactly. In Table 5.8, we see that the assembling of matrices for the y_h and w_h is three times more time-consuming in comparison to assembling the system for u_h.

8.4 Example 4: solution with a singularity in the time derivative

We now consider a solution with a singularity with respect to the time coordinate; that is, we take

$$u(x,t) = \sin \pi x |1 - t|^{\lambda}, \quad (x,t) \in \overline{Q} = [0,1] \times [0,2],$$

where the parameter is taken either as $\lambda = \frac{3}{2}$, $\lambda = 1$, or $\lambda = \frac{2}{3}$ (see Figure 5.9 illustrating u for different λ). The right-hand side $f(x,t)$ follows from the substitution of u into (2.1), and the Dirichlet boundary condition is defined as $u_D = u$ on Σ.

The solution $u(x,t)$ is smooth with respect to the spatial coordinates, whereas the regularity in time depends on the parameter λ. Then, the expected convergence in the term $h^{1/2}\|\partial_t(u - u_h)\|_Q$ is $O(h^{p-1}) \cdot O(h^{1/2}) = O(h^{p-1/2})$; see [27].

Table 5.6: *Example 2–2*. Assembling and solving time (in seconds) spent for the systems generating d. o. f. of $u_h \in S_h^2$ and $y_h, w_h \in S_{5h}^7$ ($N_{\text{ref},0} = 3$).

ref.	d. o. f.			t_{as}			t_{sol}			$\frac{t_{appr.}}{t_{er.est.}}$
	u_h	y_h	w_h	u_h	y_h	w_h	u_h	y_h	w_h	
(a) $\sigma = 0.4$										
6	30,101	225	225	$5.99 \times 10^{+1}$	2.29	2.92	3.57	8.52×10^{-3}	4.33×10^{-3}	12.14
7	86,849	529	529	$3.57 \times 10^{+2}$	9.30	9.41	$1.11 \times 10^{+1}$	5.19×10^{-2}	3.47×10^{-2}	19.58
8	141,987	1,425	1,425	$6.36 \times 10^{+2}$	$6.50 \times 10^{+1}$	$5.91 \times 10^{+1}$	$2.56 \times 10^{+1}$	3.00×10^{-1}	1.29×10^{-1}	5.31
				$t_{as}(u_h)$: $t_{as}(y_h)$: $t_{as}(w_h)$			$t_{sol}(u_h)$: $t_{sol}(y_h)$: $t_{sol}(w_h)$			
				10.76	1.10	1.00	198.84	2.32	1.00	
(b) $\sigma = 0.6$										
6	15,436	225	225	$2.61 \times 10^{+1}$	2.36	2.41	1.77	1.45×10^{-2}	3.12×10^{-3}	11.73
7	35,745	529	529	$8.99 \times 10^{+1}$	9.86	$1.01 \times 10^{+1}$	4.68	7.06×10^{-2}	4.12×10^{-2}	9.52
8	52,453	1,249	1,249	$1.05 \times 10^{+2}$	$8.03 \times 10^{+1}$	$7.08 \times 10^{+1}$	7.38	3.47×10^{-1}	1.66×10^{-1}	1.39
				$t_{as}(u_h)$: $t_{as}(y_h)$: $t_{as}(w_h)$			$t_{sol}(u_h)$: $t_{sol}(y_h)$: $t_{sol}(w_h)$			
				1.49	1.13	1.00	44.46	2.09	1.00	

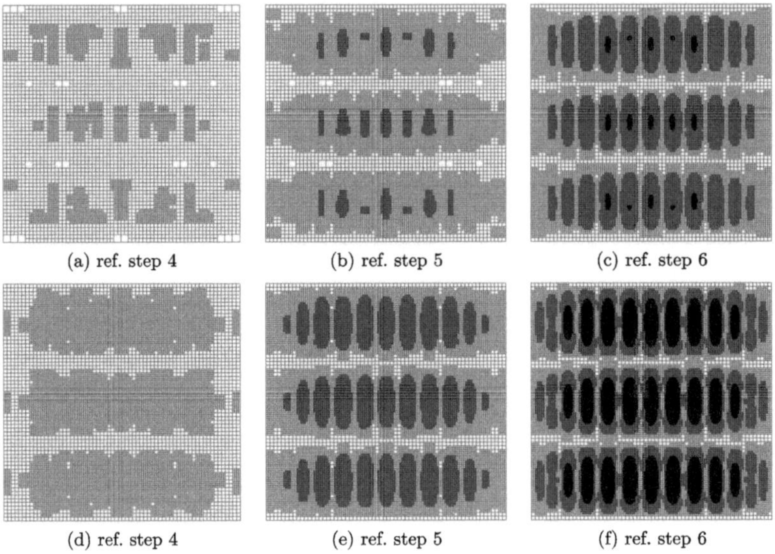

Figure 5.6: *Example 2-2.* Meshes obtained by the marking criteria $\mathbb{M}_{\text{BULK}}(0.6)$ (top row) and $\mathbb{M}_{\text{BULK}}(\sigma = 0.4)$ (bottom row) with respect to the refinement steps 4–6.

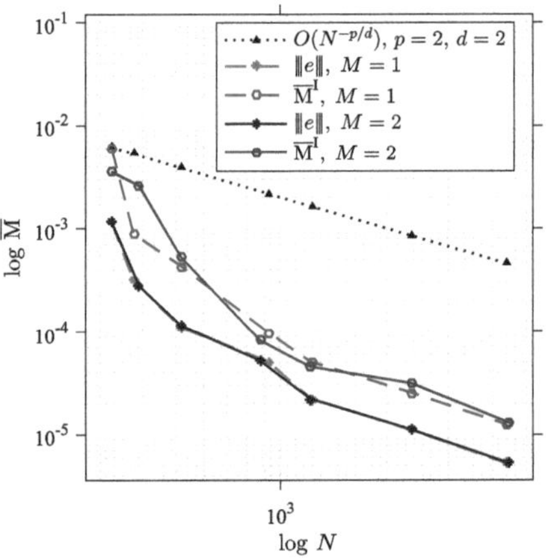

Figure 5.7: *Example 3.* The majorant $\overline{\mathrm{M}}^{\mathrm{I}}$ and e. o. c. for $u_h \in S_h^2$ and two different choices of auxiliary functions (a) $y_h, w_h \in S_h^3$ and (b) $y_h, w_h \in S_{2h}^6$.

Table 5.7: *Example 3.* Efficiency of $\overline{\mathrm{M}}^{\mathrm{I}}$, $\overline{\mathrm{M}}^{\mathrm{II}}$, $\overline{\mathrm{M}}^{\mathrm{I}}_h$, $\mathbb{E}$, and the order of convergence of $\|e\|_{\text{loc},h}$ and $\|e\|_{\mathcal{L}}$ with marking $\mathbb{M}_{\text{BULK}}(0.6)$ for $u_h \in S_h^2$ and $y_h, w_h \in S_h^3$ ($N_{\text{ref},0} = 4$).

ref.	$\|\nabla_x e\|_Q$	$I_{\text{eff}}(\overline{\mathrm{M}}^{\mathrm{I}})$	$I_{\text{eff}}(\overline{\mathrm{M}}^{\mathrm{II}})$	$\|e\|_{\text{loc},h}$	eoc	$\|e\|_{\mathcal{L}}$	eoc	$I_{\text{eff}}(\mathbb{E})$
2	3.1311×10^{-4}	2.85	1.55	3.1335×10^{-4}	17.71	5.6510×10^{-2}	8.64	1.00
3	1.0915×10^{-4}	3.93	1.73	1.0944×10^{-4}	6.49	3.1506×10^{-2}	3.60	1.00
5	2.2033×10^{-5}	2.27	1.36	2.2042×10^{-5}	5.87	1.4796×10^{-2}	3.59	1.00
7	5.2517×10^{-6}	2.38	1.22	5.2526×10^{-6}	2.41	7.2473×10^{-3}	1.27	1.00

Table 5.8: *Example 3*. Assembling and solving time (in seconds) spent for the systems generating d. o. f. of $u_h \in S_h^2$ and $y_h, w_h \in S_h^3$ for $\sigma = 0.6$ ($N_{\text{ref},0} = 4$).

ref.	d. o. f.			t_{as}			t_{sol}			$\frac{t_{appr.}}{t_{er.est.}}$
	u_h	y_h	w_h	u_h	y_h	w_h	u_h	y_h	w_h	
3	520	544	544	6.50×10^{-1}	2.39	1.80	1.09×10^{-2}	8.60×10^{-3}	1.91×10^{-2}	0.27
5	1,232	1,237	1,237	1.72	6.26	4.71	6.12×10^{-2}	8.25×10^{-2}	1.26×10^{-1}	0.28
7	4,368	4,246	4,246	6.20	$3.04 \times 10^{+1}$	$1.87 \times 10^{+1}$	6.38×10^{-1}	6.05×10^{-1}	5.26×10^{-1}	0.22
				$t_{as}(u_h) : t_{as}(y_h) : t_{as}(w_h)$			$t_{sol}(u_h) : t_{sol}(y_h) : t_{sol}(w_h)$			
				0.33	1.62	1.00	1.21	1.15	1.00	

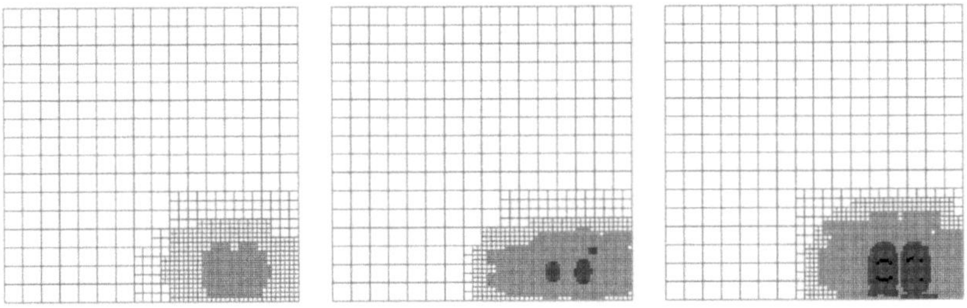

(a) ref. step 4 (b) ref. step 5 (c) ref. step 6

Figure 5.8: *Example 3.* Meshes obtained in the refinement steps 4–6, $\sigma = 0.6$ ($N_{\mathrm{ref},0} = 4$) for $u_h \in S_h^2$ and $y_h, w_h \in S_h^3$.

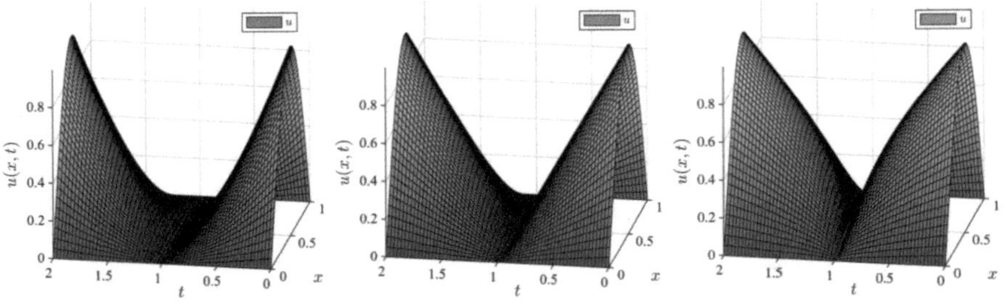

(a) $\lambda = \frac{3}{2}$ (b) $\lambda = 1$ (c) $\lambda = \frac{2}{3}$

Figure 5.9: *Example 4.* Exact solution $u(x, t) = \sin \pi x |1 - t|^{\lambda}$.

Unlike previous examples, where the bulk marking criterion was used, we consider the so-called "absolute threshold" marking criterion; that is, $\mathfrak{S}_{\mathrm{GARU}} := \sigma \max_{K \in \mathcal{K}_h} \overline{\mathrm{m}}^{\mathrm{I},2}_{\mathrm{d},K}$, $\sigma \in [0, 1]$. Theoretical convergence for similar λ were tested in [42] (for $p = 2$). Table 5.9 illustrates an improved order of convergence for $\lambda \in \{\frac{2}{3}, 1, \frac{3}{2}\}$. In particular, for the case $\lambda = 1$, we compare the error convergence of the uniform refinement to the adaptive one. We note that the cases (a) and (c) are considered in detail in the preceding paper [42]. Figure 5.10 highlights the fact that in all the cases we managed to obtain improved convergence rates. In Table 5.9, we add an additional test case for $\lambda = 1$, where, instead of C^1 basis functions, we consider C^0-continuous approximations. This allows getting an optimal rate of convergence and rather efficient error estimates. Figures 5.11–5.12 present meshes obtained in the adaptive refinement steps 5–8 for the parameters $\lambda = 3/2$ and $2/3$ and reconfirm that the functional error estimates detect the local singularities rather efficiently. The singularity at $t = 1$ is captured and very well, represented by the error indicator and resulting adaptive mesh. Moreover, for the rest of the times; that is, $(0, 1) \cup (1, 2)$, where the solution is smooth, the mesh is not over-refined. Figures 5.13 and 5.14 present the adaptive meshes obtained for the case $\lambda = 1$ for C^1-smooth B-spline basis-functions and for C^0-continuous ones, respectively. It is easy to see that in Figure 5.13 the line $t = 1$

Table 5.9: *Example 4*. Efficiency of $\overline{\mathrm{M}}^{\mathrm{I}}$, $\overline{\mathrm{M}}^{\mathrm{II}}$, $\mathbb{E}$, and order of convergence of $\|e\|_{\mathrm{loc},h}$ and $\|e\|_{\mathcal{L}}$ for $\sigma = 0.4$ ($N_{\mathrm{ref},0} = 1$).

ref.	d. o. f.	$\|\nabla_x e\|_Q$	$I_{\mathrm{eff}}(\overline{\mathrm{M}}^{\mathrm{I}})$	$I_{\mathrm{eff}}(\overline{\mathrm{M}}^{\mathrm{II}})$	$\|e\|_{\mathrm{loc},h}$	eoc
(a) $\lambda = \frac{3}{2}$						
adaptive refinement for $u_h \in S_h^2$ and $y_h, w_h \in S_h^3$					theoretical eoc $O(h^{3/2})$	
5	252	3.7652×10^{-3}	3.39	1.38	3.8038×10^{-3}	1.86
7	870	8.2312×10^{-4}	5.02	1.55	8.2588×10^{-4}	2.28
9	5,526	2.4187×10^{-4}	3.41	1.64	2.4205×10^{-4}	1.67
(b) $\lambda = 1$						
uniform refinement for $u_h \in S_h^2$ and $y_h, w_h \in S_h^3$,					theoretical eoc $O(h)$	
2	16	1.3246×10^{-1}	1.58	0.81	1.6352×10^{-1}	2.30
4	100	2.3788×10^{-2}	4.05	1.41	2.8059×10^{-2}	1.41
6	1,156	5.7163×10^{-3}	11.98	3.55	6.4975×10^{-3}	1.08
8	16,900	1.4259×10^{-3}	42.99	15.13	1.6016×10^{-3}	1.02
adaptive refinement for $u_h \in S_h^2$ and $y_h, w_h \in S_h^3$,					theoretical eoc $O(h)$	
5	172	1.9026×10^{-2}	4.60	1.56	1.9489×10^{-2}	1.26
7	662	4.1260×10^{-3}	11.19	2.20	4.1898×10^{-3}	1.91
9	3,798	1.4618×10^{-3}	20.85	2.76	1.4674×10^{-3}	1.77
C^0-continuous basis functions on the mesh–line $t = 1$,					theoretical eoc $O(h^2)$	
4	190	3.2619×10^{-3}	1.10	0.94	3.2623×10^{-3}	2.64
6	1,302	6.7686×10^{-4}	1.05	0.82	6.7687×10^{-4}	2.59
8	4,496	1.7603×10^{-4}	1.04	0.85	1.7603×10^{-4}	2.66
(c) $\lambda = \frac{2}{3}$						
adaptive refinement for $u_h \in S_h^2$ and $y_h, w_h \in S_h^3$,					theoretical eoc $O(h^{2/3})$	
5	188	3.5335×10^{-2}	6.32	1.81	3.8350×10^{-2}	1.61
7	710	1.8142×10^{-2}	12.16	2.08	1.8499×10^{-2}	0.73
9	4,106	7.5502×10^{-3}	32.74	2.97	7.6290×10^{-3}	0.63

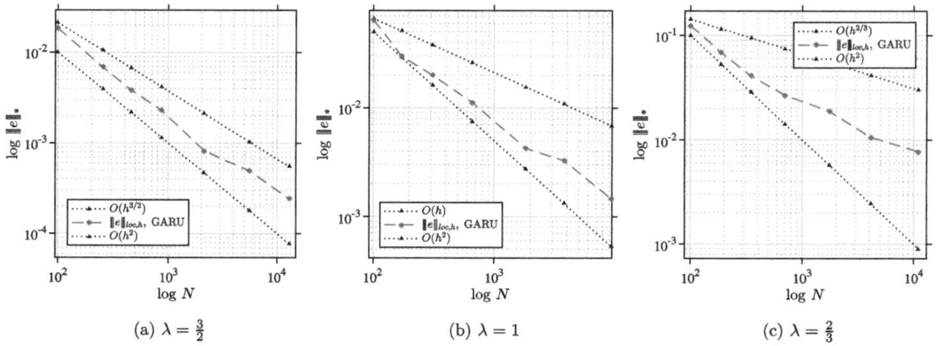

Figure 5.10: *Example 4*. The estimated order of convergence for approximations with $u \in S_h^2$ and S_h^3: (a) $\lambda = \frac{3}{2}$, (b) $\lambda = 1$, and (c) $\lambda = \frac{2}{3}$.

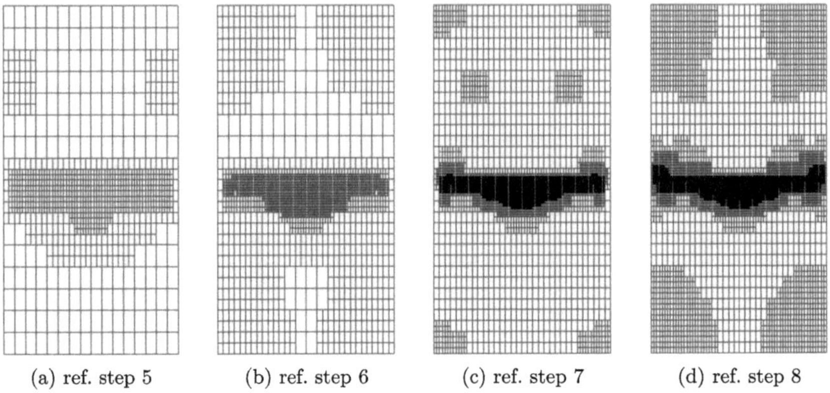

Figure 5.11: *Example 3 (case (a): $\lambda = \frac{3}{2}$).* Meshes obtained in the refinement steps 5–7 for $u_h \in S_h^2$ and $y_h, w_h \in S_h^3$.

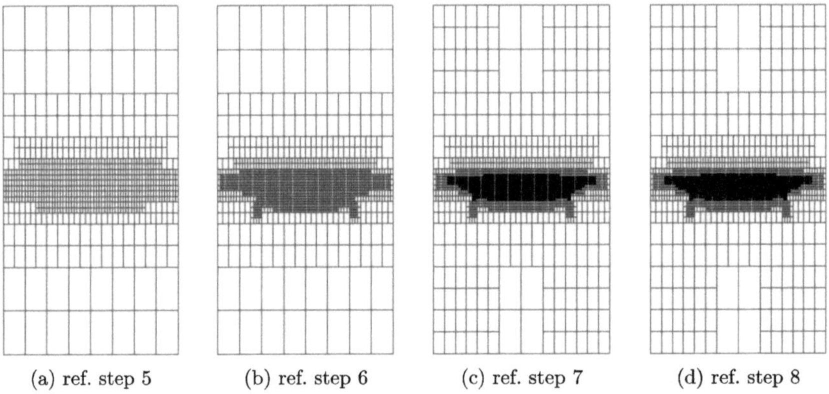

Figure 5.12: *Example 3 (case (c): $\lambda = \frac{2}{3}$).* Meshes obtained in the refinement steps 5–7 for $u_h \in S_h^2$ and $y_h, w_h \in S_h^3$.

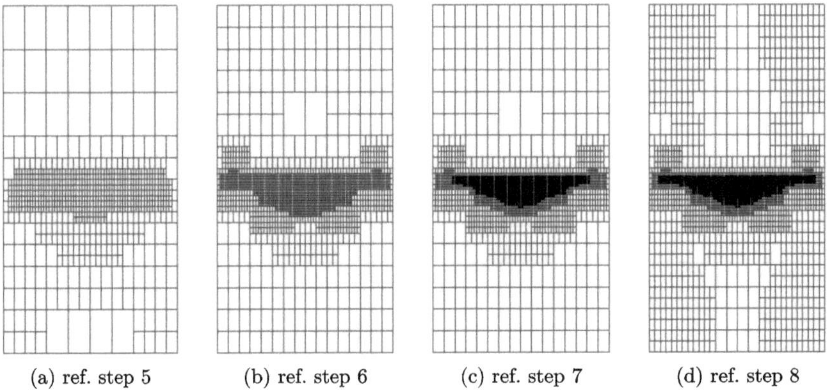

Figure 5.13: *Example 3 (case (b): $\lambda = 1$).* Meshes obtained in the refinement steps 5–7 for $u_h \in S_h^2$ and $y_h, w_h \in S_h^3$.

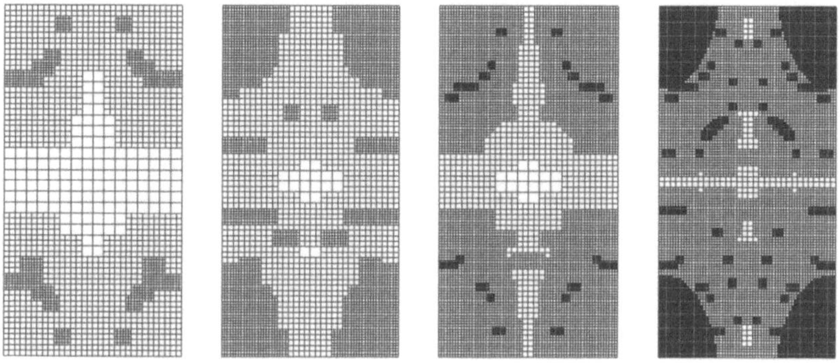

(a) ref. step 5 (b) ref. step 6 (c) ref. step 7 (d) ref. step 8

Figure 5.14: *Example 3 (case (b): $\lambda = 1$ with C^0–continuity on the mesh–line $t = 1$).* Meshes obtained in the refinement steps 5–7 for $u_h \in S_h^2$ and $y_h, w_h \in S_h^3$.

is treated as a singularity since we try to represent $|1 - t|$ with C^1-smooth B-splines. Adaptive meshes of Figure 5.14, however, refine everywhere else, but $t = 1$, since C^0 basis-functions fit the module function exactly.

8.5 Example 5: quarter-annulus domain extended in time

In the last example, we test the problem defined in the three-dimensional space-time cylinder $Q = \Omega_{\circlearrowleft} \times (0, T)$, where $\Omega_{\circlearrowleft}$ is represented by a quarter-annulus, which extended from $t = 0$ till $t = T = 1$. The exact solution is defined by

$$u(x,y,t) = (1-x)x^2(1-y)y^2(1-t)t^2, \quad (x,y,t) \in \overline{Q} := \overline{\Omega}_{\circlearrowleft} \times [0,1].$$

The right hand side $f(x,y,t)$, $(x,y,t) \in Q := \Omega_{\circlearrowleft} \times (0,1)$, is computed based on the substitution of u into the equation (2.1), and the Dirichlet boundary condition is defined as $u_D = u$ on Σ.

The initial mesh for the test is generated by one uniform refinement $N_{\text{ref},0} = 1$, the bulk marking parameter is set to $\sigma = 0.4$. The estimated order of convergence is illustrated in Figure 5.15, which corresponds to the theoretical expectation. We start the analysis by looking at Table 5.10. It is easy to see that all majorants have adequate performance, taking into account that the auxiliary functions $\mathbf{y}_h \in \oplus^2 S_{3h}^5$, and $w_h \in S_{3h}^5$. Table 5.11 confirms that assembling and solving of the systems reconstructing d. o. f. of u_h requires more time than assembling and solving routines for the systems generating $\mathbf{y}_h$ and w_h.

Figure 5.16 presents adaptive meshes generated for space-time approximations of functions in the quarter annulus Q. From the plots presented, we see that refinements are localized in the area close to the lateral surface of the quarter-annulus with the radius two. This happens because the solution is changing fast in the vicinity of this "outer"surface.

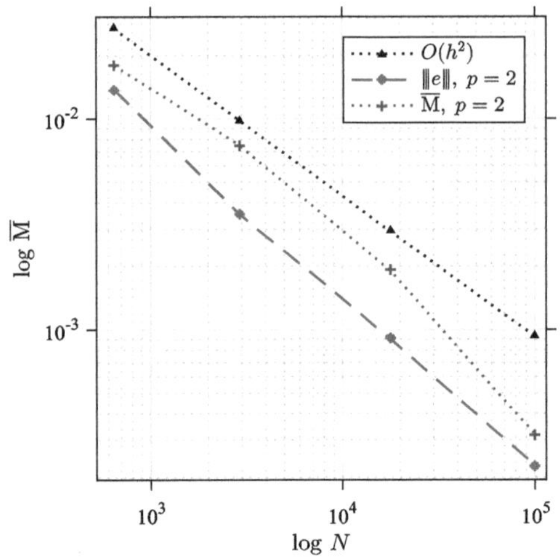

Figure 5.15: *Example 5*. The estimated order of convergence for $u \in S_h^2$.

Table 5.10: *Example 5*. Efficiency of $\overline{\mathrm{M}}^{\mathrm{I}}$, $\overline{\mathrm{M}}^{\mathrm{II}}$, and $\mathbb{E}$ for the bulk marking parameter $\sigma = 0.4$ for $u_h \in S_h^2$, $y_h \in \oplus^2 S_h^3$, and $w_h \in S_{3h}^5$ ($N_{\mathrm{ref},0} = 4$).

ref.	$\|\nabla_x e\|_Q$	$I_{\mathrm{eff}}(\overline{\mathrm{M}}^{\mathrm{I}})$	$I_{\mathrm{eff}}(\overline{\mathrm{M}}^{\mathrm{II}})$	$\|\|\|e\|\|\|_{\mathrm{loc},h}$	eoc	$\|\|\|e\|\|\|_{\mathcal{L}}$	eoc	$I_{\mathrm{eff}}(\mathbb{E})$
3	1.3711×10^{-2}	1.31	1.20	1.3722×10^{-2}	4.66	2.5548×10^{-1}	2.14	1.00
4	3.5322×10^{-3}	2.12	1.74	3.5331×10^{-3}	2.70	1.2719×10^{-1}	1.39	1.00
5	9.0289×10^{-4}	2.11	1.90	9.0425×10^{-4}	2.25	5.9632×10^{-2}	1.25	1.00
6	2.2747×10^{-4}	1.40	1.69	2.2749×10^{-4}	2.41	3.1509×10^{-2}	1.11	1.00

9 Conclusions

We derived a new locally stabilized space-time IgA scheme for parabolic I-BVPs, where global scaling h in the upwind test functions is replaced by a local scaling that depends on the local element size h_K, and supplied it with a priori error estimates that take care of locally refined meshes. However, the main purpose of this work was to study the properties of locally stabilized space-time IgA schemes in the context of mesh-adaptive computational methods. Adaptive mesh refinement is based on error indicators generated by functional-type a posteriori error estimates, which naturally account specific features and advantages of the IgA method. Since the corresponding error majorants are presented by integrals formed by element-wise contributions, they can efficiently be used for indication of the local errors and subsequent mesh refinement. We consider a fully unstructured space-time adaptive IgA scheme and use localized THB-splines for the mesh refinement. Finally, we illustrated the reliability and efficiency of the presented a posteriori error estimates in a series of examples exhibiting differ-

Table 5.11: *Example 5*. Assembling and solving time (in seconds) spent for the systems generating d. o. f. of $u_h \in S_h^2$, $y_h \in \oplus^2 S_h^3$, and $w_h \in S_{3h}^5$ with $\sigma = 0.4$ ($N_{\mathrm{ref},0} = 4$).

ref.	d. o. f.			t_{as}			t_{sol}			$\frac{t_{\mathrm{appr.}}}{t_{\mathrm{er.est.}}}$
	u_h	y_h	w_h	u_h	y_h	w_h	u_h	y_h	w_h	
3	646	686	343	7.03	$1.47 \times 10^{+1}$	6.17	1.10×10^{-2}	5.83×10^{-1}	2.13×10^{-3}	0.33
4	2,910	686	343	$4.04 \times 10^{+1}$	$1.21 \times 10^{+1}$	5.74	4.19×10^{-1}	5.45×10^{-1}	2.85×10^{-3}	2.22
5	17,881	1,458	729	$2.75 \times 10^{+2}$	$8.16 \times 10^{+1}$	$4.08 \times 10^{+1}$	$2.75 \times 10^{+1}$	4.00	4.72×10^{-2}	2.39
6	99,842	4,140	2,070	$2.90 \times 10^{+3}$	$2.33 \times 10^{+3}$	$1.51 \times 10^{+3}$	$1.26 \times 10^{+3}$	$8.88 \times 10^{+1}$	3.42×10^{-1}	1.06
				$t_{\mathrm{as}}(u_h): t_{\mathrm{as}}(y_h): t_{\mathrm{as}}(w_h)$			$t_{\mathrm{sol}}(u_h): t_{\mathrm{sol}}(y_h): t_{\mathrm{sol}}(w_h)$			
				1.91	1.54	1.00	3,683.81	259.30	1.00	1.06

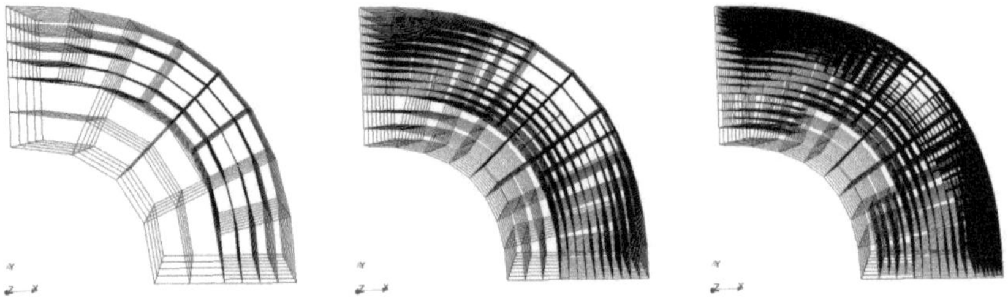

(a) ref. step 1 (b) ref. step 2 (c) ref. step 3

Figure 5.16: *Example 5*. Adaptive meshes in the refinement steps 1–3 and $\sigma = 0.6$.

ent features of exact solutions. Numerical tests performed have demonstrated high efficiency of the approach. Moreover, we also made a comparative study of the computational expenses for assembling the systems, finding an approximate solution and computing guaranteed and sufficiently accurate error bounds. In the majority of examples, error estimation required much lesser time than the reconstruction of the approximate IgA solution. Last but not least, the numerical examples have confirmed the high efficiency of the locally stabilized space-time THB-spline-based methods used in combination with suitable error indicators and mesh adaptive procedures. Of course, besides THB-spline, other local spline refinement techniques, such as mentioned in the introduction, can also be utilized in this adaptive framework. Adaptive methods are typically applied in combination with multigrid or multilevel solvers or preconditioners for the algebraic systems that we have to solve, since the adaptive procedure naturally provides a space-time hierarchy of meshes in a nested setting. Preceding experiments on massively parallel computers presented in [43] show that even algebraic multigrid preconditioners in connection with GMRES accelerations result in very efficient solvers for large systems with billions of space-time unknowns arising from (3+1)d examples. It is clear that the approach presented can be extended to a wider class of parabolic problems, such as eddy-current problems in electromagnetics.

Bibliography

[1] M. Ainsworth and J. T. Oden. *A posteriori error estimation in finite element analysis*. Wiley and Sons, New York, 2000.

[2] I. Anjam and D. Pauly. An elementary method of deriving a posteriori error equalities and estimates for linear partial differential equations. *Comput. Meth. Appl. Math.*, published online, 2016.

[3] W. Bangerth and R. Rannacher. *Adaptive finite element methods for differential equations*. Lectures in Mathematics, ETH Zürich, Birkhäuser, Basel, 2003.

[4] Y. Bazilevs, L. Beirão da Veiga, J. A. Cottrell, T. J. R. Hughes, and G. Sangalli. Isogeometric analysis: approximation, stability and error estimates for *h*-refined meshes. *Math. Models Methods Appl. Sci.*, 16 (7): 1031–1090, 2006.
[5] Y. Bazilevs, V. M. Calo, J. A. Cottrell, J. A. Evans, T. J. R. Hughes, S. Lipton, M. A. Scott, and T. W. Sederberg. Isogeometric analysis using T-splines. *Comput. Methods Appl. Mech. Engrg.*, 199 (5–8): 229–263, 2010.
[6] L. Beirão da Veiga, A. Buffa, D. Cho, G. Sangalli. Isogeometric analysis using T-splines on two-patch geometries. *Comput. Methods Appl. Mech. Engrg.*, 200 (21–22): 1787–1803, 2011.
[7] L. Beirão da Veiga, A. Buffa, G. Sangalli, and R. Vazquez. Mathematical analysis of variational isogeometric methods. *Acta Numerica*, 23: 157–287, 2014.
[8] A. Bressan. Some properties of LR-splines. *Comput. Aided Geom. Design*, 30 (8): 778–794, 2013.
[9] A. Buffa and C. Giannelli. Adaptive isogeometric methods with hierarchical splines: error estimator and convergence. *Math. Models Meth. Appl. Sciences*, 26 (01): 1–25, 2016.
[10] J. A. Cottrell, T. Hughes, and Y. Bazilevs. *Isogeometric analysis, toward integration of CAD and FEA*. John Wiley and Sons, Sussex, 2009.
[11] L. Dedè and H. A. F. A. Santos. B-spline goal-oriented error estimators for geometrically nonlinear rods. *Comput. Mech.*, 49 (1): 35–52, 2012.
[12] W. Dörfler. A convergent adaptive algorithm for Poisson's equation. *SIAM J. Numer. Anal.*, 33 (3): 1106–1124, 1996.
[13] T. Dokken, T. Lyche, and K. F. Pettersen. Polynomial splines over locally refined box-partitions. *Comput. Aided Geom. Design*, 30 (3): 331–356, 2013.
[14] M. R. Dörfel, B. Jüttler, and B. Simeon. Adaptive isogeometric analysis by local *h*-refinement with T-splines. *Comput. Methods Appl. Mech. Engrg.*, 199 (5–8): 264–275, 2010.
[15] N. Engleitner and B. Jüttler. Patchwork B-spline refinement. *Computer Aided Design*, 90: 168–179, 2017.
[16] J. A. Evans and T. J. R. Hughes. Explicit trace inequalities for isogeometric analysis and parametric hexahedral finite elements. *Numer. Math.*, 123 (2): 259–290, 2013.
[17] S. Fackler. J.–L. Lions' problem concerning maximal regularity of equations governed by non-autonomous forms. *Ann. Inst. H. Poincaré Anal. Non Linéaire*, 34 (3): 699–709, 2017.
[18] D. R. Forsey and R. H. Bartels. Hierarchical B-spline refinement. *Comput. Graph.*, 22: 205–212, 1988.
[19] A. V. Gaevskaya and S. I. Repin. A posteriori error estimates for approximate solutions of linear parabolic problems. *Differential Equations*, 41 (7): 970–983, 2005.
[20] M. Gander. 50 years of time parallel time integration. In T. Carraro, M. Geiger, S. Körkel, and R. Rannacher, editors, *Multiple shooting and time domain decomposition methods: MuS-TDD, Heidelberg, May 6–8, 2013*, pages 69–113, Springer, Cham, 2015.
[21] M. Gander and M. Neumüller. Analysis of a new space-time parallel multigrid algorithm for parabolic problems. *SIAM J. Sci. Comput.*, 38 (4): A2173–A2208, 2016.
[22] C. Giannelli, B. Jüttler, S. K. Kleiss, A. Mantzaflaris, B. Simeon, and J. Speh. THB-splines: an effective mathematical technology for adaptive refinement in geometric design and isogeometric analysis. *Comput. Methods Appl. Mech. Engrg.*, 299: 337–365, 2016.
[23] C. Giannelli, B. Jüttler, and H. Speleers. THB-splines: the truncated basis for hierarchical splines. *Comput. Aided Geom. Design*, 29 (7): 485–498, 2012.
[24] C. Giannelli, B. Jüttler, and H. Speleers. Strongly stable bases for adaptively refined multilevel spline spaces. *Adv. Comput. Math.*, 40 (2): 459–490, 2014.
[25] A. Mantzaflaris et al. G+Smo (geometry plus simulation modules), v0.8.1, http://gs.jku.at/gismo, 2015.
[26] P. Hansbo. Space-time oriented streamline diffusion methods for nonlinear conservation laws

in one dimension. *Comm. Numer. Meth. Eng.*, 10 (3): 203–215, 1994.
[27] C. Hofer, U. Langer, M. Neumüller, and I. Toulopoulos. Time-multipatch discontinuous Galerkin space-time isogeometric analysis of parabolic evolution problems. *Electr. Trans. Numer. Anal.*, 49: 126–150, 2018.
[28] B. Holm and S. Matculevich. Fully reliable error control for evolutionary problems. *Comput. Math. Appl.*, 75 (4): 1302–1329, 2018.
[29] T. Hughes, J. A. Cottrell, and Y. Bazilevs. Isogeometric analysis: CAD, finite elements, NURBS, exact geometry and mesh refinement. *Comput. Methods Appl. Mech. Engrg.*, 194: 4135–4195, 2005.
[30] K. A. Johannessen. An adaptive isogeometric finite element analysis. Master thesis, Norwegian University of Science and Technology, 2009.
[31] C. Johnson. *Numerical solution of partial differential equations by the finite element method.* Dover Publications, Inc., Mineola, NY, 1987.
[32] C. Johnson and J. Saranen. Streamline diffusion methods for the incompressible Euler and Navier–Stokes equations. *Math. Comput.*, 47 (175): 1–18, 1986.
[33] S. K. Kleiss and S. K. Tomar. Guaranteed and sharp a posteriori error estimates in isogeometric analysis. *Comput. Math. Appl.*, 70 (3): 167–190, 2015.
[34] C. Koutschan, M. Neumüller, and S. Radu. Inverse inequality estimates with symbolic computation. *Adv. Appl. Math.*, 80: 1–23, 2016.
[35] R. Kraft. Adaptive and linearly independent multilevel B-splines. In *Surface fitting and multiresolution methods (Chamonix–Mont–Blanc, 1996)*, pages 209–218, Vanderbilt Univ. Press, Nashville, TN, 1997.
[36] M. Kumar, T. Kvamsdal, and K. A. Johannessen. Simple a posteriori error estimators in adaptive isogeometric analysis. *Comput. Math. Appl.*, 70 (7): 1555–1582, 2015.
[37] G. Kuru. Goal-adaptive isogeometric analysis with hierarchical splines. Master thesis, Mechanical Engineering, Eindhoven University of Technology, 2013.
[38] G. Kuru, C. V. Verhoosel, K. G. van der Zee, and E. H. van Brummelen. Goal-adaptive isogeometric analysis with hierarchical splines. *Comput. Methods Appl. Mech. Engrg.*, 270: 270–292, 2014.
[39] O. A. Ladyzhenskaya. On solvability of classical boundary value problems for equations of parabolic and hyperbolic types. *Dokl. Akad. Nauk SSSR*, 97 (3): 395–398, 1954.
[40] O. A. Ladyzhenskaya. *The boundary value problems of mathematical physics*. Springer, New York, 1985.
[41] U. Langer, S. Matculevich, and S. Repin. A posteriori error estimates for space-time IgA approximations to parabolic initial boundary value problems. arXiv:math.NA/1612.08998, 2016.
[42] U. Langer, S. Matculevich, and S. Repin. Guaranteed error control bounds for the stabilised space–time IgA approximations to parabolic problems. arXiv:math.CS/1712.06017v2, 2017.
[43] U. Langer, S. Moore, and M. Neumüller. Space-time isogeometric analysis of parabolic evolution equations. *Comput. Methods Appl. Mech. Engrg.*, 306: 342–363, 2016.
[44] U. Langer, M. Neumüller, and A. Schafelner. Space-time finite element methods for parabolic evolution problems with variable coefficients. In T. Apel, U. Langer, A. Meyer, and O. Steinbach, editors, *Advanced finite element methods with applications. Selected papers of the 30th Chemnitz FEM Symposium 2017*, Volume 128 of Lecture Notes in Computational Science and Engineering, pages 247–275, Springer, 2019.
[45] O. Mali, P. Neittaanmäki, and S. Repin. *Accuracy verification methods*. Volume 32 of Computational Methods in Applied Sciences, Springer, Dordrecht, 2014.
[46] S. Matculevich. Functional approach to the error control in adaptive IgA schemes for elliptic boundary value problems. *J. Comput. Appl. Math.*, 344: 394–423, 2018.

[47] S. Matculevich and S. Repin. Computable estimates of the distance to the exact solution of the evolutionary reaction–diffusion equation. *Appl. Math. Comput.*, 247: 329–347, 2014.

[48] N. Nguyen–Thanh, H. Nguyen–Xuan, S. P. Bordas, and T. Rabczuk. Isogeometric analysis using polynomial splines over hierarchical T-meshes for two-dimensional elastic solids. *Comput. Methods Appl. Mech. Engrg.*, 200 (21–22): 1892–1908, 2011.

[49] L. Piegl and W. Tiller. *The NURBS book*. Springer, Berlin, Heidelberg, 1997.

[50] S. I. Repin. A posteriori error estimation for nonlinear variational problems by duality theory. *Zapiski Nauchnych Seminarov POMIs*, 243: 201–214, 1997.

[51] S. I. Repin. A posteriori error estimates for approximate solutions to variational problems with strongly convex functionals. *J. Math. Sci.*, 97: 4311–4328, 1999.

[52] S. I. Repin. Estimates of deviations from exact solutions of initial–boundary value problem for the heat equation. *Rend. Mat. Acc. Lincei*, 13 (9): 121–133, 2002.

[53] S. I. Repin. *A posteriori estimates for partial differential equations*. Volume 4 of Radon Series on Computational and Applied Mathematics, de Gruyter, Berlin, 2008.

[54] S. I. Repin and S. K. Tomar. A posteriori error estimates for approximations of evolutionary convection–diffusion problems. *J. Math. Sci.*, 170 (4): 554–566, 2010.

[55] C. Schwab. *p- and hp-finite element methods, theory and applications in solid and fluid mechanics*. The Clarendon Press, Oxford University Press, New York, 1998.

[56] M. A. Scott, M. J. Borden, C. V. Verhoosel, T. W. Sederberg, and T. J. R. Hughes. Isogeometric finite element data structures based on Bézier extraction of T-splines. *Internat. J. Numer. Meth. Engrg.*, 88 (2): 126–156, 2011.

[57] M. A. Scott, X. Li, T. W. Sederberg, and T. J. R. Hughes. Local refinement of analysis-suitable T-splines. *Comput. Methods Appl. Mech. Engrg.*, 213/216: 206–222, 2012.

[58] T. W. Sederberg, J. Zheng, A. Bakenov, and A. Nasri. T-splines and t-nurccs. *ACM Trans. Graphics*, 22 (3): 477–484, 2003.

[59] T. W. Sederberg, D. C. Cardon, G. T. Finnigan, N. N. North, J. Zheng, and T. Lyche. T-splines simplification and local refinement. *ACM Trans. Graphics*, 23 (3): 276–283, 2004.

[60] O. Steinbach and H. Yang. Space-time finite element methods for parabolic evolution equations: Discretization, a posteriori error estimation, adaptivity and solution. In U. Langer and O. Steinbach, editors, *Space-Time Methods: Application to partial differential equations*, Volume 25 of Radon Series on Computational and Applied Mathematics, pages 207–248, de Gruyter, Berlin, 2019.

[61] K. Takizawa and T. E. Tezduyar. Multiscale space–time fluid–structure interaction techniques. *Comput. Mech.*, 48 (3): 247–267, 2011.

[62] K. Takizawa and T. E. Tezduyar. Space–time computation techniques with continuous representation in time. *Comput. Mech.*, 53 (1): 91–99, 2014.

[63] K. G. van der Zee and C. V. Verhoosel. Isogeometric analysis-based goal-oriented error estimation for free-boundary problems. *Finite Elem. Anal. Des.*, 47 (6): 600–609, 2011.

[64] A.-V. Vuong, C. Giannelli, B. Jüttler, and B. Simeon. A hierarchical approach to adaptive local refinement in isogeometric analysis. *Comput. Methods Appl. Mech. Engrg.*, 200 (49–52): 3554–3567, 2011.

[65] P. Wang, J. Xu, J. Deng, and F. Chen. Adaptive isogeometric analysis using rational pht-splines. *Computer Aided Design*, 43 (11): 1438–1448, 2011.

Martin Neumüller and Elias Karabelas

6 Generating admissible space-time meshes for moving domains in $(d+1)$ dimensions

Abstract: In this paper, we present a discontinuous Galerkin finite element method for the solution of the transient Stokes equations on moving domains. For the discretization, we use an interior penalty Galerkin approach in space, and an upwind technique in time. The method is based on a decomposition of the space-time cylinder into finite elements. Our focus lies on three-dimensional moving geometries, thus we need to triangulate four dimensional objects. For this, we will present an algorithm to generate $(d+1)$-dimensional simplex space-time meshes, and we show under natural assumptions that the resulting space-time meshes are admissible. Further, we will show how one can generate a four-dimensional object resolving the domain movement. First numerical results for the transient Stokes equations on triangulations generated with the newly developed meshing algorithm are presented.

Keywords: finite elements, moving domains, four-dimensional mesh generation, parabolic PDE, space-time, discontinuous Galerkin

MSC 2010: 65N30, 65L50

1 Introduction

The finite element approximation of transient partial differential equations is, in most cases, based on explicit or implicit time discretization schemes. In particular, the simultaneous consideration of different time steps requires an appropriate interpolation to couple the solutions at different time levels. Especially for spatial domains with a moving boundary, one encounters various numerical difficulties. One usually relies on an arbitrary Lagrangian–Eulerian formulation. See, for example, the article [16] and references therein for an overview of the ongoing discussion. In this paper, we consider the application of finite elements in the whole space-time cylinder Q. We decompose the space-time domain Q into simplicial elements and apply stable discretization schemes. This strategy replaces the problem of time discretization by a meshing problem, and we can resolve the possible movement of the domain Ω directly. We focus

Acknowledgement: This research was supported by the grant F3210–N18 from the Austrian Science Fund (FWF).

Martin Neumüller, Institute of Computational Mathematics, Johannes Kepler University Linz, Altenberger Strasse 69, 4040 Linz, Austria, e-mail: martin@neumuller.at
Elias Karabelas, Institute of Biophysics, Medical University of Graz, Neue Stiftingtalstraße 6/D04, 8010 Graz, Austria, e-mail: elias.karabelas@medunigraz.at

https://doi.org/10.1515/9783110548488-006

on simplicial space-time meshes, since it is easier to decompose complex space-time meshes by using those kind of elements.

Space-time finite element methods have been applied to several parabolic model problems. Least square methods for convection-diffusion problems are considered, for example, in [5, 7] and for flow problems, for example, in [21, 26, 30, 31, 32, 33]. Discontinuous Galerkin finite element methods have been applied to solve transient convection-diffusion problems in [29]; for fluid dynamics see [35], and for problems from solid mechanics, see [1, 2, 13, 22]. The generation of 4D simplicial meshes from a sequence of 3D MRI data has been considered in [15]. Rather recently, the X-FEM method has been considered in the space-time setting, see [18]. In most cases, the time dependent equation is discretized in the space-time domain on *space-time slabs*. These all-at-once discretization concepts allow for local mesh refinement in the space-time domain, see for example [28, 30].

The construction of simplicial space-time elements is considered in [7, 8, 17] and [36, 37]. These algorithms allow the construction of complex space-time domains, which are interesting, especially for moving spatial domains. In these articles, no theory for the presented algorithms is provided. That is, it may not be guaranteed that these algorithms terminate, or if they result in an admissible decomposition. Moreover, these algorithms use random perturbations to ensure an admissible decomposition, which will lead to different space-time meshes for different runtimes. In this work, we present a very simple and new algorithm, which works in any dimension, and in addition we provide an analysis and show that the resulting space-time meshes are admissible under some natural conditions. Moreover, this algorithm is suited for parallel meshing, since no communication is needed during the extrusion process.

We will consider Stokes flow as a motivating model problem. For the approximation of the transient Stokes equations in the space-time cylinder, we consider a discontinuous Galerkin finite element method. In particular, we apply an interior penalty approach in space [3, 6, 11, 27], and an upwind technique in time [24, 34].

This paper is organized as follows: In Section 2, we describe the discontinuous Galerkin finite element method to solve the transient Stokes equations as a model problem. The core part of this paper and the main results are given in Section 3, where we describe our algorithm to generate a four-dimensional triangulation out of a given three-dimensional one. In Section 4, we present some numerical results, which underline the applicability of the proposed approach. We close the paper with some conclusions and comments on further work.

2 Space-time discontinuous Galerkin method

For any $t \in (0,T)$, let $\Omega(t) \subsetneq \mathbb{R}^d$ with $d = 1,2,3$ be a bounded Lipschitz domain with boundary $\Gamma(t) := \partial\Omega(t)$. We assume that the boundary $\Gamma(t)$ admits the following decomposition for every $t \in (0, T)$:

$$\Gamma(t) = \Gamma_D(t) \cup \Gamma_R(t). \tag{2.1}$$

We assume that the movement of the domain $\Omega(t)$ is known for every $t \in [0, T]$. We define the space-time cylinder Q as

$$Q := \{(\boldsymbol{x}, t) \in \mathbb{R}^{d+1} : \boldsymbol{x} \in \Omega(t), t \in (0, T)\},$$

and the space-time mantle Σ as

$$\Sigma := \{(\boldsymbol{x}, t) \in \mathbb{R}^{d+1} : \boldsymbol{x} \in \Gamma(t), t \in (0, T)\}.$$

The decomposition (2.1) induces

$$\Sigma = \Sigma_D \cup \Sigma_R.$$

The model problem we intend to study is governed by the transient Stokes equations. It reads as: find $(\boldsymbol{u}, p)$ such that

$$\begin{aligned} \frac{\partial}{\partial t}\boldsymbol{u} - \nu\Delta\boldsymbol{u} + \nabla p = \boldsymbol{f} &\quad \text{in } Q,\\ \operatorname{div}\boldsymbol{u} = 0 &\quad \text{in } Q,\\ \boldsymbol{u} = \boldsymbol{g}_D &\quad \text{on } \Sigma_D,\\ \nabla\boldsymbol{u}\cdot\boldsymbol{n} + \alpha_R\boldsymbol{u} - p\boldsymbol{n} = \boldsymbol{g}_R &\quad \text{on } \Sigma_R,\\ \boldsymbol{u} = \boldsymbol{u}_0 &\quad \text{on } \Sigma_0 := \Omega(0). \end{aligned} \tag{2.2}$$

Remark 2.1. In the case of a nonmoving domain, the definition of Q and Σ simplifies to

$$Q := \Omega \times (0, T), \quad \Sigma := \partial\Omega \times (0, T).$$

For deriving a discrete variational formulation, we need to decompose the space-time cylinder Q into simplicial elements, see [25]. Let $\mathcal{T}_h$ be a sequence of decompositions

$$\overline{Q} = \overline{\mathcal{T}_h} = \bigcup_{k=1}^{N} \tau_k$$

into (closed) finite elements of mesh size h_k. For $d = 1$, we have triangles; for $d = 2$, we use tetrahedrons, and for $d = 3$, pentatopes are chosen. The generation of such triangulations from a given triangulation of $\Omega(0)$ is not trivial. We will address this topic in Section 3.

Definition 2.1 (Admissible decomposition). A decomposition $\mathcal{T}_h$ is called *admissible* iff the nonempty intersection of two finite elements is either an edge (for $d = 1, 2, 3$), a triangle (for $d = 2, 3$), or a tetrahedron (for $d = 3$).

It is worth noting that discontinuous Galerkin methods are not restricted to admissible decompositions. However, additional assumptions are required, see [10].

Definition 2.2 (Interior facet). Let $\mathcal{T}_h$ be a decomposition of Q into finite elements τ_k. For two neighboring elements $\tau_k, \tau_l \in \mathcal{T}_h$, we call

$$\Gamma_{kl} := \tau_k \cap \tau_l$$

an *interior facet* iff Γ_{kl} is a d-dimensional manifold. The set of all interior facets will be defined as $\mathcal{I}_h$.

Any interior element Γ_{kl} has an exterior normal vector $\boldsymbol{n}_{kl}$ with a nonunique direction. We fix the direction of the normal vector such that $\boldsymbol{n}_{kl}$ is the exterior normal vector of the element τ_k when $k < l$. So the direction of the normal vector $\boldsymbol{n}_{kl}$ depends on the ordering of the finite elements, but the variational formulation, which we are going to use, will be independent of this ordering.

Definition 2.3. Let $\Gamma_{kl} \in \mathcal{I}_h$ be an interior facet with outer normal $\boldsymbol{n}_k = (\boldsymbol{n}_{\boldsymbol{x},k}, n_{t,k})^\top \in \mathbb{R}^{d+1}$ for τ_k and $\boldsymbol{n}_l = -\boldsymbol{n}_k$ for τ_l. For a given function ϕ smooth enough restricted to either τ_k or τ_l, we define on Γ_{kl}:

- The *jump* across Γ_{kl} as

$$[\![\phi]\!]_{kl} := \phi_{|\tau_k} \boldsymbol{n}_k + \phi_{|\tau_l} \boldsymbol{n}_l.$$

- The *space jump* across Γ_{kl} as

$$[\![\phi]\!]_{\boldsymbol{x},kl} := \phi_{|\tau_k} \boldsymbol{n}_{\boldsymbol{x},k} + \phi_{|\tau_l} \boldsymbol{n}_{\boldsymbol{x},l}.$$

- The *time jump* across Γ_{kl} as

$$[\![\phi]\!]_{t,kl} := \phi_{|\tau_k} n_{t,k} + \phi_{|\tau_l} n_{t,l}.$$

- The *average* of ϕ on Γ_{kl} as

$$\langle \phi \rangle_{kl} := \frac{1}{2}(\phi_{|\tau_k} + \phi_{|\tau_l}).$$

- The *upwind* in time direction of ϕ is defined as

$$\{\phi\}_{kl}^{\mathrm{up}} := \begin{cases} \phi_{|\tau_k} & \text{if } n_{k,t} > 0, \\ 0 & \text{if } n_{k,t} = 0, \\ \phi_{|\tau_l} & \text{if } n_{k,t} < 0. \end{cases}$$

Let $p, q \in \mathbb{N}_0$. Then we define the spaces of piecewise polynomials:

$$V_h^p := \{\boldsymbol{v}_h \in [L^2(Q)]^d : \boldsymbol{v}_{h|\tau_l} \in [\mathbb{P}_p(\tau_l)]^d \text{ for all } \tau_l \in \mathcal{T}_h, \boldsymbol{v}_{h|\Sigma_D} = \boldsymbol{0}\},$$
$$Q_h^q := \{q_h \in L^2(Q) : q_{h|\tau_l} \in \mathbb{P}_q(\tau_l) \text{ for all } \tau_l \in \mathcal{T}_h\}.$$

Inspired by works in [11, 24], we will use the following bilinear form defined for $\boldsymbol{u}_h, \boldsymbol{v}_h \in V_h^p(\mathcal{T}_h)$:

$$a(\boldsymbol{u}_h, \boldsymbol{v}_h) := a_t(\boldsymbol{u}_h, \boldsymbol{v}_h) + a_x(\boldsymbol{u}_h, \boldsymbol{v}_h).$$

The individual components read as follows:

$$\begin{aligned} a_x(\boldsymbol{u}_h, \boldsymbol{v}_h) := {} & \nu \sum_{l=1}^{N} \int_{\tau_l} \nabla_{\boldsymbol{x}} \boldsymbol{u}_h : \nabla_{\boldsymbol{x}} \boldsymbol{v}_h \, dq - \nu \sum_{\Gamma_{kl} \in \mathcal{I}_h} \int_{\Gamma_{kl}} \langle \nabla_{\boldsymbol{x}} \boldsymbol{u}_h \rangle_{kl} \boldsymbol{v}_{h,kl,\boldsymbol{x}} \, ds_q \\ & - \nu \sum_{\Gamma_{kl} \in \mathcal{I}_h} \int_{\Gamma_{kl}} \langle \nabla_{\boldsymbol{x}} \boldsymbol{v}_h \rangle_{kl} \boldsymbol{u}_{h,kl,\boldsymbol{x}} \, ds_q \\ & + \sum_{\Gamma_{kl} \in \mathcal{I}_h} \frac{\sigma_u}{\overline{h}_{kl}} \int_{\Gamma_{kl}} \boldsymbol{u}_{h,kl,\boldsymbol{x}} \boldsymbol{v}_{h,kl,\boldsymbol{x}} \, ds_q + \int_{\Sigma_R} \alpha_R(\boldsymbol{x}, t) \boldsymbol{u}_h \cdot \boldsymbol{v}_h \, ds_q, \end{aligned}$$

and

$$\begin{aligned} a_t(\boldsymbol{u}_h, \boldsymbol{v}_h) := {} & -\sum_{l=1}^{N} \int_{\tau_l} \boldsymbol{u}_h \cdot \frac{\partial}{\partial t} \boldsymbol{v}_h \, dq + \int_{\Sigma_T} \boldsymbol{u}_h \cdot \boldsymbol{v}_h \, ds_q \\ & + \sum_{\Gamma_{kl} \in \mathcal{I}_h} \int_{\Gamma_{kl}} \{\boldsymbol{u}_h\}^{\mathrm{up}} \boldsymbol{v}_{h,kl,\boldsymbol{x}} \, ds_q, \end{aligned}$$

for a given velocity stabilization parameter $\sigma_u > 0$. Furthermore, we define the following pressure bilinear forms for $\boldsymbol{v}_h \in V_h^p$ and $p_h, q_h \in Q_h^q$:

$$\begin{aligned} b(\boldsymbol{v}_h, p_h) &:= \sum_{l=1}^{N} \int_{\tau_l} p_h \operatorname{div} \boldsymbol{v}_h \, dq - \sum_{\Gamma_{kl}} \int_{\Gamma_{kl}} \langle p_h \rangle_{kl} \boldsymbol{v}_{h,kl,\boldsymbol{x}} \, ds_q, \\ c(p_h, q_h) &:= \sum_{\Gamma_{kl} \in \mathcal{I}_h} \sigma_p \overline{h}_{kl} \int_{\Gamma_{kl}} p_{h,kl,\boldsymbol{x}} q_{h,kl,\boldsymbol{x}} \, ds_q, \end{aligned}$$

for a given pressure stabilization parameter σ_p. In all the bilinear forms defined above, we have used $\overline{h}_{kl} := \frac{1}{2}(h_k + h_l)$. Hence, we have to find $\boldsymbol{u}_h^0 \in V_h^p$ and $p_h \in Q_h^q$, such that

$$a(\boldsymbol{u}_h^0, \boldsymbol{v}_h) - b(\boldsymbol{v}_h, p_h) = \langle \boldsymbol{f}, \boldsymbol{v}_h \rangle_Q + \langle \boldsymbol{u}_0, \boldsymbol{v}_h(0) \rangle_\Omega - a(\boldsymbol{u}_g^h, \boldsymbol{v}_h), \tag{2.3}$$

$$b(\boldsymbol{u}_h^0, q_h) + c(p_h, q_h) = -b(\boldsymbol{u}_g^h, q_h). \tag{2.4}$$

Here we used a discrete extension $\boldsymbol{u}_g^h$ of the given Dirichlet data, whereas in our experiments we will use a $L^2(\Sigma_D)$ projection. The stability of the bilinear form $a(\cdot,\cdot)$ has been shown in the PhD thesis [24]. The inf-sup condition for $b(\cdot,\cdot)$ is still an open problem, but all the numerical experiments indicate that using $p = q + 1$, $q \in \mathbb{N}_0$ results in a stable element.

3 Triangulations in $d + 1$ dimensions

In this section, we will introduce an algorithm to decompose a hyperprism into simplices to generate a $(d + 1)$ simplex space-time mesh. Moreover, we will show that the resulting mesh is admissible if the nodes of the simplices from the initial mesh are sorted in a special way.

3.1 Tensor product extensions

A simple idea for constructing a space-time mesh for a given three-dimensional simplicial spatial mesh is to extrude the mesh in time direction by a tensor product extension; see also Figure 6.1. Afterwards, we decompose the resulting prisms or so-called hyperprisms into simplicial elements. For moving domains, the idea is to move the shifted nodes, accordingly, to the given movement.

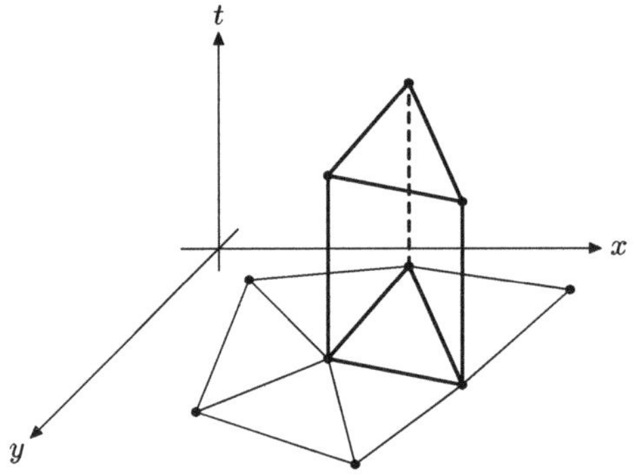

Figure 6.1: Tensor extension of a two-dimensional simplex.

Before we can start, we need a precise definition of a d-dimensional simplex.

Definition 3.1 (d-dimensional simplex). Let

$$\{\boldsymbol{p}_1, \dots, \boldsymbol{p}_{d+1}\} \subset \mathbb{R}^d,$$

$d \in \mathbb{N}$, be a set of nodes; then a d-dimensional simplex S_d is defined as

$$S_d := [\boldsymbol{p}_1, \dots, \boldsymbol{p}_{d+1}] := \operatorname{conv}(\{\boldsymbol{p}_1, \dots, \boldsymbol{p}_{d+1}\}),$$

where $\operatorname{conv}(\cdot)$ is the convex hull of a set of nodes. Note that we also fix the ordering of the nodes in the definition of a d-dimensional simplex.

Now we can extrude one simplex in time direction, and we obtain the following definition of a hyperprism:

Definition 3.2 (Hyperprism). For a given simplex $S_d = [\boldsymbol{p}_1, \dots, \boldsymbol{p}_{d+1}]$, the tensor product extension in time direction for a given time interval $[0, \tau]$, or the so called hyperprism H_{d+1}, is given by

$$P_{d+1} := [\boldsymbol{p}_1, \dots, \boldsymbol{p}_{d+1}; \tau] := \operatorname{conv}(\{\boldsymbol{p}'_1, \dots, \boldsymbol{p}'_{d+1}, \boldsymbol{p}''_1, \dots, \boldsymbol{p}''_{d+1}\}) \subset \mathbb{R}^{d+1},$$

with

$$\boldsymbol{p}'_i := (\boldsymbol{p}_i^\top, 0)^\top, \quad \boldsymbol{p}''_i := (\boldsymbol{p}_i^\top, \tau)^\top,$$

for $i = 1, \dots, d+1$.

3.2 Decomposing hyperprisms

In this section, we will give an algorithm to decompose the hyperprisms given in Definition 3.2 into simplices.

Definition 3.3 (Decomposed hyperprism). Let S_d be a given simplex and P_{d+1} the hyperprism with respect to the simplex S_d and $\tau > 0$. Then we define the following simplices:

$$\begin{aligned}
S^1_{d+1} &:= [\boldsymbol{p}'_1, \boldsymbol{p}'_2, \boldsymbol{p}'_3, \dots, \boldsymbol{p}'_{d+1}, \boldsymbol{p}''_1],\\
S^2_{d+1} &:= [\boldsymbol{p}'_2, \boldsymbol{p}'_3, \dots, \boldsymbol{p}'_{d+1}, \boldsymbol{p}''_1, \boldsymbol{p}''_2],\\
S^3_{d+1} &:= [\boldsymbol{p}'_3, \dots, \boldsymbol{p}'_{d+1}, \boldsymbol{p}''_1, \boldsymbol{p}''_2, \boldsymbol{p}''_3],\\
&\ \vdots\\
S^{d+1}_{d+1} &:= [\boldsymbol{p}'_{d+1}, \boldsymbol{p}''_1, \boldsymbol{p}''_2, \boldsymbol{p}''_3, \dots, \boldsymbol{p}''_{d+1}].
\end{aligned} \tag{3.1}$$

Furthermore, we define the set of simplices $\mathcal{T}_P(S_d, \tau) := \{S^1_{d+1}, \dots, S^{d+1}_{d+1}\}$.

Note that the ordering of the nodes of a hyperprism P_{d+1} is essential for the resulting decomposition (3.1). To ensure that the simplices $S^1_{d+1}, \dots, S^{d+1}_{d+1}$, defined in (3.1), decompose the hyperprism P_{d+1}, we need the following lemma:

Lemma 3.1. *Let P_{d+1} be some given hyperprism with respect to the simplex S_d and $\tau > 0$. Then the set of simplices*

$$\mathcal{T}_P(S_d, \tau) = \{S^1_{d+1}, \dots, S^{d+1}_{d+1}\},$$

defined in (3.1), is an admissible decomposition of the hyperprism P_{d+1}.

Proof. By construction, the set of simplices $\mathcal{T}_P(S_d, \tau) = \{S^1_{d+1}, \dots, S^{d+1}_{d+1}\}$ is admissible. Furthermore, every simplex S^i_{d+1} for $i = 1, \dots, d+1$ is contained in the hyperprism P_{d+1} since P_{d+1} is convex. It remains to show that the union of all simplices $\mathcal{T}_P(S_d, \tau)$

is equal to the hyperprism. That is, we have to show that the volume of the union of all simplices $\mathcal{T}_P(S_d, \tau)$ coincides with the volume of the hyperprism. To do so, we transform the hyperprism P_{d+1} to a reference hyperprism $\hat{P}_{d+1}$, where we easily can compute all the volume terms. For this, we define the reference simplex $\hat{S}_d \subset \mathbb{R}^d$ as

$$\hat{S}_d := [\boldsymbol{e}_0, \boldsymbol{e}_1, \dots, \boldsymbol{e}_d] = \operatorname{conv}(\{\boldsymbol{e}_0, \boldsymbol{e}_1, \dots, \boldsymbol{e}_d\}),$$

with

$$\begin{aligned} \boldsymbol{e}_0 &:= (0, 0, \dots, 0, 0)^\top, \\ \boldsymbol{e}_1 &:= (1, 0, \dots, 0, 0)^\top, \\ \boldsymbol{e}_2 &:= (0, 1, \dots, 0, 0)^\top, \\ &\vdots \\ \boldsymbol{e}_d &:= (0, 0, \dots, 0, 1)^\top. \end{aligned}$$

Then we define the reference hyperprism $\hat{P}_{d+1}$ as

$$\hat{P}_{d+1} := [\boldsymbol{e}_0, \dots, \boldsymbol{e}_{d+1}; 1].$$

With the standard affine transformation, we have a bijective mapping between the reference hyperprism $\hat{P}_{d+1}$ and the hyperprism P_{d+1}. This affine transformation consists of the standard transformation for d-dimensional simplices and a scaling in time direction. So we only have to compare the volume for the reference hyperprism. Now, the volume of the reference simplex $\hat{S}_d$ is given by $|\hat{S}_d| = \frac{1}{d!}$. Hence, the volume of the reference hyperprism is

$$|\hat{P}_{d+1}| = \frac{1}{d!}.$$

The simplices of our decomposition in the reference domain are given by

$$\begin{aligned} \hat{S}^1_{d+1} &:= [\boldsymbol{e}'_0, \boldsymbol{e}'_1, \boldsymbol{e}'_2, \dots, \boldsymbol{e}'_d, \boldsymbol{e}''_0], \\ \hat{S}^2_{d+1} &:= [\boldsymbol{e}'_1, \boldsymbol{e}'_2, \dots, \boldsymbol{e}'_d, \boldsymbol{e}''_0, \boldsymbol{e}''_1], \\ \hat{S}^3_{d+1} &:= [\boldsymbol{e}'_2, \dots, \boldsymbol{e}'_d, \boldsymbol{e}''_0, \boldsymbol{e}''_1, \boldsymbol{e}''_2], \\ &\vdots \\ \hat{S}^{d+1}_{d+1} &:= [\boldsymbol{e}'_d, \boldsymbol{e}''_0, \boldsymbol{e}''_1, \boldsymbol{e}''_2, \dots, \boldsymbol{e}''_d]. \end{aligned}$$

It is easy to see that these simplices have the same volume; That is,

$$|\hat{S}^i_{d+1}| = \frac{1}{(d+1)!}, \qquad \text{for } i = 1, \dots, d+1.$$

Hence, we have

$$\left| \bigcup_{i=1}^{d+1} \hat{S}^i_{d+1} \right| = (d+1)\frac{1}{(d+1)!} = \frac{1}{d!} = |\hat{P}_{d+1}|,$$

which completes the proof. □

3.3 Admissible tensor product triangulations

For a given d-dimensional triangulation $\mathcal{T}_h$, we now want to construct a tensor product extension by applying the algorithm (3.1) for every simplex of the simplicial mesh $\mathcal{T}_h$. With Lemma 3.1, we know that every hyperprism can be admissibly decomposed into simplex elements. In this section, we want to formulate conditions such that the overall space-time mesh is admissible. To this end, a special ordering of the nodes is required.

Definition 3.4 (Consistently numbered). Let

$$\mathcal{T}_h = \{S_d^i : S_d^i = [\boldsymbol{p}_1^i, \ldots, \boldsymbol{p}_{d+1}^i]\}$$

be an admissible d-dimensional simplex mesh. Then $\mathcal{T}_h$ is called *consistently numbered*, iff for any two simplices $S_d^i, S_d^j \in \mathcal{T}_h$ with nonempty intersection; that is, $S_d^i \cap S_d^j \neq \emptyset$, there exist indices $k_1 < \cdots < k_n$ and $\ell_1 < \cdots < \ell_n$ with $n \in \mathbb{N}$, $n \leq d+1$, such that

$$S_d^i \cap S_d^j = [\boldsymbol{p}_{k_1}^i, \ldots, \boldsymbol{p}_{k_n}^i] \equiv [\boldsymbol{p}_{\ell_1}^j, \ldots, \boldsymbol{p}_{\ell_n}^j].$$

Here "=" means that the two sets are the same and "≡" means that the two sets are equal, and that also the numbering of the nodes is the same. That is, we have $\boldsymbol{p}_{k_1}^i = \boldsymbol{p}_{\ell_1}^j, \ldots, \boldsymbol{p}_{k_n}^i = \boldsymbol{p}_{\ell_n}^j$.

The definition of a consistently numbered triangulation can also be found in [9], and it is important for the refinement of d-dimensional simplices, especially for $d \geq 4$. If an admissible mesh is consistently numbered, we can prove the following Theorem:

Theorem 3.1. *Let $\mathcal{T}_h$ be a consistently numbered, admissible d-dimensional triangulation. Given $\tau > 0$, let*

$$\mathcal{T}_{h,\tau} := \{\mathcal{T}_P(S_d, \tau) : S_d \in \mathcal{T}_h\}$$

be the $(d+1)$–dimensional simplex mesh, resulting by decomposing every hyperprism with the algorithm given in (3.1). *Then the space-time mesh $\mathcal{T}_{h,\tau}$ is admissible.*

Proof. With Lemma 3.1, we know that every hyperprism is decomposed admissible into simplices. To obtain a global admissible mesh, we have to prove that the tensor product triangulations $\mathcal{T}_P(S_d^i, \tau)$ and $\mathcal{T}_P(S_d^j, \tau)$ for each neighboring elements $S_d^i, S_d^j \in \mathcal{T}_h$ are matching. Let

$$S_d^i = [\boldsymbol{p}_1^i, \ldots, \boldsymbol{p}_{d+1}^i] \quad \text{and} \quad S_d^j = [\boldsymbol{p}_1^j, \ldots, \boldsymbol{p}_{d+1}^j]$$

with $S_d^i, S_d^j \in \mathcal{T}_h$ be some neighboring simplices and

$$P_d := P_{d+1}^i \cap P_{d+1}^j,$$

with

$$P^i_{d+1} := [\boldsymbol{p}^i_1, \ldots, \boldsymbol{p}^i_{d+1}; \tau] \quad \text{and} \quad P^j_{d+1} := [\boldsymbol{p}^j_1, \ldots, \boldsymbol{p}^j_{d+1}; \tau]$$

be the intersecting hyperprisms, and

$$\mathcal{T}^i_P := \{S_{d+1} \cap P_d : S_{d+1} \in \mathcal{T}_P(S^i_d, \tau)\},$$
$$\mathcal{T}^j_P := \{S_{d+1} \cap P_d : S_{d+1} \in \mathcal{T}_P(S^j_d, \tau)\}$$

be the corresponding triangulations of P_d, obtained by $\mathcal{T}_P(S^i_d, \tau)$ and $\mathcal{T}_P(S^j_d, \tau)$. It remains to show that the intersecting hyperprism P_d is decomposed in the same way from both sides, meaning that $\mathcal{T}^i_P = \mathcal{T}^j_P$. Since $\mathcal{T}_h$ is consistently numbered, there exists indices $k_1 < \cdots < k_n$ and $\ell_1 < \cdots < \ell_n$ with $n = d$, such that

$$S^i_d \cap S^j_d = S^i_{d-1} := [\boldsymbol{p}^i_{k_1}, \ldots, \boldsymbol{p}^i_{k_n}] \equiv S^j_{d-1} := [\boldsymbol{p}^j_{\ell_1}, \ldots, \boldsymbol{p}^j_{\ell_n}]. \tag{3.2}$$

Therefore, the intersecting simplex $S^i_d \cap S^j_d$ is obtained by simply removing the nodes from S^i_d or S^j_d, which are not shared together and furthermore they have the same ordering of the nodes. For the intersecting hyperprism P_d, the decompositions from both sides $\mathcal{T}^i_P$ and $\mathcal{T}^j_P$ are given by removing the nodes, which are not shared together from the formula (3.1) and, with (3.2), we have

$$\mathcal{T}^i_P = \mathcal{T}_P(S^i_{d-1}, \tau) \quad \text{and} \quad \mathcal{T}^j_P = \mathcal{T}_P(S^j_{d-1}, \tau).$$

Since in equation (3.2) also the node ordering of S^i_{d-1} and S^j_{d-1} is the same, we also obtain

$$\mathcal{T}_P(S^i_{d-1}, \tau) = \mathcal{T}_P(S^j_{d-1}, \tau),$$

which implies that $\mathcal{T}^i_P = \mathcal{T}^j_P$. □

Remark 3.1. To obtain an admissible space-time mesh $\mathcal{T}_{h,\tau}$, we only have to ensure that the nodes of the spatial mesh $\mathcal{T}_h$ are consistently numbered. This can be achieved by sorting the local nodes of each simplex $S_d \in \mathcal{T}_h$ with respect to the global node numbers.

3.4 Tensor product triangulations for moving domains

If the movement of a computational domain is known in advance, we can generate admissible space-time meshes by applying the methods from above. The idea is to move the points at the top of the tensor-product extension. Assuming that the displacement of points on the boundary $\Gamma(t)$ is governed by a function

$$\boldsymbol{g}_{\text{mov}}(\boldsymbol{X}, t) \colon \Gamma(0) \times (0, T) \to \mathbb{R}^d,$$

then a point $\boldsymbol{x} \in \Gamma(t)$ can be written as

$$\boldsymbol{x} = \boldsymbol{X} + \boldsymbol{g}_{\text{mov}}(\boldsymbol{X}, t),$$

where $\boldsymbol{X} \in \Gamma(0)$. Here, we assume that $\boldsymbol{g}_{\text{mov}}$ is continuous to be able to extend $\boldsymbol{g}_{\text{mov}}$ to the whole space-time domain Q. Recall the definition of a hyperprism in Definition 3.2. Instead of using $\boldsymbol{p}_i'' := (\vec{p}_i^{\top}, \tau)^{\top}$ on the surface, we can apply the displacement and use $\boldsymbol{p}_i'' := (\boldsymbol{p}_i^{\top} + \boldsymbol{g}_{\text{mov}}(\boldsymbol{p}_i, \tau)^{\top}, \tau)^{\top}$ for all boundary points of the simplex mesh that are subject to a movement. The remaining generation of the 4D mesh stays untouched. For boundary movements that are of small magnitude and do not change the topology of the initial geometry, this can be sufficient. For stronger, yet topology-preserving, movements, this concept would create degenerating simplex elements. A remedy to this is to use the movement $\boldsymbol{g}_{\text{mov}}$ as Dirichlet datum for a vector Laplacian or a linear elasticity problem. Then the resulting displacement is applied to all simplex points in the domain. For more on mesh smoothing, we refer to [19, 20].

In the case of stronger displacements or even topology changes, remeshing would be required, and we need further meshing algorithms to connect different spatial domains in space and time. Especially for four-dimensional space-time meshes, this remains a future research topic.

If the movement of the computational domain is not known in advance, we can solve the problem on a coarse spatial grid with coarse time steps to obtain a coarse approximation for the movement. Afterwards, we can construct the coarse space-time mesh with the methods given in this work. By using adaptive schemes in space and time, we further can refine the space-time domain adaptively and move the points in the space-time domain by the computed finer approximations. Note that the movement of the points has to be only done in the range of the approximation error, which is usually small. Of course this is also considered as a further research topic.

3.5 Visualization

Here we want to address the issue of visualizing results for four-dimensional triangulations $\mathcal{T}_h$. In applications, it is desired to visualize results at given time instances $t_k \in [0, T]$. The main idea is to cut the decomposition $\mathcal{T}_h$ with several planes to obtain a finite number of three-dimensional manifolds. For this, we need to have a hyperplane to calculate the intersections with the decomposition.

Definition 3.5 (Hyperplane). Let $\boldsymbol{p}_0 \in \mathbb{R}^4$ be arbitrary, and let $\boldsymbol{p}_1, \boldsymbol{p}_2, \boldsymbol{p}_3$, and $\boldsymbol{p}_4 \in \mathbb{R}^4$ be linearly independent. Then the set

$$H_4 := \{\boldsymbol{x} \in \mathbb{R}^4 : \boldsymbol{x} = \boldsymbol{p}_0 + \mu_1 \boldsymbol{p}_1 + \mu_2 \boldsymbol{p}_2 + \mu_3 \boldsymbol{p}_3 \text{ for } \mu_1, \mu_2, \mu_3 \in \mathbb{R}\}$$

is called a *hyperplane*.

To cut a given decomposition $\mathcal{T}_h$ with a hyperplane H_4, we have to cut every element $\tau_k \in \mathcal{T}_h$ with the hyperplane. For this, we have to calculate for every edge $e_i = (\boldsymbol{x}_{i_1}, \boldsymbol{x}_{i_2})$, $i = 1, \ldots, 10$ of τ_k, the intersection with the hyperplane. A point $\boldsymbol{x} \in e_i$ can be written as

$$\boldsymbol{x} = \boldsymbol{x}_{i_1} + \lambda(\boldsymbol{x}_{i_2} - \boldsymbol{x}_{i_1})$$

for a given $\lambda \in [0,1]$. Hence, an intersection point $\boldsymbol{\xi}_i$ of the edge e_i with the hyperplane H_4 has to satisfy

$$\boldsymbol{x}_{i_1} + \lambda(\boldsymbol{x}_{i_2} - \boldsymbol{x}_{i_1}) = \boldsymbol{p}_0 + \mu_1\boldsymbol{p}_1 + \mu_2\boldsymbol{p}_2 + \mu_3\boldsymbol{p}_3,$$

or in matrix notation

$$\underbrace{\begin{pmatrix} \boldsymbol{p}_1 & \boldsymbol{p}_2 & \boldsymbol{p}_3 & \boldsymbol{x}_{i_1} - \boldsymbol{x}_{i_2} \end{pmatrix}}_{=:\mathbf{A}_i} \begin{pmatrix} \mu_1 \\ \mu_2 \\ \mu_3 \\ \lambda \end{pmatrix} = \boldsymbol{x}_{i_1} - \boldsymbol{p}_0.$$

The matrix $\mathbf{A}_i$ is invertible iff the vector $\boldsymbol{x}_{i_1} - \boldsymbol{x}_{i_2}$ is linearly independent to the vectors $\boldsymbol{p}_1$, $\boldsymbol{p}_2$, $\boldsymbol{p}_3$. In fact, the matrix $\mathbf{A}_i$ is not invertible if the edge e_i is parallel to the hyperplane H_4. In this case, there exists either no intersection point, or infinitely many. If the matrix is invertible, we can calculate the coefficients μ_1, μ_2, μ_3, and $\lambda \in \mathbb{R}$ uniquely. Let D_k denote the set of all intersection points of the element $\tau_k \in \mathcal{T}_h$ with the hyperplane H_4. We distinguish two relevant cases:

1. If $|D_k| = 4$, then the intersection points form a tetrahedron.
2. If $|D_k| = 6$, then the intersection points form a general irregular prism.

If we use the special vectors

$$\boldsymbol{p}_0 := t_* \boldsymbol{e}_t, \quad \boldsymbol{p}_1 := \boldsymbol{e}_x, \quad \boldsymbol{p}_2 := \boldsymbol{e}_y, \quad \boldsymbol{p}_3 := \boldsymbol{e}_z$$

for a given $t_* \in [0, T]$, we can now calculate a three-dimensional object, which can be visualized with existing software tools, for example, [4].

4 Numerical results

In this section, we will present first numerical examples. Starting point is the discrete variational formulation (2.3)–(2.4). This can be equivalently written as the following block system:

$$\begin{pmatrix} \mathbf{A}_h & -\mathbf{B}_h^\top \\ \mathbf{B}_h & \mathbf{C}_h \end{pmatrix} \begin{pmatrix} \boldsymbol{U} \\ \boldsymbol{P} \end{pmatrix} = \begin{pmatrix} \boldsymbol{F}_1 \\ \boldsymbol{F}_2 \end{pmatrix}. \tag{4.1}$$

It is worth noting that, due to the discretization of the time derivative, we have that $\mathbf{A}_h \neq \mathbf{A}_h^\top$. The four-dimensional computational geometries and the resulting linear systems in the subsequent numerical examples were solved with the software package NESHMET, developed by the authors. In particular, we used a preconditioned GMRes method. As preconditioner, we use the following block diagonal preconditioner:

$$\mathbf{P} := \begin{pmatrix} \overline{\mathbf{A}}_h & \\ & \overline{\mathbf{S}}_h \end{pmatrix}, \tag{4.2}$$

where $\overline{\mathbf{A}}_h$ is chosen as a component-wise algebraic multigrid with respect to the diagonal blocks of $\mathbf{A}_h$. For $\overline{\mathbf{S}}_h$, we chose an ILU(2)-factorization of

$$\mathbf{D}_h + \mathbf{B}_h \mathrm{diag}(\mathbf{A}_h)^{-1} \mathbf{B}_h^\top.$$

These preconditioners were taken from the HYPRE library [14].

4.1 Convergence studies

To demonstrate the correctness of the presented algorithms, we present a convergence study for a simple academic test problem. The spatial geometry under consideration is given by the three-dimensional unit cube $\Omega := (0,1)^3$. We prescribe a constant displacement of the $z = 1$ plane by the function $g(t) := t(1-t)$. The time interval is chosen as $[0,1]$. The triangulation of the resulting 4D geometry was accomplished using the tools described in Section 3. The initial triangulation consisted of 48 pentatopes. Some snapshots of the triangulation of the moving domain are depicted in Figure 6.2. These snapshots were generated by slicing through the refined 4D mesh along the time axis as described in Section 3.5. Here, one can clearly see, that the resulting slices do not consist of tetrahedrons only.

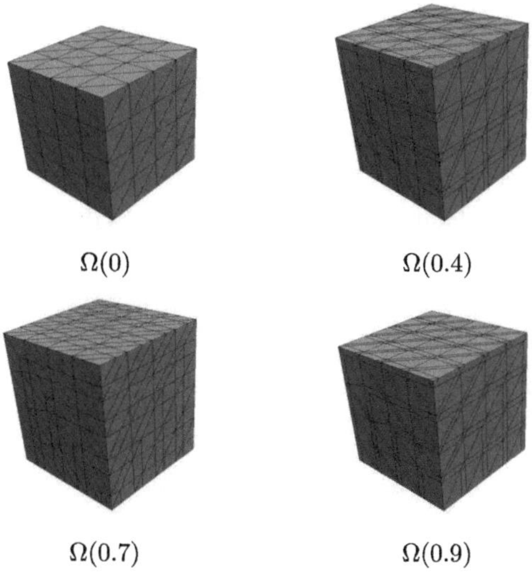

Figure 6.2: Snapshots of the triangulations $\Omega(t)$.

For our convergence study we define the following boundaries:

$$\Sigma_R := \{(\boldsymbol{x}, t) \in \Sigma : \boldsymbol{x}[1] = 1\}, \quad \Sigma_D := \overline{\Sigma \backslash \Sigma_R}.$$

On Σ_R, we set $\alpha_R = 0$, thus having a Neumann boundary. The given data $\boldsymbol{f}$, $\boldsymbol{g}_D$, $\boldsymbol{g}_R$, and $\boldsymbol{u}_0$ are chosen, such that the solutions $\boldsymbol{u}$ and p of (2.2) are given by the regular functions (taken from [12]):

$$\boldsymbol{u}(\boldsymbol{x}, t) = t(1-t)\begin{pmatrix} x_0 + x_0^2 + x_0x_1 + x_0^3x_1 \\ x_1 + x_0x_1 + x_1^2 + x_0^2x_1^2 \\ -2x_2 - 3x_0x_2 - 3x_1x_2 - 5x_0^2x_1x_2 \end{pmatrix},$$

$$p(\boldsymbol{x}, t) = t(1-t)\left(x_0x_1x_2 + x_0^3x_1^3x_2 - \frac{5}{32}\right).$$

To study the convergence behavior of the numerical approach (2.3)–(2.4), we applied several uniform refinement steps. Furthermore, we used the finite element pairing with $p = 1$ and $q = 0$. The stabilization parameters were set to $\sigma_u = 10$ and $\sigma_p = 1$. For the measurement of the error, we used the following norm for $\boldsymbol{u}_h$ taken from [24]:

$$\|\boldsymbol{u}_h\|_{\mathbf{DG}}^2 := \sum_{i=0}^{d-1} \|\boldsymbol{u}_h[i]\|_{\mathrm{DG}}^2,$$

where

$$\|u\|_{\mathrm{DG}}^2 := \sum_{\tau_l \in \mathcal{T}_h} \|\nabla_x u\|_{L_2(\tau_l)}^2 + \sum_{\Gamma_{kl} \in \mathcal{I}_h} \frac{\sigma_u}{\overline{h}_{kl}} \|u_{kl,x}\|_{L_2(\tau_l)}^2$$
$$+ \sum_{\tau_l \in \mathcal{T}_h} h_l \|\partial_t u\|_{L_2(\tau_l)}^2 + \sum_{\Gamma_{kl} \in \mathcal{I}_h} \|u_{kl,t}\|_{L_2(\tau_l)}^2 + \|u\|_{L_2(\Sigma_0)}^2 + \|u\|_{L_2(\Sigma_T)}^2.$$

For the pressure p_h, we take the standard $L_2(Q)$ norm. In Table 6.1, we depicted the results from our convergence study. For the theoretical estimates, we refer to [23, Remark 5.1.4]. The $L_2(Q)$ error for the velocity $\boldsymbol{u}_h$ is depicted for comparison. However, there is no theoretical bound for the estimated order of convergence for this norm. The other errors depicted behave as theoretically expected.

Table 6.1: Numerical results for the finite element pairing $(V_h^1(\mathcal{T}_h), Q_h^0(\mathcal{T}_h))$.

level	elements	dof $\boldsymbol{u}_h$	dof p_h	$\|\boldsymbol{u} - \boldsymbol{u}_h\|_{\mathrm{DG}}$	eoc	$\|\boldsymbol{u} - \boldsymbol{u}_h\|_{L_2(Q)}$	eoc	$\|p - p_h\|_{L_2(Q)}$	eoc
0	48	168	48	1.84×10^{0}	–	1.23×10^{-1}	–	2.30×10^{-1}	–
1	768	6,456	768	8.94×10^{-1}	0.82	3.65×10^{-2}	1.38	1.32×10^{-1}	0.63
2	12,288	141,072	12,288	4.16×10^{-1}	1.00	1.03×10^{-2}	1.65	5.94×10^{-2}	1.05
3	196,608	2,591,904	196,608	1.96×10^{-1}	1.04	2.90×10^{-3}	1.75	2.24×10^{-2}	1.34
Theory:					1.00				1.00

4.2 Robin boundary conditions for simulating valves

In the subsequent examples, we want to simulate opening and closing valves. This means, that we need to switch between an inflow and a no-slip condition. To this end, we used the following configuration for (2.2): We set $\boldsymbol{g}_R \equiv \boldsymbol{0}$. Furthermore, we use the following Robin coefficients for outflow:

$$\alpha_R(x,t) := \begin{cases} 10^6 & \text{if } t \in [0, \frac{1}{2}), \\ 0 & \text{if } t \in [\frac{1}{2}, 1], \end{cases}$$

and the following for inflow:

$$\alpha_R(x,t) := \begin{cases} 0 & \text{if } t \in [0, \frac{1}{2}), \\ 10^6 & \text{if } t \in [\frac{1}{2}, 1]. \end{cases}$$

4.3 First example: diaphragm pump

In the first example, we consider Stokes flow in a diaphragm pump. The geometry consists of the intersection of two cylinders. The first one has its main axis aligned with the z-axis with a radius of 0.8, and ranges between $z = -0.4$ and $z = 0.4$. The second cylinder has its main axis aligned with the x-axis with a radius of 0.2, and ranges from $x = -1$ to $x = 1$. A front view of the geometry is depicted in Figure 6.3.

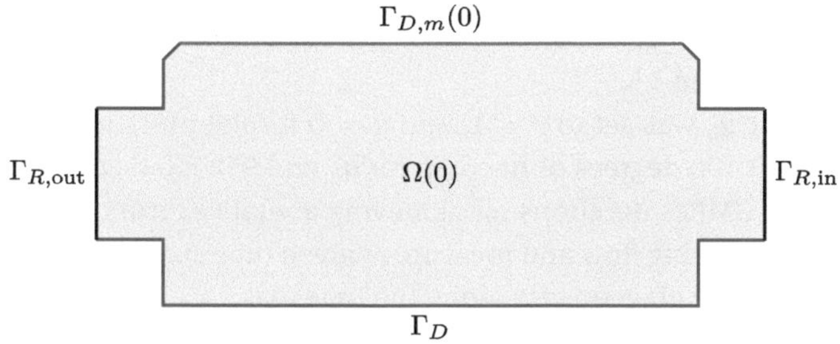

Figure 6.3: Front view of the initial geometry $\Omega(0)$. Red boundaries belong to Γ_D.

The movement of the boundary $\Gamma(t)$ was prescribed as follows:

$$\boldsymbol{g}_{\text{mov}}(t, \boldsymbol{X}) := \left(0.4 + \sin^2(\pi t)\left(1 - \frac{[X(0)]^2 + [X(1)]^2}{0.75^2}\right)\right)\boldsymbol{e}_z - \boldsymbol{X},$$

for $\boldsymbol{X} \in \Gamma_{D,m}(0)$ and $\boldsymbol{0}$ else. The following boundary conditions are used:

- Dirichlet boundary condition $\boldsymbol{u} = \boldsymbol{0}$ on Γ_D.
- Dirichlet boundary condition $\boldsymbol{u} = \frac{\partial}{\partial t}\boldsymbol{g}_{\text{mov}}$ on $\Gamma_{D,m}(t)$.

- On $\Gamma_{R,\mathrm{in}}$ and $\Gamma_{R,\mathrm{out}}$, we used the Robin boundary conditions discussed in Section 4.2.
- The initial condition for $\boldsymbol{u}$ was set to $\boldsymbol{u}(\boldsymbol{x}, 0) = \boldsymbol{0}$.

The triangulation of the resulting 4D geometry was accomplished using the tools described in Section 3. The resulting mesh consisted of 951,360 pentatopes.

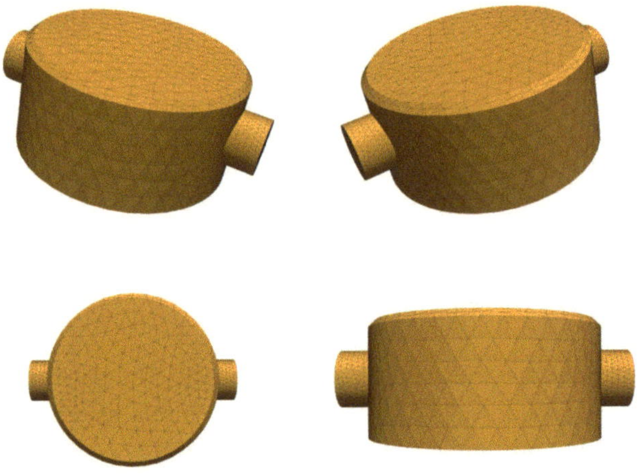

Figure 6.4: Initial Triangulation $\Omega(0)$.

Some snapshots of the triangulation of the moving domain are depicted in Figure 6.4 and Figure 6.5. These snapshots were generated by slicing through the 4D mesh along the time axis as described in Section 3.5.

The polynomial degree for $\boldsymbol{u}_h$ was set to $p = 1$, and $q = 0$ for the pressure variable p_h. This resulted in 14,270,400 degrees of freedom for $\boldsymbol{u}_h$ and 951,360 degrees of freedom for p_h. We needed 95 GMRes iterations for achieving a relative error of 10^{-5}. In Figure 6.6, one can see the resulting flow and pressure at given time stamps, which were again produced by slicing the 4D geometry along the time axis.

4.4 Second example: Y-shaped pipe

For the second example, we considered a Y-shaped pipe. A schematic view is depicted in Figure 6.7. We prescribed the following movement of $\Gamma(t)$:

$$\boldsymbol{g}_{\mathrm{mov}}(\boldsymbol{X}, t) := \begin{cases} \boldsymbol{0} & \text{for } \boldsymbol{X} \notin \Gamma_{D,m} \cup \widetilde{\Gamma}_{D,m}, \\ 4\frac{|X(2)+3|}{7} \sin^2(\pi t)\boldsymbol{e}_z & \text{for } \boldsymbol{X} \in \Gamma_{D,m} \cup \widetilde{\Gamma}_{D,m}. \end{cases}$$

Some snapshots of the domain movement are depicted in Figure 6.8 and Figure 6.9. The boundary conditions were set as follows:

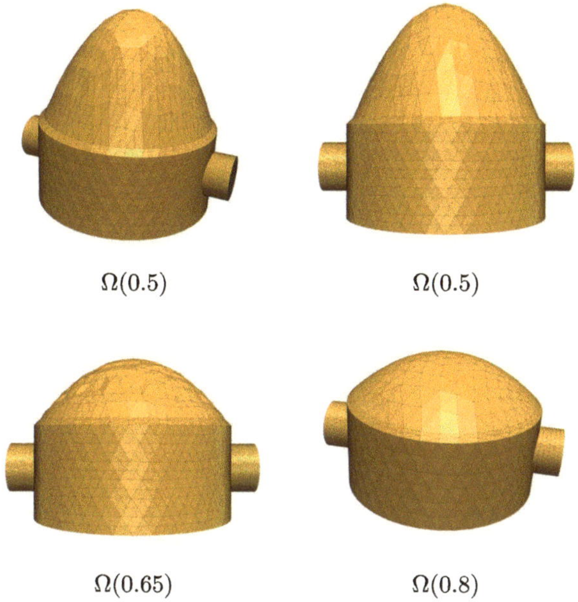

Figure 6.5: Snapshots of the Triangulations $\Omega(t)$.

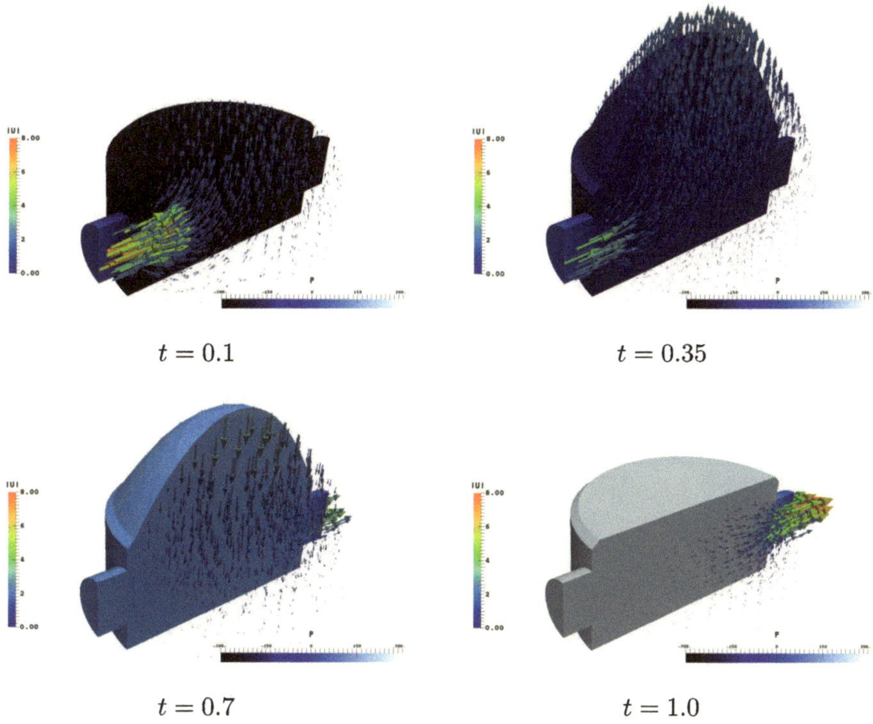

Figure 6.6: Snapshots of the solution. Additionally, we have cut along the y–axis.

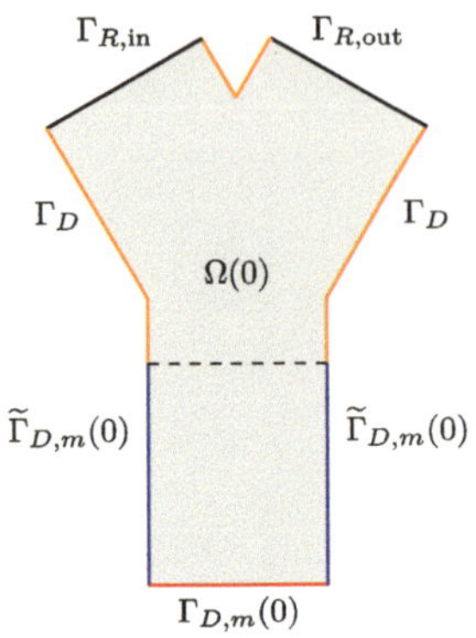

Figure 6.7: Front view of the initial geometry $\Omega(0)$. Orange boundaries belong to Γ_D. The dashed line represents the plane $z = -3$. The height from top to bottom is 17. The base of the pipe is located at $z = -10$. The radius of the pipe is 3.

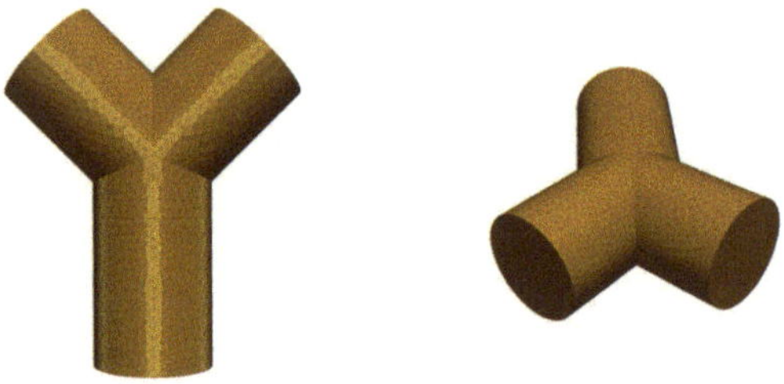

Figure 6.8: Initial triangulation $\Omega(0)$.

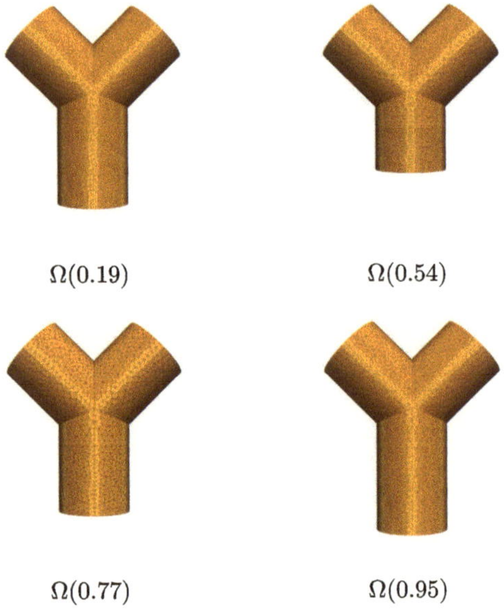

Figure 6.9: Snapshots of the triangulations $\Omega(t)$.

- Dirichlet boundary condition $\boldsymbol{u} = \mathbf{0}$ on $\Gamma_D \cup \widetilde{\Gamma}_{D,m}(t)$.
- Dirichlet boundary condition $\boldsymbol{u} = \frac{\partial}{\partial t}\boldsymbol{g}_{\text{mov}}$ on $\Gamma_{D,m}(t)$.
- On $\Gamma_{R,\text{in}}$ and $\Gamma_{R,\text{out}}$, we used the Robin boundary conditions discussed in Section 4.2.
- The initial condition for $\boldsymbol{u}$ was set to $\boldsymbol{u}(\boldsymbol{x}, 0) = \mathbf{0}$.

The resulting 4D mesh consisted of 2,618,880 pentatopes. With the same ansatz spaces, as used for Example 1, we have 39,283,200 degrees of freedom for $\boldsymbol{u}_h$ and 2,618,880 degrees of freedom for p_h. We needed 107 GMRes iterations for achieving a relative error of 10^{-5}. In Figure 6.10, we have depicted some results.

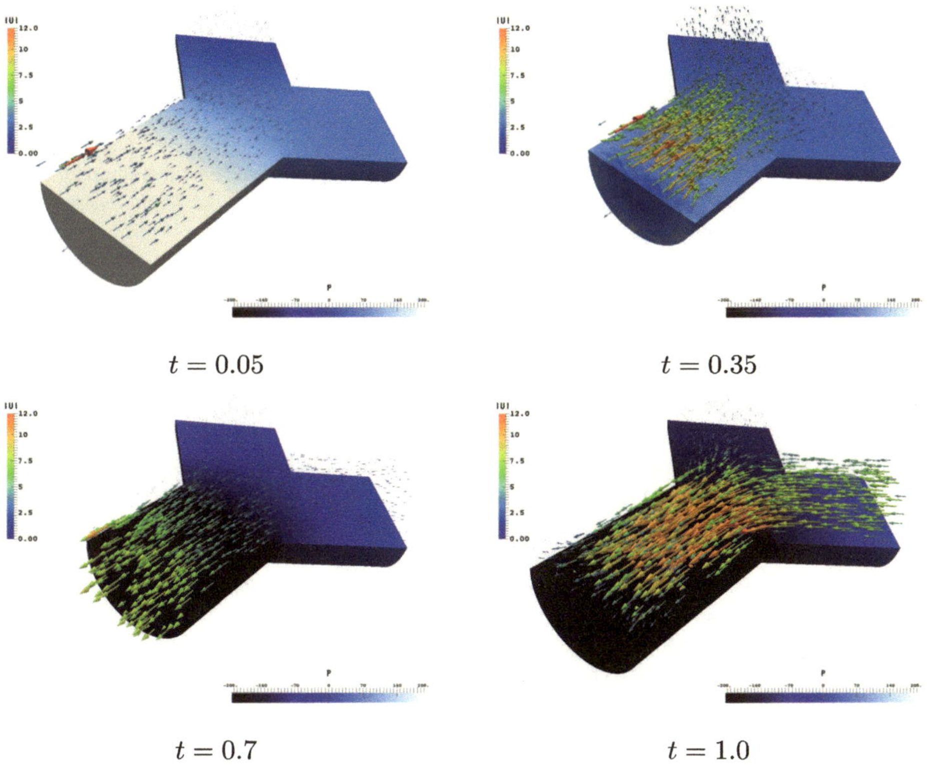

Figure 6.10: Snapshots of the solution. Additionally, we have cut along the y-axis.

5 Conclusions

In this paper, we have presented a novel approach to construct $(d + 1)$-dimensional triangulations for moving domains. The main focus in the presented experiments was devoted to the four-dimensional case. This was done by extending the elements of the space triangulation into hyperprisms. Assuming a consistent numbered spatial triangulation, we were able to prove that our algorithm produces admissible space-time meshes. We implemented the presented algorithm and applied it to solve the transient Stokes equations with a space-time discontinuous Galerkin finite element method. Other interesting candidates for applying space-time methods are given by optimal control problems with time-dependent partial differential equations as constraints. In this case, one has to solve a forward and backward problem, which are both coupled in space and time. Another attractive aspect of general space-time meshes is the possibility to apply adaptive refinement strategies to resolve local behaviors in space and time.

Bibliography

[1] R. Abedi, B. Petracovici, and R. B. Haber. A space-time discontinuous Galerkin method for linearized elastodynamics with element-wise momentum balance. *Comput. Methods Appl. Mech. Engrg.*, 195 (25): 3247–3273, 2006.

[2] R. Abedi, M. Hawker, R. Haber, and K. Matouš. An adaptive space-time discontinuous Galerkin method for cohesive models of elastodynamic fracture. *Internat. J. Numer. Methods Engrg.*, 81: 1207–1241, 2010.

[3] D. Arnold, F. Brezzi, B. Cockburn, and L. Marini. Unified analysis of discontinuous Galerkin methods for elliptic problems. *SIAM J. Numer. Anal.*, 39: 1749–1779, 2002.

[4] U. Ayachit. *The paraview guide: A parallel visualization application*. Kitware Inc., Clifton Park, 2015.

[5] R. Bank and M. Metti. An error analysis of some higher order space-time moving finite elements. *Comput. Visual. Sci.*, 16 (5): 219–229, 2013.

[6] C. Baumann and J. Oden. A discontinuous hp finite element method for convection–diffusion problems. *Comput. Methods Appl. Mech. Engrg.*, 175: 311–341, 1999.

[7] M. Behr. Simplex space-time meshes in finite element simulations. *Internat. J. Numer. Methods Fluids*, 57: 1421–1434, 2008.

[8] M. Behr and V. Karyofylli. Simplex space-time meshes in mold filling simulations. *Oberwolfach report on space-time methods for time-dependent partial differential equations*, 14: 872–874, 2017.

[9] J. Bey. Simplicial grid refinement: On Freudenthal's algorithm and the optimal number of congruence classes. *Numer. Math.*, 85: 1–29, 2000.

[10] A. Buffa and C. Ortner. Compact embeddings of broken Sobolev spaces and applications. *IMA J. Numer. Anal.*, 29 (4): 827–855, 2009.

[11] D. Di Pietro and A. Ern. *Mathematical aspects of discontinuous Galerkin methods*. Springer, Berlin, Heidelberg, 2012.

[12] C. R. Dohrmann and P. B. Bochev. A stabilized finite element method for the Stokes problem based on polynomial pressure projections. *Int. J. Numer. Meth. Fl.*, 46 (2): 183–201, 2004.

[13] S. Dumont and F. Jourdan. A space-time finite element method for elastodynamics problems with a moving loading zone. In *CST 2012*, 2012.

[14] R. Falgout, J. Jones, and U. Yang. The design and implementation of Hypre, a library of parallel high performance preconditioners. *Lect. Notes Comput. Sci. Eng.*, 51: 267–294, 2006.

[15] P. Foteinos and N. Chrisochoides. 4D Space-time Delaunay meshing for medical images. In *Proceedings of the 22nd International Meshing Roundtable*, pages 223–240. Springer, 2014.
[16] E. S. Gawlik and A. J. Lew. High-order finite element methods for moving boundary problems with prescribed boundary evolution. *Comput. Methods Appl. Mech. Engrg.*, 278: 314–346, 2014.
[17] V. Karyofylli, M. Frings, S. Elgeti, and M. Behr. Simplex space-time meshes in two-phase flow simulations. *Internat. J. Numer. Methods Fluids*, 86 (3): 218–230, 2018.
[18] C. Lehrenfeld. The Nitsche XFEM–DG space-time method and its implementation in three space dimensions. *SIAM J. Sci. Comput.*, 37 (1): A245–A270, 2015.
[19] P. I. Liakopoulos and K. C. Giannakoglou. Unstructured remeshing using an efficient smoothing scheme approach. In *ECCOMAS CFD 2006: Proceedings of the European Conference on computational fluid dynamics, Egmond aan Zee, The Netherlands, September 5–8, 2006*. Delft University of Technology; European Community on Computational Methods in Applied Sciences (ECCOMAS), 2006.
[20] E. J. López, N. M. Nigro, and M. A. Storti. Simultaneous untangling and smoothing of moving grids. *Internat. J. Numer. Methods Engrg.*, 76 (7): 994–1019, 2008.
[21] A. Masud and T. Hughes. A space-time Galerkin/least-squares finite element formulation of the Navier–Stokes equations for moving domain problems. *Comput. Methods Appl. Mech. Engrg.*, 146: 91–126, 1997.
[22] S. T. Miller, B. Kraczek, R. B. Haber, and D. D. Johnson. Multi-field spacetime discontinuous Galerkin methods for linearized elastodynamics. *Comput. Methods Appl. Mech. Engrg.*, 199 (1): 34–47, 2009.
[23] M. Neumüller. Eine Finite Elemente Methode für optimale Kontrollprobleme mit parabolischen Randwertaufgaben. Master's thesis, TU Graz, 2010.
[24] M. Neumüller. *Space-time Methods: Fast solvers and applications*. Volume 20 of Monographic Series TU Graz: Computation in Engineering and Science, 2013.
[25] M. Neumüller and O. Steinbach. A DG space-time domain decomposition method. In *Domain decomposition methods in science and engineering XX* R. Bank, M. Holst, O. Widlund, and J. Xu, editors. Volume 91 of Lecture Notes in Computational Science and Engineering, pages 623–630, Springer, Heidelberg, 2013.
[26] T. Rendall, C. B. Allen, and E. D. Power. Conservative unsteady aerodynamic simulation of arbitrary boundary motion using structured and unstructured meshes in time. *Int. J. Numer. Meth. Fl.*, 70 (12): 1518–1542, 2012.
[27] B. Rivière. *Discontinuous Galerkin methods for solving elliptic and parabolic equations: Theory and implementation*. Cambridge University Press, 2008.
[28] O. Steinbach and H. Yang. Comparison of algebraic multigrid methods for an adaptive space-time finite-element discretization of the heat equation in 3d and 4d. *Numer. Linear Algebra Appl.*, 25 (3): e2143, 2018.
[29] J. Sudirham, J. van der Vegt, and R. van Damme. Space–time discontinuous Galerkin method for advection–diffusion problems on time-dependent domains. *Appl. Num. Math.*, 56: 1491–1518, 2006.
[30] T. Tezduyar and S. Sathe. Enhanced-discretization space–time technique (EDSTT). *Comput. Methods Appl. Mech. Engrg.*, 193: 1385–1401, 2004.
[31] T. Tezduyar, M. Behr, and J. Liou. A new strategy for finite element computations involving moving boundaries and interfaces—The deforming-spatial-domain/space-time procedure. I: The concept and the preliminary numerical tests. *Comput. Methods Appl. Mech. Engrg.*, 94 (3): 339–351, 1992.
[32] T. Tezduyar, M. Behr, S. Mittal, and J. Liou. A new strategy for finite element computations involving moving boundaries and interfaces—The deforming-spatial-domain/space-time

procedure: II. Computation of free-surface flows, two-liquid flows, and flows with drifting cylinders. *Comput. Methods Appl. Mech. Engrg.*, 94 (3): 353–371, 1992.

[33] T. E. Tezduyar. Interface-tracking and interface-capturing techniques for finite element computation of moving boundaries and interfaces. *Comput. Methods in Appl. Mech. Eng.*, 195 (23): 2983–3000, 2006.

[34] V. Thomée. *Galerkin finite element methods for parabolic problems.* Springer, New York, 2006.

[35] J. van der Vegt and J. Sudirham. A space–time discontinuous Galerkin method for the time-dependent Oseen equations. *Appl. Numer. Math.*, 58 (12): 1892–1917, 2008.

[36] L. Wang and P.–O. Persson. A discontinuous Galerkin method for the Navier–Stokes equations on deforming domains using unstructured moving space–time meshes. In *21st AIAA Computational Fluid Dynamics Conference*, pages 2833–2847, 2013.

[37] L. Wang and P.–O. Persson. A high-order discontinuous Galerkin method with unstructured space-time meshes for two-dimensional compressible flows on domains with large deformations. *Comput. & Fluids*, 118: 53–68, 2015.

Olaf Steinbach and Huidong Yang

7 Space-time finite element methods for parabolic evolution equations: discretization, a posteriori error estimation, adaptivity and solution

Abstract: In this work, we present an overview on the development of space-time finite element methods for the numerical solution of some parabolic evolution equations with the heat equation as a model problem. Instead of using more standard semidiscretization approaches, such as the method of lines or Rothe's method, our specific focus is on continuous space-time finite element discretizations in space and time simultaneously. Whereas such discretizations bring more flexibility to the space-time finite element error analysis and error control, they usually lead to higher computational complexity and memory consumptions in comparison with standard time-stepping methods. Therefore, progress on a posteriori error estimation and respective adaptive schemes in the space-time domain is reviewed, which aims to save a number of degrees of freedom, and hence reduces complexity, and recovers optimal-order error estimates. Further, we provide a summary on recent advances in efficient parallel space-time iterative solution strategies for the related large-scale linear systems of algebraic equations that are crucial to make such all-at-once approaches competitive with traditional time-stepping methods. Finally, some numerical results are given to demonstrate the benefits of a particular adaptive space-time finite element method, the robustness of some space-time algebraic multigrid methods, and the applicability of space-time finite element methods for the solution of some parabolic optimal control problem.

Keywords: Space-time finite elements, a priori and a posteriori error estimates, h–adaptivity, robust space-time iterative solvers, parabolic optimal control

MSC 2010: 65F08, 65F10, 65M15, 65M22, 65M55, 65M60

Acknowledgement: This work has been supported by the Austrian Science Fund (FWF) under the Grant SFB Mathematical Optimization and Applications in Biomedical Sciences. The authors would like to thank the anonymous referees for their valuable suggestions and remarks.

Olaf Steinbach, Institute of Applied Mathematics, TU Graz, Steyrergasse 30, 8010 Graz, Austria, e-mail: o.steinbach@tugraz.at
Huidong Yang, Johann Radon Institute for Computational and Applied Mathematics, Altenberger Strasse 69, 4040 Linz, Austria, e-mail: huidong.yang@oeaw.ac.at

https://doi.org/10.1515/9783110548488-007

1 Introduction

Throughout this paper, we mainly focus on a space-time finite element solution for the following parabolic model problem:

$$\begin{aligned} \partial_t u(x,t) - \Delta_x u(x,t) &= f(x,t) \quad \text{for } (x,t) \in Q := \Omega \times (0,T], \\ u(x,t) &= g(x,t) \quad \text{for } (x,t) \in \Sigma := \partial\Omega \times (0,T], \\ u(x,0) &= u_0(x) \quad \text{for } x \in \Omega, \end{aligned} \tag{1.1}$$

where $\Omega \subset \mathbb{R}^d$, $d = 2,3$, is a convex polygonal or polyhedral bounded Lipschitz domain, and T is a given final time. Note that f, g, and u_0 are given data, which are specified later. For the ease of presentation, modifications of the model problem (1.1), including nonlinear terms, will be stated explicitly in the context.

Classical numerical methods for solving the model problem (1.1) are the method of lines, that is, discretize first in space and then in time (see, for example, [210], or Rothe's method, that is, discretize first in time and then in space; see, for example, [139]. However, here we review alternative approaches performing Galerkin-type finite element discretizations simultaneously in space and time.

Galerkin-type finite element methods, continuous in space, and possibly discontinuous in time, on unstructured space-time meshes for the solution of general linear parabolic equations in a moving domain have been considered already in the 1970s [120], but without optimal-order error estimates. In this approach, time is considered as another variable. Later, related error estimates were further improved when using discontinuous in time Galerkin methods, see [78, 79, 80, 81, 82, 161].

In the 1980s, a time-discontinuous Galerkin least-squares formulation for elastodynamics in mixed displacement-velocity form, along with an optimal convergence rate in a norm stronger than the total energy norm, was derived in [118]. The basis functions were chosen to be piecewise polynomials on each space-time finite element at a time slab, with no separation in spatial and temporal variables. The concept of this method was recently extended in [202] for continuous space-time finite element discretizations of linear parabolic problems. Therein, a Petrov–Galerkin finite element method was derived, using continuous and piecewise linear finite elements simultaneously in space and time on arbitrary admissible simplicial meshes, along with optimal error estimates in the respective energy norm. In a similar spirit, discontinuous Galerkin space-time finite element discretizations, also allowing hanging nodes, and related optimal error estimates in the respective DG-norms have been consecutively investigated in [122, 123, 124, 172, 173, 174]. Current areas of interests in space-time finite element methods include hp–approximations [72], isogeometric analysis [140, 145], Petrov–Galerkin streamline diffusion and edge average finite elements [22], finite element exterior calculus [11], fractional diffusion equations [180], discontinuous Petrov–Galerkin [102], to name a few.

Whereas the above-mentioned space-time finite element methods are based on variational formulations in the Bochner space $L^2(0,T;H_0^1(\Omega)) \cap H^1(0,T;H^{-1}(\Omega))$, an alternative approach is based on variational formulations in anisotropic Sobolev spaces $H^{1,1/2}(Q)$. This goes back to the pioneering work [156] using Fourier analysis in space and time with $T = \infty$. Related variational formulations were then analyzed in [96], and a corresponding stability and error analysis of wavelet discretizations is given in [71, 72, 148], where the Hilbert transformation is used to construct optimal test functions. This approach can be generalized to the case of a finite time interval $(0,T]$; see [206] for more details.

Meanwhile, a posteriori error estimates and related adaptive schemes for parabolic problems have been considerably studied since the early works [77, 78, 79, 80, 81, 83] in the 1980s–1990s. As it is well known, the general aim of a posteriori error estimates is to obtain computable upper (reliability) and lower (efficiency) bounds for the error with respect to a certain norm in terms of local error indicators that are used to drive a local mesh refinement to achieve optimality in the error control. Most of the techniques to obtain a posteriori error estimates for parabolic problems are borrowed and adapted from the ones for elliptic problems. Usually, the full error is split into spatial and temporal contributions, and others; for example, the error from data approximations, each of which can be bounded separately. Well studied approaches are parabolic duality methods [78, 177], using energy arguments [59, 60, 199], reconstruction techniques [24, 69, 160], functional-type estimates [142, 164, 186], residual-type estimates [181, 218, 220], flux reconstruction methods [84, 87], and using recovered gradients [137, 152]. However, many of these methods demand adaptive refinements in space and in time separately, which often results in complications of the link between the adaptive mesh refining/coarsening in space and in time. Recently, we have considered a residual type a posteriori error indicator [203, 204, 205] in space and time simultaneously, which shows reliability in our numerical experiments. A posteriori error estimates are also applicable to more involved parabolic type evolution problems, for example, for parabolic variational inequalities [1, 171, 178], the Allen–Cahn equation [94, 127], the Schrödinger equation [126], the p-Laplacian [56, 132], interface problems [198, 199] with jumping coefficients [35, 36], the Navier–Stokes system [34, 37, 184], and parabolic optimal control problems [158, 168, 182].

With the rapid growth in hardware development, parallel space-time solution methods become more feasible running on high-performance computers [91, 105, 147], since we usually face a large-scale system of algebraic equations arising from space-time finite element discretizations. In comparison with space parallel methods, the time-parallel approach has only a relatively short history [104] due to the naturally sequential feature of conventional time-stepping methods. Very recent advances in parallel space-time solution methods concern space-time multigrid with concurrency [92, 93, 99, 105] and space-time domain decomposition by constraints [17]. We have recently considered space-time algebraic multigrid (AMG) [204, 205] methods as black-box-type solvers for the linear system of algebraic equations arising from the space-time finite element discretization [202] of the heat equation. In comparison with space-time geometric multigrid methods [105, 173], coarsening in algebraic multigrid requires special care, since the spatial and temporal directions are not easily detected on the pure algebraic level, and the strength of connections may need to be taken into account [48, 50, 191].

The remainder of this paper is organized in the following way: Section 2 mainly deals with Galerkin space-time finite element methods for the discretization of related parabolic problems. In Section 3, we review a posteriori error estimates and space-time adaptive schemes mainly for linear parabolic equations and their variants. In Section 4, we describe recent developments on space-time solution methods. Some numerical results are presented in Section 5. Finally, we draw some conclusions in Section 6.

2 Space-time finite element discretization

In this section, we present an overview on space-time Galerkin finite element methods for an approximate solution of the model problem (1.1). The main focus is on discontinuous approximations, either in time, or both in space and time (see Subsection 2.1), and on approximations continuous in space and time (see Subsection 2.2). A Petrov–Galerkin space-time finite element method [202] that we used in our numerical experiments will be discussed in Subsection 2.3. Subsection 2.4 provides a short report on recent developments in space-time finite element methods, for example, on solving the related parabolic evolution equations, the exterior calculus, fractional diffusion equations, and discontinuous Petrov–Galerkin methods.

2.1 Discontinuous space-time finite element methods

A general class of Galerkin-type methods, which are based on a weak formulation similar to the one in [155], for the solution of general linear parabolic equations in a given time-dependent domain were discussed in [120]. The used approximations are continuous in space and, possibly, discontinuous in time. Although the method, in principle, allows using quite flexible space-time finite element meshes, the total space-time domain is decomposed into time slabs, and each time slab is decomposed into simplicial (or prismatic) elements. It is possible to have hanging nodes on the interface between two time slabs. In most cases, the consideration of time slabs allows the interpretation of discontinuous finite elements in time as time-stepping schemes. The finite element functions restricted to a simplicial element are polynomials of degree k with respect to the spatial and temporal variables, and they are, in general, discontinuous at a time level t^n. In this method, time is treated as another variable, and the discretization is performed in space and time simultaneously. The convergence of the space-time finite element solution is considered in $L^2(0,T;H^1(\Omega))$, see [120, Theorem 5.1]. Under proper assumptions on the structure of the mesh, improved error estimates were considered in [78, 79, 80, 81, 82, 161].

A time-discontinuous Galerkin least-squares method for a mixed displacement-velocity formulation in elastodynamics and an optimal convergence result with respect to a special norm, which is stronger than the energy norm, were derived in [118]. The finite element mesh for this approach is obtained by a decomposition of tensor-structured time slabs into triangular elements. The finite element basis functions for the displacement and for the velocity are piecewise polynomials on each space-time finite element, where no separation in spatial and temporal variables is considered. Optimal error estimates were numerically confirmed for arbitrary combinations of polynomial basis functions for the displacement and the velocity.

For the heat equation, the stability for a class of discontinuous Galerkin methods on tensor-structured meshes, which are constructed as a product of separate partitions in space and time, was analyzed in [161]. The finite element spaces consist of piecewise polynomial functions on each tensor-structured space-time slab, which are the product of continuous piecewise polynomial functions in space, and piecewise polynomial and possible discontinuous functions in time. The methods were shown to be stable with respect to a mesh-dependent norm, a discrete analogue to the L^2 norm in space and time. An optimal error estimate in $L^2(0,T;L^2(\Omega))$ was derived, see [161, Theorem 1.2]. The main tools for proving the error estimates are the well known finite element interpolation properties in space [61], the approximation property of the interpolation operator on the time slab [82], and the stability properties of the L^2 projection in time.

Recently, in a series of papers [123, 172, 173, 174], discontinuous Galerkin discretizations in the space-time domain have been analyzed for parabolic problems. There, the jumping operator in spatial and in the temporal directions, and the upwinding operator in time have been exploited. In fact, time is considered as another spatial coordinate, and the discrete finite element spaces of piecewise polynomials are designed accordingly on each space-time finite element. Optimal error estimates in the respective DG-norms were derived. The method allows to use arbitrary admissible simplex meshes in the space-time domain, also allowing hanging nodes. For an application in coupled cardiac electromechanics, see [122, 124].

In [150], a global best approximation and an interior best approximation of fully discrete Galerkin finite element solutions of second-order parabolic problems on convex polygonal and polyhedral domains were shown with respect to the L^∞-norm, see [150, Theorems 1,2]. The space-time finite element space for the discrete approximation consists of a product of continuous piecewise polynomial functions in space and discontinuous piecewise polynomial functions in time on each space-time slab. The main tools for proving the best approximation results are corresponding elliptic estimates in weighted norms, weighted resolvent estimates, and maximal parabolic [151] and smoothing estimates.

2.2 Continuous space-time finite element methods

A novel variational method for approximating the heat equation using continuous spatial and temporal finite element functions was analyzed in [16]. The tensor-product-based finite element space consists of continuous piecewise polynomials of order p on the spatial mesh, and continuous piecewise polynomials of order q on the temporal partition. In the discrete variational formulation, the test function has been differentiated with respect to the time variable. The solution of the proposed discrete variational formulation can be computed by successively marching through the time partition, whereas the test function is of one order less in temporal direction. This explains that the proposed method can be viewed as a Petrov–Galerkin method with trial functions continuous in space and time, and test functions continuous in space, but discontinuous in time. Error estimates in different norms have been derived using the properties of the constructed spatial elliptic projection and the one-dimensional temporal projection from the continuous to discrete spaces.

The stability of space-time Petrov–Galerkin discretizations applied to parabolic evolution problems was discussed in [169]. To obtain a mesh independent positive lower bound in the discrete inf-sup condition, different discretization levels for the trial and test spaces have to be chosen. In particular, this requirement can be fulfilled by using suitable hierarchical families of discrete spaces. The method is applicable to both finite element and wavelet discretizations in space and time [170]. The analysis of the method is based on some regularity assumption on the spatial partial differential operator involved, and on the so-called Jackson- and Bernstein estimates and a reverse Cauchy–Schwarz inequality with respect to the discrete trial and test spaces that can be easily fulfilled with properly chosen spaces.

Recently, a new stable single patch space-time isogeometric analysis method [119] for the numerical solution of parabolic evolution equations in both spatially fixed and moving computational domains were derived in [140]. Starting from the standard weak space-time variational formulation [134, 135], a stable discrete weak formulation is achieved by using a time-upwind test function. The space-time finite element spaces consist of tensor-product multivariate B-spline basis functions. Optimal a priori error estimates for such a stable discretization were shown with respect to a corresponding mesh-dependent norm. This approach has been extended to (time-) multipatch space-time isogemetric analysis [113, 147]. Furthermore, the stabilization term has been localized in [146, 193]. In [211], the approach has been extended to stabilized space-time finite element methods using bubble spaces.

Two stable space-time variational formulations in weighted Bochner spaces for the heat equation on the unbounded temporal interval were devised in [8]. The discrete weak formulations were shown to be stable with respect to suitable weighted space-time norms. The space-time finite element spaces were taken as space-time tensor-product spaces of Laguerre polynomials in time and arbitrary nontrivial finite-dimensional subspaces in space.

In [22], a standard Petrov–Galerkin streamline diffusion method [51] and the edge average finite element method [229] were applied to a time-dependent partial differential equation that is embedded into a convection-diffusion-type equation with singularity. These schemes provide proper discretizations for convection-diffusion problems with suitable stability and approximation properties. The methods allow using arbitrary simplex meshes in high dimensions, which usually demand extensive memory usage in space-time simulations. To cure this drawback, accurate dimension reduction algorithms on tetrahedral space-time meshes have been proposed in, for example, [149].

2.3 A Petrov–Galerkin space-time finite element method

In [202], a Petrov–Galerkin finite element method for the approximate solution of parabolic evolution equations was proposed, in which stability conditions and a priori error estimates were derived for the space-time finite element approximations. Since we have used this approach in our numerical experiments, we will discuss this method in more detail. Note that similar weak formulations have been considered in [6, 7, 196, 216], but using wavelet discretization techniques. Since time is considered as another spatial coordinate, we employ piecewise linear finite elements in both space and time simultaneously, where arbitrary admissible simplex meshes are allowed.

Since the initial condition is, as the Dirichlet boundary condition, considered as an essential boundary condition, we introduce a suitable $\overline{u}_0 \in L^2(0,T;H^1(\Omega)) \cap H^1(0,T;H^{-1}(\Omega))$ as an arbitrary, but fixed, extension of the given Dirichlet and initial data. Then the Petrov–Galerkin variational formulation for the heat equation (1.1) is to find $\overline{u} \in X := \{v \in L^2(0,T;H_0^1(\Omega)) \cap H^1(0,T;H^{-1}(\Omega)), v(x,0) = 0 \text{ for } x \in \Omega\}$, such that

$$a(\overline{u}, v) = \langle f, v\rangle - a(\overline{u}_0, v) \tag{2.2}$$

is satisfied for all $v \in Y := L^2(0,T;H_0^1(\Omega))$, where

$$a(u,v) := \int_0^T\int_\Omega [\partial_t u(x,t)v(x,t) + \nabla_x u(x,t)\cdot\nabla_x v(x,t)]\,dx\,dt,$$

$$\langle f, v\rangle := \int_0^T\int_\Omega f(x,t)v(x,t)\,dx\,dt.$$

Under proper assumptions, the uniqueness of the solution to the variational problem (2.2) can be shown; see also [156, 196, 216].

Theorem 2.1 ([202]). *Let $\overline{u}_0 \in L^2(0,T;H^1(\Omega)) \cap H^1(0,T;H^{-1}(\Omega))$ be an extension of the initial datum $u_0 \in L^2(\Omega)$ and of the Dirichlet datum $g \in H^{1/2,1/4}(\Sigma)$, and assume $f \in L^2(0,T;H^{-1}(\Omega))$. The bilinear form $a(\cdot,\cdot)$ is bounded,*

$$a(u,v) \leq \sqrt{2}\|u\|_{L^2(0,T;H_0^1(\Omega))\cap H^1(0,T;H^{-1}(\Omega))}\|v\|_{0,T;H_0^1(\Omega)} \tag{2.3}$$

for all $u \in L^2(0,T;H_0^1(\Omega)) \cap H^1(0,T;H^{-1}(\Omega))$ and $v \in L^2(0,T;H_0^1(\Omega))$. In addition, it satisfies the stability condition

$$\frac{1}{2\sqrt{2}}\|u\|_{L^2(0,T;H_0^1(\Omega))\cap H^1(0,T;H^{-1}(\Omega))} \leq \sup_{0\neq v\in L^2(0,T;H_0^1(\Omega))} \frac{a(u,v)}{\|v\|_{L^2(0,T;H_0^1(\Omega))}}. \tag{2.4}$$

Then there exists a unique solution $\overline{u} \in X$ of the variational formulation (2.2), satisfying

$$\begin{aligned}&\|\overline{u}\|_{L^2(0,T;H_0^1(\Omega))\cap H^1(0,T;H^{-1}(\Omega))}\\ &\quad\leq 2\sqrt{2}\|f\|_{L^2(0,T;H^{-1}(\Omega))} + 4\|\overline{u}_0\|_{L^2(0,T;H^1(\Omega))\cap H^1(0,T;H^{-1}(\Omega))}.\end{aligned} \tag{2.5}$$

Proof. See the proof for Theorem 2.1 and Corollary 2.3 in [202]. □

The related discrete Galerkin–Petrov problem is to find $\overline{u}_h \in X_h \subset X$ such that

$$a(\overline{u}_h, v_h) = \langle f, v_h\rangle - a(\overline{u}_0, v_h) \tag{2.6}$$

is satisfied for all $v_h \in Y_h \subset Y$, where we assume $X_h \subset Y_h$. The discrete stability condition is shown in the following theorem, where we use a discrete norm for $H^1(0,T;H^{-1}(\Omega))$,

$$\|u\|_{X_h}^2 := \|w_h\|_{L^2(0,T;H_0^1(\Omega))}^2 + \|u\|_{L^2(0,T;H_0^1(\Omega))}^2,$$

with $w_h \in Y_h$ being the unique finite element solution of the quasistatic variational formulation

$$\int_0^T\int_\Omega \nabla_x w_h(x,t)\cdot\nabla_x v_h(x,t)\,dx\,dt = \int_0^T\int_\Omega \alpha\partial_t u(x,t) v_h(x,t)\,dx\,dt$$

for all $v_h \in Y_h$.

Theorem 2.2 ([202]). *Assuming $X_h \subset X$, $Y_h \subset Y$, and $X_h \subset Y_h$, there holds the stability condition*

$$\frac{1}{2\sqrt{2}}\|u_h\|_{X_h} \leq \sup_{0\neq v_h\in Y_h} \frac{a(u_h,v_h)}{\|v_h\|_{L^2(0,T;H_0^1(\Omega))}} \tag{2.7}$$

for all $u_h \in X_h$.

Proof. See the proof of Theorem 3.1 in [202]. □

We then have the following a priori error estimate:

Theorem 2.3 ([202]). *Let $\overline{u} \in X$ and $\overline{u}_h \in X_h$ be the unique solutions of the variational problems (2.2) and (2.7), respectively. Then there holds the a priori error estimate*

$$\|\overline{u} - \overline{u}_h\|_{X_h} \leq 5 \inf_{z_h \in X_h} \|\overline{u} - z_h\|_X. \tag{2.8}$$

Proof. See the proof of Theorem 3.3 in [202]. □

In particular, the space-time cylinder $Q = \Omega \times (0, T)$ is decomposed into admissible and shape regular finite elements q_ℓ, i. e. $Q_h = \cup_{\ell=1}^N \overline{q}_\ell$. For simplicity, we assume that Ω is polygonal or polyhedral bounded, i. e., $\overline{Q} = Q_h$. The finite element spaces are given by $X_h = S_h^1(Q_h) \cap X$ and $Y_h = X_h$ with $S_h^1(Q_h) = \text{span}\{\varphi_i\}_{i=1}^M$ being the span of piecewise linear and continuous basis functions φ_i. The following energy error estimate is shown in [202].

Theorem 2.4 ([202]). *Let $\overline{u} \in X$ and $\overline{u}_h \in X_h$ be the unique solutions of the variational problem (2.2) and (2.6), respectively. Assuming $\overline{u} \in H^2(Q)$, then there holds the energy error estimate*

$$\|\overline{u} - \overline{u}_h\|_{L^2(0,T;H_0^1(\Omega))} \leq ch\|\overline{u}\|_{H^2(Q)}, \tag{2.9}$$

Proof. See the proof of Theorem 3.3 in [202]. □

We mention that on the discrete level, we obtain a nonsymmetric, but positive, definite linear system of algebraic equations that will be solved by algebraic multigrid (AMG) methods as discussed in Subsection 5.2. The adaptivity related to this method will be discussed in Subsection 3.8, and corresponding numerical experiments are presented in Section 5.

2.4 Some other space-time finite element methods

As an extension of the finite element exterior calculus for linear elliptic problems in mixed variational formulations [13, 14] to parabolic problems, a Galerkin method for a model Hodge heat equation was considered in [11]. Both semidiscrete and fully-discrete numerical schemes, which are based on a mixed formulation, were analyzed therein. The wellposedness of the mixed variational formulation was shown using the Hille–Yosida–Phillips theory. Error estimates for the finite element approximation to the evolution equation were obtained by a comparison with a corresponding elliptic projection of the exact solution into the finite element space. In a special case, this mixed form reduces to the standard weak formulation for the heat equation that has been considered in [73, 227].

Recently, another extension of the finite element exterior calculus has been considered in [107], namely, to parabolic and hyperbolic problems. A priori error estimates for Galerkin finite element approximations in the natural Bochner space norms were derived therein by combining recent results on the finite element exterior calculus for elliptic problems with a classical approach in [210]. The method has been recently extended to parabolic evolution problems on Riemannian manifolds [116] by using the framework developed in [115].

Numerical solution techniques for parabolic equations with fractional diffusion and the Caputo fractional time derivative were studied in [180]. Therein, the evolution problem was written as a quasistationary elliptic problem with a dynamic boundary condition. The spatial fractional diffusion is treated as the Caffarelli–Silvestre extension problem on a semiinfinite cylinder in one more spatial dimension [54]. The finite difference scheme proposed in [154] was employed to discretize the fractional time derivative. First-degree tensor-product finite elements for the truncation problem with exponential decay adapted from [179] were used for the spatial discretization. Unconditional stability and error estimates for the fully discrete scheme were shown therein. Using a similar discretization scheme, a convergence analysis for the discretization of a space-time fractional optimal control problem was discussed in the recent work [9].

A time-stepping discontinuous Petrov–Galerkin method with optimal test functions for the heat equation was discussed in [102]. The stability for the semidiscrete and fully-discrete schemes based on a backward Euler time-stepping and an ultraweak variational formulation [65] at each time step was shown. We mention that a more detailed discussion on discontinuous Petrov–Galerkin methods with optimal test functions for elliptic and fluid problems can be found in, for example, [65, 66, 67, 75, 76].

3 A posteriori error estimates and adaptivity

In this section, we discuss a posteriori error estimates and corresponding adaptive schemes for parabolic problems. Well established methods for deriving a posteriori error estimates and devising respective adaptive refinement strategies are reviewed, namely,

(i) parabolic duality – Subsection 3.1,
(ii) energy arguments – Subsection 3.2,
(iii) reconstruction – Subsection 3.3,
(iv) functional type – Subsection 3.4,
(v) residual type – Subsection 3.5,
(vi) flux reconstruction – Subsection 3.6, and
(vii) recovered gradient – Subsection 3.7.

Furthermore, we provide some details on our space-time adaptive method relying on a residual-type error indicator with conforming local mesh refinements (see Subsection 3.8) that have been recently developed [204, 205], and which drive the adaptive refinement in space and time simultaneously.

A posteriori error estimates in different applications are reported in Subsection 3.9. We mention that the given overview is not restricted to a posteriori error estimates using space-time finite element approximations, that means, time-stepping methods may also be considered.

3.1 Parabolic duality

Adaptive finite element methods for linear parabolic problems using a discontinuous Galerkin approach in time, and on each time interval, a continuous finite element approximation in space were considered in [78]. The discrete space is a tensor-product space of polynomials in space and time. This separation offers some flexibility on the spatial and temporal mesh adaptivity. The a posteriori error estimates in $L^\infty(0,T;L^2(\Omega))$ were derived using duality techniques involving both continuous and discrete dual problems, and strong stability properties of the dual problems. In [79], error estimates have been extended to $L^\infty(0,T;L^\infty(\Omega))$ and $L^\infty(0,T;L^2(\Omega))$. The a posteriori estimates have been further generalized to nonlinear parabolic problems in [80], for the error control in $L^\infty(0,T;L^2(\Omega))$. Using similar techniques from [78], adaptive finite element methods for long-time integration of parabolic problems have been discussed in [81] for time discontinuous Galerkin methods.

The duality technique has been applied to a posteriori error estimates for a degenerate parabolic problem in [177], where the related dual problem corresponds to a nonstrictly parabolic equation in nondivergence form with a vanishing rough diffusion coefficient [101, 176]. Moreover, goal-oriented a posteriori error estimates for space-time discretizations based on dual weighted residual approaches are considered, for example, in [195].

3.2 Energy arguments

Using so the so-called direct energy estimate argument [60], that is, a coupled system of one parabolic equation with one variational inequality, the error indicator, which consists of the time and space error indicators, which are designed for linear parabolic problems discretized by backward Euler in time and continuous finite elements in space, has been shown to be an upper bound of the error in the respective norm; see [59, Theorem 2.1]. A new refinement/coarsening strategy based on a bisection algorithm (see, for example, [23]), has been proposed. In addition, a coarsening error indicator was also provided therein. The time and space adaptive algorithms were further developed; see Algorithm 3.2 in [59]. The algorithm includes time and space refining, and space and time coarsening with respect to a prescribed spatial and temporal tolerance. The adaptive algorithm has been further improved in [133] so that it always reaches the final time for a given tolerance.

3.3 Reconstruction methods

Using the so-called elliptic reconstruction of the finite element solution of a spatially semidiscretized equation, which is considered as an "a posteriori dual" to the elliptic projection in the classical a priori error analysis for semidiscrete linear parabolic problems [106, 111, 112, 210, 227], a posteriori error estimates for parabolic problems in $L^\infty(0,T;L^2(\Omega))$ were derived in [160]. A similar idea, the so-called postprocessed Galerkin approximation, has been used to derive a posteriori error estimates for nonlinear parabolic problems [63, 64, 106]. A posteriori analysis of evolution problems based on both spatial and temporal reconstructions can be found in [58].

As an extension of [160], and using a combination of the elliptic reconstruction and other properly related techniques, for example, main parabolic error estimates, heat kernel estimates [15], pointwise boundedness of the spatial derivatives of the Green's function for the parabolic problem [74], a posteriori error estimates for fully discrete linear parabolic problems, were derived in various norms. In particular, the following spaces were considered: $L^\infty(0,T;L^2(\Omega))$, see [24, 136, 138]; $L^\infty(0,T;H_0^1(\Omega))$, see [136], $L^\infty(0,T;H^1(\Omega))$; $H^1(0,T;L_2(\Omega))$, see [136]; $L^\infty(0,T;L^\infty(\Omega))$, see [41, 69, 129, 130], and $L^\infty(D \subset Q)$, see [68].

3.4 Functional-type estimates

As an extension of functional-type a posteriori error estimates derived for elliptic problems in [162, 185, 187], error bounds for the heat equation were derived in [186]. These functional estimates are upper bounds (majorants) for the difference in a certain norm between the exact solution of the heat equation and any admissible approximation from the associated function space. Error majorants have been further derived for evolutionary convection-diffusion problems in [188]. In [165], both the error majorant and error minorant (lower bound) were derived for evolutionary reaction-diffusion problems with mixed Dirichlet–Neumann boundary conditions.

Functional-type a posteriori error estimates for time-periodic parabolic boundary value problems have been derived in [142], which provide guaranteed and fully computable upper bounds (majorants) for the error in $H^{1,1/2}(Q)$. As an extension, functional-type a posteriori error estimates for the state and adjoint errors in distributed time-periodic parabolic optimal control problems were derived in [141]. Furthermore, guaranteed and computable upper bounds for the cost functionals, and their sharpness, were derived therein, using a posteriori estimates for the state equation. Similar results for elliptic optimal control problems can be found in [103, 187].

Recently, following the approaches [165, 186], error majorants for the stabilized space-time weak formulation of the parabolic problem using isogeometric analysis [140] have been derived in [143, 144]. A comparison of efficiency of the functional-type a posteriori error estimates applied to a class of parabolic problems, using both the time-marching and space-time approaches, was discussed in [114].

3.5 Residual-type methods

Based on a general framework [218], the work [220] derives a residual-type posteriori error estimate in $L^r(0, T; W^{1,\rho}(\Omega))$ for a space-time finite element discretization of a nonlinear parabolic boundary value problem and nonstationary incompressible Navier–Stokes equations. With additional regularity assumptions, a residual-type a posteriori error estimate in the weaker space $L^r(0, T; L^\rho(\Omega))$ with $1 < r, \rho < \infty$ has been shown in [219].

Under the assumption that the triangulations are nested, residual a posteriori error indicators with respect to $L^2(0, T; H_0^1(\Omega))$ for a standard discretization of the heat equation have been reported in [181].

Residual a posteriori error indicators with respect to a certain norm were derived in [221] for a discretization of the heat equation by A-stable θ-schemes ($\theta \in [0.5, 1]$) in time and conforming finite elements in space.

Residual a posteriori error indicators in $L^\infty(0, T; L^2(\Omega))$ for a discretization of the heat equation using Euler's implicit scheme in time and continuous, piecewise polynomial finite elements in space were constructed in [32].

3.6 Flux reconstruction methods

A posteriori error estimates in a broken norm for an approximate solution of the heat equation were derived in [84]. The estimates are based on H^1-conforming reconstructions of the potential, which is continuous and piecewise affine in time, and a locally conservative $H(\mathrm{div})$-conforming reconstruction of the flux, which is piecewise constant in time. Such a method is inspired by a posteriori error estimates for elliptic problems in [86, 89, 224, 225]. Furthermore, the $H(\mathrm{div})$-conforming flux reconstruction can be found in [2, 32, 44, 125], whereas the potential reconstruction is discussed in [53, 125, 221].

In [87], a posteriori error estimates in $L^2(0, T; H^1(\Omega)) \cap H^1(0, T; H^{-1}(\Omega))$ for the error and for the temporal jumps of the numerical solution of parabolic problems were considered. This technique can be viewed as a natural extension of flux reconstructions for elliptic problems [43, 44, 70, 85]. Such estimators have been shown to be unconditionally locally space-time-efficient with respect to local errors, with constants independent of both the spatial and temporal approximation orders. They also allow very general refinement and coarsening strategies between the time steps.

The equilibrated flux reconstruction has been used in [88] to obtain a posteriori estimates in $L^2(0,T;H^1(\Omega))$ of the error and of the temporal jumps. Under the so-called one-side parabolic condition $h^2 \lesssim \tau$, it was shown that the constants in the bounds are robust with respect to the mesh and time step sizes h and τ, respectively, the spatial polynomial degrees, and the refinement and coarsening strategies between time steps.

3.7 Recovered gradient approach

In [152], gradient recovery a posteriori error estimators [4, 20, 55, 95, 153] have been extended from elliptic equations to the linear heat equation, under the condition that the time-stepping error must be strictly smaller than the space discretization error. As improvement, in [137], a posteriori error estimates in the energy norm using gradient recovery approaches have been investigated for the full discretization of the linear heat equation, without such restrictive assumptions on the time step size. Here, gradient recovery is a local weighted averaging with gradient sampled from neighboring elements.

3.8 A residual-type error indicator with conforming space-time local mesh refinements

3.8.1 A space-time local error indicator

In our recent work [204, 205], a residual-based a posteriori error indicator for the space-time Petrov–Galerkin finite element method [202] was proposed.

Let $u_h \in X_h$ be the space-time finite element solution of the variational problem (2.6), where X_h is a properly defined finite element space, see Subsection 2.3. Then we can define the local residuals

$$R_{q_\ell}(u_h) := f + \Delta_x u_h - \partial_t u_h$$

on each space-time finite element q_ℓ, and the jumps

$$J_\gamma(u_h) := [n_x \cdot \nabla_x u_h]_{|\gamma}$$

of the normal flux in the spatial direction across the inner boundaries γ between q_ℓ and its neighboring elements. Then, the local error indicator on each element q_ℓ is given as

$$\eta_{q_\ell} = \{c_1 h_{q_\ell}^2 \|R_{q_\ell}\|_{L^2(q_\ell)}^2 + c_2 h_{q_\ell} \|J_\gamma\|_{L^2(\partial q_\ell)}^2\}^{\frac{1}{2}}, \tag{3.10}$$

with suitably chosen positive constants c_1, c_2, which may depend on the model problem and the shape of the considered domain. In all our numerical examples, we have used $c_1 = c_2 = 1$ for simplicity. For more details, we refer to our recent work [203]. It is obvious that this method allows performing spatial and temporal adaptivity simultaneously.

3.8.2 An adaptive space-time finite element loop

The adaptive loop in the space-time finite element method follows the standard adaptive finite element approach, see, for example, [223], which consists of the following four main steps: Given a conforming decomposition Q_0 at the initial mesh level $k = 0$,

1. SOLVE: Solve the discrete problem (2.6) on the adaptive mesh level k;
2. ESTIMATE: Compute the local error indicators (3.10) on each element q_ℓ and the global error indicator; stop if the solution is accurate enough;
3. MARK: Mark the elements for refinement using a proper marking strategy;
4. REFINE: Perform the local mesh refinement using octasection or bisection (see the following description), increase the level $k := k + 1$, obtain the conforming decomposition Q_k, and go to Step 1.

For the module MARK, we use the maximal marking strategy: For a given parameter $\vartheta \in [0, 1]$, mark all elements q_k that fulfill

$$\eta_{q_k} \geq \vartheta \max_{\ell=1,\dots,N_k} \eta_{q_\ell}, \tag{3.11}$$

where N_k denotes the total number of space-time finite elements on the current level k. Those marked and the affected neighboring elements will be refined on the next level $k + 1$. In our numerical experiments, we use $\vartheta = 0.5$ for the adaptive mesh refinement.

3.8.3 An octasection-based adaptive mesh refinement method

For the adaptive local mesh refinement, we first adopt a method mainly following the idea in [38], that is, the so-called octasection, which is strongly connected to the *Red–Green* refinement in two dimensions [19, 21, 201]. In the software package UG [27], a similar method has been implemented, whereas in [109], a parallel version of this method has been developed. Here, we have only considered a refining procedure without coarsening. Whereas in the following the focus is to discuss a Galerkin–Petrov space-time finite element method on simplicial meshes, space-time adaptivity can be considered for more general situations as well, for example, for higher-order tensor-product spaces, or hexahedral meshes.

Starting from an initial mesh with shape regular tetrahedral elements, we mark the tetrahedra that shall be refined on the next level. Each of these marked tetrahedra is divided into four congruent tetrahedra and one octahedron by connecting middle edge points on each face of the tetrahedron. Following the shortest-interior-edge strategy [231], we further divide the remaining octahedron into four tetrahedra.

During the regular refining procedure, hanging nodes appear; that is, some edges of the tetrahedron are divided, but the tetrahedron itself is not yet divided, which has to be closed by a combined regular and irregular refinement strategy. According to how many, that is, maximal five, and which edges of a tetrahedron are divided, there exist 62 possible cases for the irregular refinement. Due to a symmetry argument, the number of cases is reduced to nine. In [38, 42], only four types of irregular refinement are considered for simplicity.

By building proper connections among the tetrahedral vertex and the middle edge points on the face of a tetrahedron locally, we obtain a hybrid mesh without hanging nodes, but with a mixture of different elements: tetrahedra, pyramids, and triangular prisms; see Figure 7.1 for an illustration.

The remaining task is to further divide the pyramids and triangular prisms into tetrahedra in a conformal way, which does not introduce any additional nodes in the mesh, and, at the same time, subdivide each rectangular face shared by two elements into two triangles in a conforming way. This is realized in the same manner as detailed in [230].

It is important that on the next refinement levels, the tetrahedra from the irregular refinement will never be refined again. If such an element is marked for further refinement, we will return to its "parent", which is regular, and make a regular refinement for this element and, meanwhile, remark all neighboring elements that are affected. In this way, we avoid a mesh degeneration during the adaptive refinement procedure.

For high dimensions, the regular (*Red*) refinement has been considered in [39] using Freudenthal's algorithm [100]. The conforming *Red-Green* refinement of simplicial meshes in arbitrary dimensions was recently investigated in [108] using the placing triangulation technique. A special refinement strategy to decompose pentatopes into smaller ones for the four-dimensional space-time cylinder has been introduced in [174].

3.8.4 A bisection-based adaptive mesh refinement method

We further consider a bisection-based mesh refinement method [12] in three dimensions, which is strongly related to the newest vertex bisection method proposed and developed in [23, 131, 157]. The local mesh refinement is summarized as follows: For each marked tetrahedron, we choose one edge as the so-called refinement edge, and the two faces intersecting at this edge as the refinement faces. For the other two nonrefinement faces, we choose a particular edge on each face as the so-called marked edge.

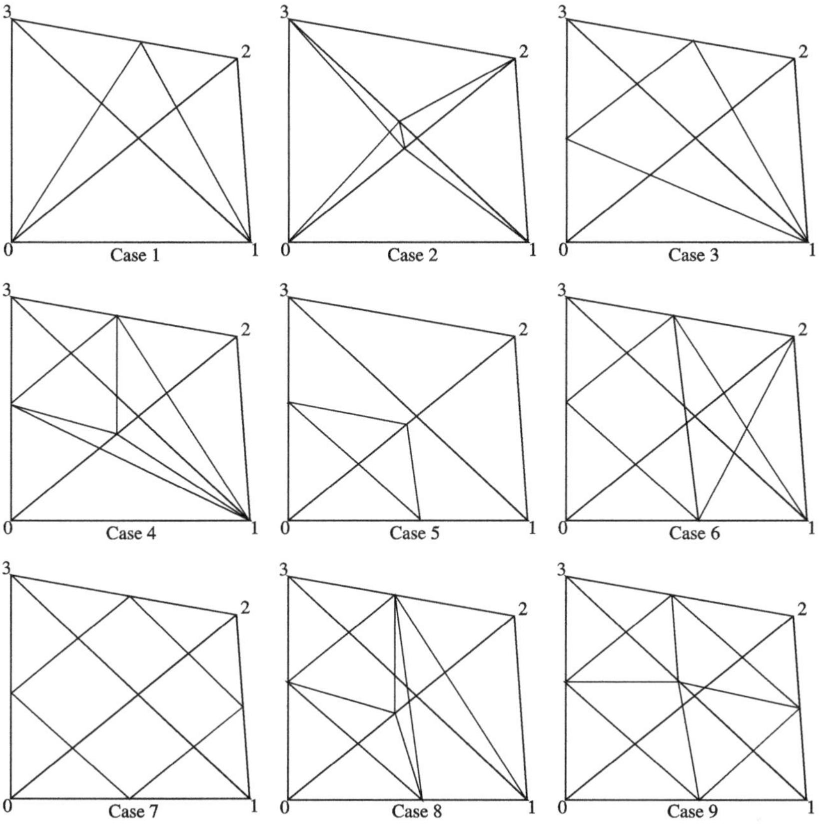

Figure 7.1: Irregular refinements of a tetrahedron with local vertex numbering $0, \ldots, 3$ into hybrid elements: tetrahedra, pyramids and triangular prisms: Case 1 (two tetrahedra), Case 2 (four tetrahedra), Case 3 (one tetrahedron and one pyramid), Case 4 (four tetrahedra), Case 5 (one tetrahedron and one triangular prism), Case 6 (one tetrahedron and two triangular prisms), Case 7 (two triangular prisms), Case 8 (two tetrahedra and two pyramids), Case 9 (two tetrahedra, one triangular prism and one pyramid).

Once a tetrahedron is marked for refinement, we will divide this tetrahedron into two smaller ones by connecting the mid point on the refinement edge with the other two tetrahedral vertices that are not lying on the refinement edge. This simplifies the local refinement patterns.

In fact, according to the relative position between the marked and refinement edges, the marked tetrahedra can be grouped into four types: Type P (planar, where the marked edges and the refinement edge are coplanar); Type A (adjacent, where each marked edge shares a common vertex with the refinement edge, but they are not coplanar); Type O (opposite, where the marked edges have no intersection with the refinement edge), and Type M (mixed, one marked edge does not intersect the refinement edge, and the other does), see Figure 7.2 for an illustration.

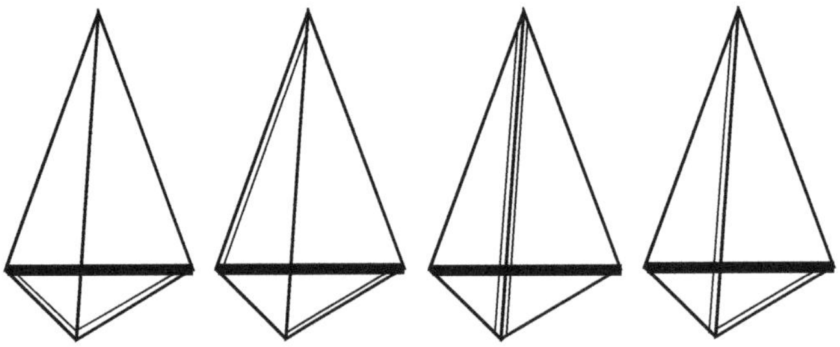

Figure 7.2: The four types of marked tetrahedra (from left to right): *P*, *A*, *O*, and *M*. The refinement edge is indicated by the thick solid line, and the marked edge is highlighted by a double line. The two faces sharing the refinement edge are the refinement faces, and the other remaining two faces, the nonrefinement faces.

In addition, a flag $f_\tau \in \{0, 1\}$ is attached to the type P so that P is classified as type P_f ($f_\tau = 1$), or P_u ($f_\tau = 0$). The local refinement follows the following rules:

$$P_u \longrightarrow P_f, \quad A \longrightarrow P_u, \quad M \longrightarrow P_u, \quad O \longrightarrow P_u, \quad P_f \longrightarrow A. \tag{3.12}$$

It is sufficient and necessary to obtain conforming refinements in three dimensions, provided that the refinement edges of two neighboring elements have to coincide when they are on the common sharing face [175]. This condition is easily fulfilled for any initial conforming triangulation by, example, choosing the longest edge of each tetrahedral element as the refinement edge [12]. A conforming mesh is obtained by a recursive calling of the local mesh refinement until no hanging nodes exist, see [12, Theorem 3.1].

We mention that bisection methods have been extended to any dimension in [166, 207, 212], which is very useful for the space-time adaptivity in four dimensions. In addition, it has been shown in [207] that the newest vertex bisection refinement stays local for any dimension, which is extended from the result [40] in two dimensions.

Compared with the octasection-based method, the refinement pattern for each tetrahedron has been fixed from the very beginning in the bisection-based approach. During the adaptive refinement procedure, the tetrahedron chosen for refinement will be continuously refined according to the strict rules (3.12) specified above. Some related numerical results concerning adaptive space-time finite element methods for both linear and nonlinear parabolic problems are shown in Subsection 5.1.

3.9 Some other topics of a posteriori error estimates in space-time FEM

In [171], a residual-type a posteriori error analysis was derived for a fully discrete method using piecewise linear finite elements in space and backward Euler discretization in time of parabolic variational inequalities in the pricing of American options for baskets. A posteriori error estimators were derived for the error in $L^2(0,T;H^1(\Omega))$. Two residual-type error indicators have been considered in [1] for a posteriori error estimates for parabolic variational inequalities. In [178], an a posteriori error analysis was investigated for a class of integral equations and variational inequalities in the pricing of European or American options under Lévy processes, discretized by piecewise linear finite elements in space, and the implicit Euler method in time. A residual-type a posteriori error estimator was derived for the error in H^s, $s \in (0,2]$.

A posteriori error estimates for space-time discretizations with discontinuous finite elements in time for a parabolic obstacle problem were derived in [25].

Using the so-called energy argument and a topological continuation argument, a posteriori error control in $L^\infty(0,T;L^2(\Omega))$ for the Allen–Cahn's problem was derived in [127], which only has a low-order polynomial dependence in ε^{-1} with ε being the interface thickness parameter. A similar result has been also achieved in [94]. These results have been improved from the old results, which have an exponential dependence on ε^{-2}. Similarly, in [26], quasioptimal a posteriori error estimates in $L^\infty(0,T;L^2(\Omega))$ for the finite element approximation of Allen–Cahn equations were derived, with a low-order polynomial dependence in ε^{-1}.

In [126], optimal-order residual-type a posteriori error estimates for the fully discrete linear Schrödinger-type equations in $L^\infty(0,T;L^2(\Omega))$ were derived, using a Crank–Nicolson method in time, and a finite element method in space, which may change in time. In addition, a practical space-time adaptive algorithm was realized, which guarantees rigorously that the total error remains below a given tolerance as long as the algorithm converges.

As an extension of the a posteriori estimates for the linear heat equation considered in [221], a reliable and efficient a posteriori estimate for the fully discrete implicit Euler Galerkin finite element scheme of the nonlinear p-Laplacian problem was derived in [132]. In an earlier work [56], a posteriori error estimates for the finite element approximation of the nonlinear p-Laplacian have also been derived.

Following the reconstruction approaches studied in [5, 24, 136], residual-type a posteriori error estimates for the linear parabolic problems with two transmission conditions on the common space-time interface were derived in [198]. In [199], using an energy argument, residual-based a posteriori error estimates for linear parabolic interface problems were derived with optimal-order convergence in $L^2(0,T;H^1(\Omega))$, and with an almost optimal order in $L^\infty(0,T;L^2(\Omega))$.

Following the approach in [181, 221], residual-type a posteriori error estimates were derived in [35] for the heat equation with a diffusion coefficient, which is constant in time and piecewise constant within the subdomains. In [36], as an extension of the results in [35], a residual-based a posteriori error estimator was further derived for a new discretization method, that is, Crank–Nicolson in time and a conforming finite element method in space, of the heat equation with jumping diffusion coefficients. This method allows varying time step sizes on different elements at the same time.

A fully combined spatial and temporal adaptive scheme was applied to the unsteady Navier–Stokes equations in [37], using the results of a posteriori error estimates and adaptive algorithms for the steady Navier–Stokes system [34, 184], the heat equation [32, 35, 181, 221], an unsteady reaction-convection-diffusion equation [222], and the unsteady Stokes system [33]. In particular, local-in-space error indicators [34] and local-in-time error indicators [32, 33, 35, 181, 221] were adopted in the adaptive algorithm in space and in time.

Following the approach in [31], an anisotropic a posteriori error estimate for controlling the error between the true and the computed cost functional in an optimal control problem governed by a parabolic equation was derived in [182], which is discretized by a Crank–Nicolson scheme in time, and continuous piecewise linear finite elements in space. In the error analysis, a space-time interpolation operator was built as a combination of a Clément [62] or Scott–Zhang [197]-type quasiinterpolation operator on strongly anisotropic meshes [10, 97, 98] in space, and the standard Lagrange interpolation in time. With the help of the simple Zienkiewicz–Zhu-type error estimator [3, 190], and assuming that the time step size is small enough, an anisotropic a posteriori error estimate for the cost functional was designed.

In [158], a posteriori error estimates for both the state and the control approximation of a quadratic optimal control problem governed by a linear parabolic equation were derived. In [200], equivalent a posteriori error estimators of residual-type with lower and upper bounds for both the state and control approximations of a constrained optimal control problem governed by a parabolic integro-differential equation on multimeshes were considered.

As an extension of the work [28, 29, 30], a posteriori error estimates were derived for a cost functional of parabolic optimization problems in [168]. In [195], as an extension of an optimal control approach to a posteriori error estimation in finite element methods proposed in [18, 28], an a posteriori error estimator and an adaptive space-time algorithm for parabolic equations that allow dynamic locally refined meshes and nonuniform time discretizations have been considered.

4 Solution methods

For a general discussion on parallel solution methods for time-dependent problems, we refer to the recent review work [104] and the references therein, where four main types of space-time solution methods have been discussed in detail: shooting-type time parallel methods, space-time domain decomposition methods, geometric multigrid methods in space and time, and direct solvers. Here, we are mainly concerned with very recent developments of space-time geometric multigrid methods, algebraic multigrid methods, and space-time domain decomposition by constraints as discussed in Subsections 4.1–4.3, respectively.

We mention that there are many other techniques for designing parallel space-time solvers. For example, in [91, 92, 93], optimal-scaling parallel multigrid-reduction-in-time and multigrid methods with space-time concurrency were developed for solving linear and nonlinear parabolic model problems with both implicit and explicit time discretizations, which are mainly based on multigrid reduction [189]. In [110], so-called semi-geometric multigrid methods for a new continuous space-time finite element discretization of transient problems in continuum mechanics have been developed.

4.1 Space-time geometric multigrid

In [105, 173], a new space-time parallel geometric multigrid method was developed for solving fully discrete parabolic equations, that is, using an arbitrarily high-order discontinuous Galerkin discretization in time, and a finite element method in space. Using exponential local Fourier mode analysis [45, 47, 117, 226], block Jacobi smoothing factors and two-grid convergence factors for arbitrary discontinuous Galerkin time discretization schemes were investigated, which lead to a precise criterion, that is, a restriction on the discretization parameter $\mu = \tau h^{-2}$ with respect to the polynomial degree in time, for determining semicoarsening in time, or full coarsening in space and time. It is concluded that semicoarsening in time within the two-grid cycle always converges to the exact solution, whereas full space-time coarsening can be applied when the discretization parameter μ is large enough in comparison with the critical value.

Instead of an adaptive coarsening as proposed in, for example, [105, 117], a space-time multigrid method, using an adaptive smoothing strategy in combination with standard coarsening in both temporal and spatial domains, was developed in [99]. According to a critical value, the adaptive strategy determines a choice of smoothers between zebra line-in-time relaxation/read-black line-in-time relaxation and zebra line-in-space relaxation/zebra plane-in-space relaxation in one/two space dimensions, respectively. This critical value is obtained by performing a local Fourier analysis [45, 47, 209, 214, 226, 228]. The proposed multigrid method is robust for both first-order Euler and second-order Crank–Nicolson temporal discretization schemes.

4.2 Space-time algebraic multigrid

In our recent work [204, 205], considering local mesh refinement and using arbitrary simplex meshes in three and four space-time dimensions in a space-time finite element discretization method [202] for the heat equation, we have compared algebraic multigrid methods [48, 50, 191, 208], using different coarsening for solving the arising linear system of algebraic equations:

$$Ax = b. \tag{4.13}$$

It is clear from [202] that the matrix A is nonsymmetric, but positive definite. The linear system (4.13) is solved by a preconditioned GMRES method with different algebraic multigrid preconditioners. In particular, we use two $V(1,1)$-cycles with one pre- and post-smoothing step as a preconditioner in the GMRES method. We have considered a pure matrix-graph [128], a greedy coarse-grid selection [159], compatible relaxation [49, 90], and Petrov–Galerkin smoothed aggregation [217].

The simple pure matrix-graph coarsening strategy yields a very aggressive coarsening method, and a rather low operator and grid complexities. It treats all connections equally, without taking into account the strength of a connection in the classical Ruge–Stüben algebraic multigrid [50, 191], which is actually important in our space-time algebraic multigrid methods.

In the greedy strategy, the strength of a connection has been taken into account, and only strong connections are to be considered in the definition of the interpolation operator, by using a dynamic measure to determine the diagonal dominance of a row among those rows already selected as fine degrees of freedom or undesignated.

In the compatible relaxation algebraic multigrid, the coarse grid set is selected by performing compatible relaxation restricted to the fine degrees of freedom only.

In the smoothed aggregation algebraic multigrid method, the interpolation operator is constructed by smoothing a tentative interpolation operator, for example, using piecewise constant basis functions, on the decomposition of degrees of freedom into small disjoint subsets. More details are reported in our recent work [204] and the related references therein.

Furthermore, we employ one sweep of the Kaczmarz relaxation scheme [121] as presmoother and postsmoother for the nonsymmetric and positive definite system on multigrid levels: Let x^0 be a given initial guess, for $i = 1, \ldots, n$, compute

$$x^i = x^{i-1} + \frac{b_i - \langle A_i, x^{i-1} \rangle}{\|A_i\|_{l_2}^2} A_i, \tag{4.14}$$

with A_i being the ith row of A presented as a column vector, and b_i the ith component of b. The algebraic multigrid smoothing property of the Kaczmarz relaxation scheme for even more general nonsymmetric matrices has been discussed in [46, 183].

All methods have shown relative robustness with respect to the mesh discretization parameters in space and time, the heat capacity constant, and local mesh adaptivity. Some numerical results concerning algebraic multigrid performance will be demonstrated in Subsection 5.2.

4.3 Space-time balancing domain decomposition by constraints

In [17], weakly scalable space-time preconditioners based on nonoverlapping multilevel balancing domain decomposition by constraints (BDDC) methods have been developed for solving both linear and nonlinear parabolic problems discretized using finite elements in space, and backward Euler schemes in time. The essential components in the space-parallel BDDC method [163, 215], namely, subassembled spaces and operators, coarse degrees of freedom, and transfer operators, have been extended to space-time. In this method, the space domain is decomposed into fine space elements, and coarse space subdomain partitions, and the time domain is decomposed into a fine-time interval and coarse-time subdomain partitions. The space-time subdomain partition is then defined as Cartesian product of the space and time subdomains. A subassembled problem involving independent subdomain corrections within the subassembled space is then defined on the space-time subdomain partition, with introduced perturbation terms on inner time interfaces, that is, the first and last time steps of the time subinterval. The coarse degrees of freedom are associated with the geometrical objects, namely, vertices, edges, and faces, among space-time subdomains. Every coarse degree of freedom is enforced to be continuous among subdomains by respective constraints. The space-time transfer operator is then constructed as a combination of the so-called space-time weighting operator and space-time "harmonic" extension operator. Finally, using all these components, the space time BDDC preconditioner is built as in the additive Schwarz method.

5 Numerical experiments

5.1 Space-time finite element adaptivity

5.1.1 A linear model problem

We first consider some numerical results for the adaptive solution of the linear model equation (1.1). For this purpose, we consider the exact solution

$$u(x_1,x_2,t) = (x_1^2 - x_1)(x_2^2 - x_2)(t^2 - t)e^{-100((x_1-t)^2+(x_2-t)^2)} \tag{5.15}$$

for $(x_1,x_2) \in (0,1)^2$ and $t \in (0,1]$, i. e. $Q = (0,1)^3$, and the given data are defined accordingly. The mesh information is prescribed in Table 7.1: number of degrees of freedom (#Dofs), number of tetrahedral elements (#Tets), and spatial and temporal mesh size (h/τ). The estimated order of convergence (eoc) for the absolute errors in $L^2(0,T;H_0^1(\Omega))$ on five mesh levels L_1–L_5 are given in Table 7.2.

A comparison of the convergence history using uniform and adaptive refinements is shown in Figure 7.3. We mention that, for the adaptive mesh refinements, the octasection [38] and bisection [12] methods have been used. We use the a posteriori error

Table 7.1: Mesh information on five uniformly refined levels: L_1–L_5.

Level	#Dofs	#Tets	h/τ
L_1	125	384	0.25
L_2	729	3,072	0.125
L_3	4,913	24,586	0.0625
L_4	35,937	196,608	0.03125
L_5	274,625	1,572,864	0.015625

Table 7.2: The estimated order of convergence (eoc) for the linear model problem on the mesh levels L_1-L_5.

Level	$\|e\|_{L_2(0,T;H_0^1(\Omega))}$	eoc
L_1	1.64×10^{-2}	–
L_2	1.49×10^{-2}	0.13
L_3	1.08×10^{-2}	0.46
L_4	6.14×10^{-3}	0.82
L_5	3.17×10^{-3}	0.96

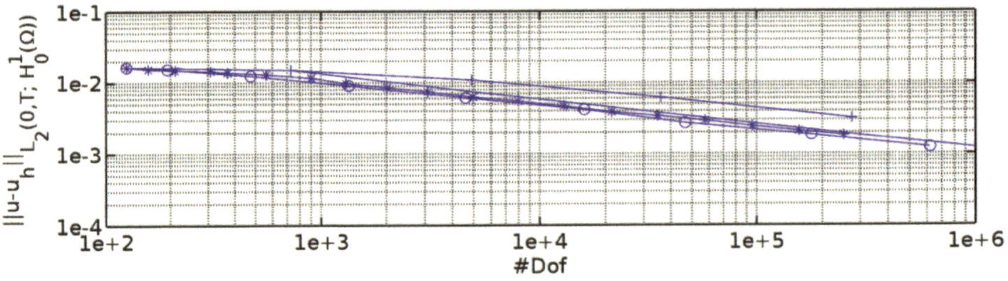

Figure 7.3: Convergence history of the space-time finite element methods for the linear model problem: uniform (– + –), octasection (– ∘ –), bisection (– ∗ –), and linear (–).

estimators and the adaptive method, as discussed in Subsection 3.8. From the results, we observe a linear order of convergence with both the uniform and two adaptive mesh refinements. The adaptive methods show more efficiency than the uniform one, in particular, in saving a number of degrees of freedom.

In Figure 7.4, we visualize the adaptive meshes using the octasection and bisection at the time levels $t = 0.25k$, $k = 0, \ldots, 4$. In Figure 7.5, we show the numerical solution and adaptive space-time meshes on three planes: $x_1 = 0.5$, $x_2 = 0.5$, $t = 0.5$. Our adaptive methods can effectively capture the moving interface in the space-time domain and make the corresponding adaptive mesh refinements in space-time. More results concerning space-time adaptivity can be found in [203, 204].

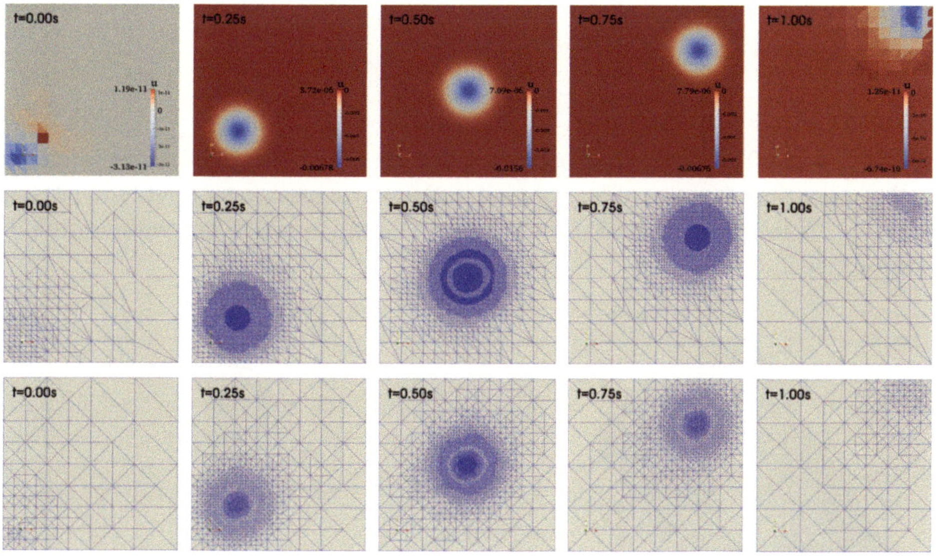

Figure 7.4: Visualization of numerical solutions and adaptive meshes at time levels $t = 0.25k$, $k = 0, \ldots, 4$, for the linear model problem: numerical solution (top), adaptive meshes using octasection at the 9th refinement level (middle), and bisection at the 19th refinement level (bottom).

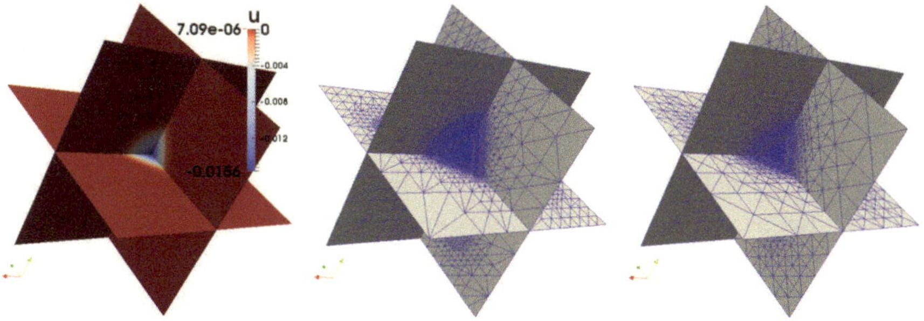

Figure 7.5: Visualization of the numerical solution and adaptive space-time meshes at $x_1 = 0.5$, $x_2 = 0.5$, $t = 0.5$ for the linear model problem: numerical solution (left), adaptive meshes using octasection at the 9th refinement level (middle), and bisection at the 19th refinement level (right).

5.1.2 A nonlinear model problem

As an extension of the linear heat equation (1.1), we also consider the following nonlinear parabolic equation with a third-order reaction term and a positive constant ε,

$$\partial_t u(x,t) - \Delta_x u(x,t) + \frac{1}{\varepsilon^2}(u^3(x,t) - u(x,t)) = 0, \tag{5.16}$$

which is the so-called Schlögl model [194] or the Nagumo equation [167], with applications in optimal control [52, 57]. We now consider an example with the exact solution

$$u(x_1, x_2, t) = \frac{1}{2}\left(1 - \tanh\frac{x_1 - st}{2\sqrt{2\varepsilon}}\right), \tag{5.17}$$

in the space-time domain $Q := (17, 19)^2 \times (0, 5]$, and with $\varepsilon = 0.38$, $s = \frac{3}{\sqrt{2\varepsilon}}$. Note that the spatial domain $\Omega = (17, 19)^2$ is chosen such that we can observe the moving interface in the space-time domain Q. The solution and uniform meshes on level 1 are plotted in Figure 7.6. It is easy to see that the solution changes from 0 to 1 smoothly within a narrow time interval.

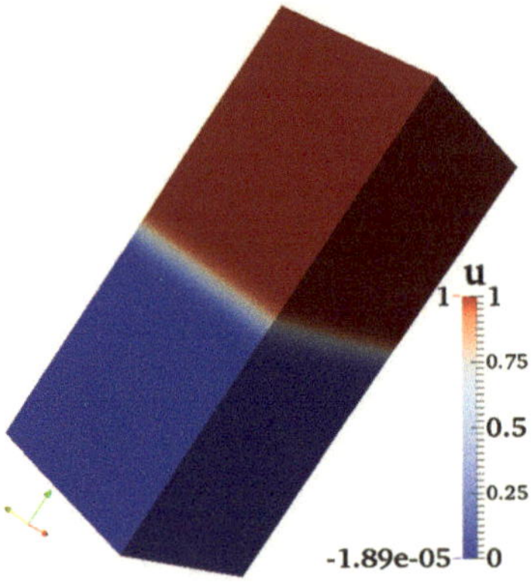

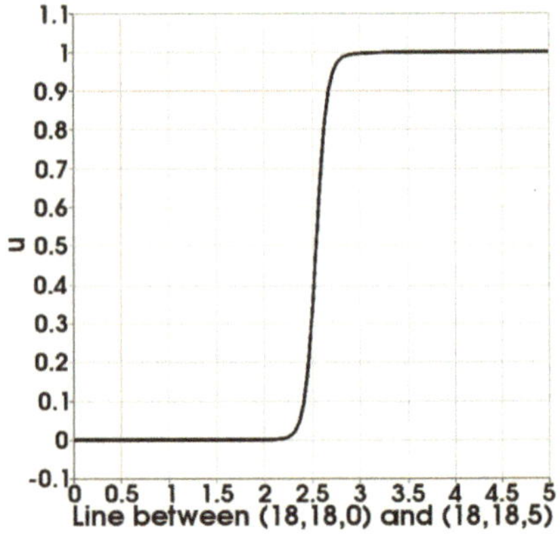

Figure 7.6: Visualization of the solution (left) for the nonlinear problem, the plot of the solution along the line (time) with the starting point (18, 18, 0), and end point (18, 18, 5) (right).

For this nonlinear model problem, we observe a linear order of convergence of the numerical solution as shown in Table 7.3. A comparison of adaptive and uniform refinements is demonstrated in Figure 7.7, where the adaptive ones show a better efficiency. As for the linear model problem, we use the residual-based local error indicator (3.10) on each element to drive the adaptive mesh refinements, where the local residual is replaced by the residual for the nonlinear problem. Note that we observe a better efficiency of the bisection method than the octasection one in this particular example.

Table 7.3: The estimated order of convergence for the nonlinear model problem on five mesh levels.

Level	#Dofs	$h(\tau)$	$\|e\|_{L_2(0,T;H_0^1(\Omega))}$	eoc
1	225	0.5	1.35×10^{-0}	–
2	1,377	0.25	8.23×10^{-1}	0.71
3	9,537	0.125	4.78×10^{-1}	0.78
4	70,785	0.0625	2.48×10^{-1}	0.95
5	545,025	0.03125	1.19×10^{-1}	1.06

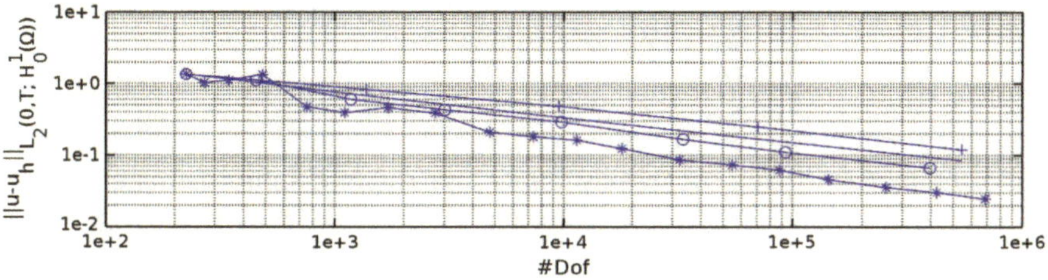

Figure 7.7: Convergence history of the space-time finite element for the nonlinear model problem: uniform (– + –), octasection (– ∘ –), bisection (– ∗ –), and linear (–).

We visualize the numerical solution on the planes $x_1 = 18$, $x_2 = 18$, and $t = 3.0$, and the adaptive space-time meshes on the boundary and on the plane $x_2 = 18$ as given in Figure 7.8. From the results, we see that the transient interface of the solution in the space-time domain can be captured by the two adaptive methods for this nonlinear model problem.

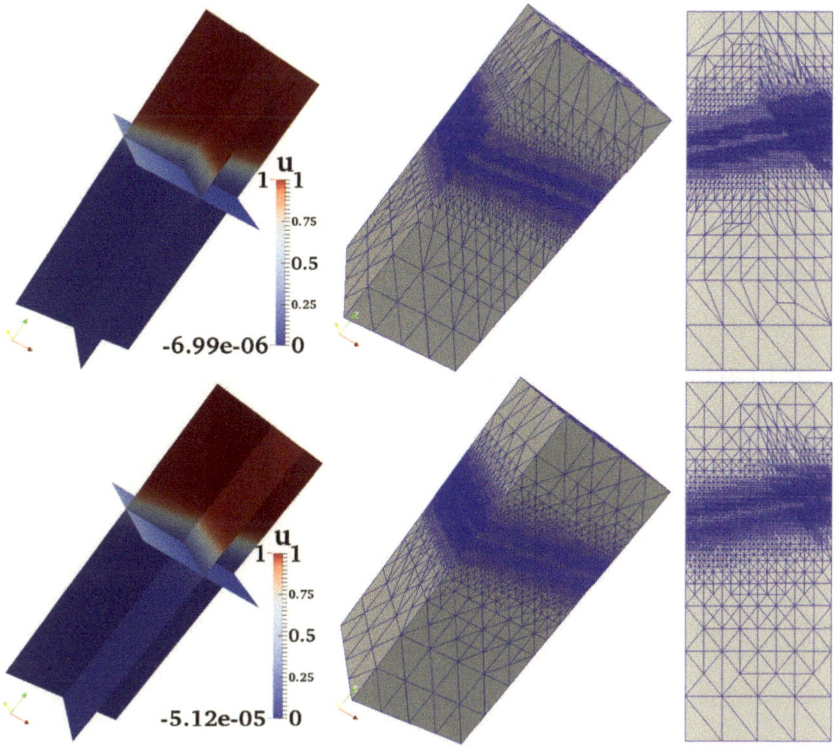

Figure 7.8: Visualization of the numerical solution on the planes $x_1 = 18$, $x_2 = 18$ and $t = 3.0$ (left), adaptive space-time meshes on the boundary (middle) and on the plane $x_2 = 18$ (right), using octasection at the 8th refinement level (top) and bisection at the 18th refinement level (bottom) for the nonlinear model problem.

5.2 Space-time algebraic multigrid methods

To study the performance of the algebraic multigrid methods, we focus on a comparison of algebraic multigrid preconditioned GMRES methods for solving the linear model problem (1.1), using the greedy (AMG_Greedy) and smoothed aggregation (AMG_SA) coarsening, as discussed in Subsection 4.2. That is because of the pure matrix-graph coarsening leads to a poor performance, and the compatible relaxation scheme leads to a similar performance as the greedy scheme. More numerical results in three and four dimensions are reported in [204].

We use the relative residual reduction tol $= 10^{-8}$ as a stopping criterion for the GMRES method with one V(1, 1)-cycle as a preconditioner. The number of AMG_Greedy and AMG_SA preconditioned GMRES iterations and costs in seconds (s) with one V(1, 1)-cycle are compared in Table 7.4. We mention that, for this example, we use uniform refinements from a bisection method [207]. The preconditioners show relatively good robustness and performance with respect to mesh refinements.

Table 7.4: Comparison of GMRES iterations and costs in seconds using one V(1, 1)-cycle preconditioner on the uniform meshes.

#Dofs	AMG_Greedy		AMG_SA	
	It	sec	It	sec
961	8	0.01	15	0.04
2,881	10	0.06	18	0.1
11,457	11	0.3	16	0.5
53,569	17	2.4	43	7.6
168,577	27	13.6	53	27.8

5.3 An application to a parabolic optimal control problem

In this example, we apply the Petrov–Galerkin space-time finite element method [202] to a parabolic optimal control problem. We consider the following model problem: Minimize the cost functional

$$\mathcal{J}(u,z) := \frac{1}{2}\int_{\Omega} [u(x,T) - \bar{u}(x)]^2 dx + \frac{1}{2}\rho\|z\|^2_{L^2(Q)} \tag{5.18}$$

with respect to the state u and control z, subject to the following parabolic problem:

$$\begin{aligned} \partial_t u(x,t) - \Delta_x u(x,t) &= z(x,t) && \text{for } (x,t) \in Q := \Omega \times (0,T), \\ u(x,t) &= g(x,t) && \text{for } (x,t) \in \Sigma := \Gamma \times (0,T), \\ u(x,0) &= u_0(x) && \text{for } x \in \Omega, \end{aligned} \tag{5.19}$$

where $\bar{u}(x)$ denotes a desired final temperature distribution, $z = z(x,t)$ is a control acting on the space-time cylinder Q, and $\rho > 0$ is a regularization parameter; see similar problems considered in [213].

The related optimality system consists of the following three parts:
(a) the primal problem:

$$\begin{aligned} \partial_t u(x,t) - \Delta_x u(x,t) &= z(x,t) && \text{for } (x,t) \in Q := \Omega \times (0,T), \\ u(x,t) &= g(x,t) && \text{for } (x,t) \in \Sigma := \Gamma \times (0,T), \\ u(x,0) &= u_0(x) && \text{for } x \in \Omega, \end{aligned} \tag{5.20}$$

(b) the adjoint problem:

$$\begin{aligned} -\partial_t p(x,t) - \Delta_x p(x,t) &= 0 && \text{for } (x,t) \in Q, \\ p(x,t) &= 0 && \text{for } (x,t) \in \Sigma, \\ p(x,T) &= u(x,T) - \bar{u}(x) && \text{for } x \in \Omega, \end{aligned} \tag{5.21}$$

and (c) the optimality condition:

$$p(x,t) + \rho z(x,t) = 0 \quad \text{for } (x,t) \in Q. \tag{5.22}$$

We apply the space-time finite element discretization method [202] to this optimality system. We consider $\Omega = (0,1)^2$, $T = 1$, i. e. $Q = (0,1)^3$, and the exact solutions:

$$\begin{aligned} u(x,t) &= \frac{3}{4}\zeta e^{2\pi^2 t} \sin(\pi x_1)\sin(\pi x_2), \\ p(x,t) &= -3\rho\pi^2\zeta e^{2\pi^2 t} \sin(\pi x_1)\sin(\pi x_2), \\ z(x,t) &= 3\pi^2\zeta e^{2\pi^2 t} \sin(\pi x_1)\sin(\pi x_2) \end{aligned}$$

with $\zeta = 10^{-9}$. The initial condition for the primal variable is then given by

$$u_0(x) = \frac{3}{4}\zeta \sin(\pi x_1)\sin(\pi x_2),$$

and the target is

$$\bar{u}(x) = \left(\frac{3}{4} + 3\rho\pi^2\right)\zeta e^{2\pi^2} \sin(\pi x_1)\sin(\pi x_2).$$

We run simulations on the five mesh levels as provided for the linear model problem in Section 5.1. The estimated order of convergence for u, p, and z in $L^2(0,T;H_0^1(\Omega))$ are given in Table 7.5 and Table 7.6 for $\rho = 1$ and $\rho = 0.001$, respectively. We observe a linear convergence rate as expected. Moreover, in $L^2(Q)$, we observe a quadratic order of convergence; see Table 7.7 and Table 7.8 for $\rho = 1$ and $\rho = 0.001$, respectively.

To solve the discrete optimality system, we utilize a monolithic AMG method using a blockwise ILU smoother [192], and a simple blockwise coarsening strategy [128]. We observe a quite robust AMG performance with respect to the mesh refinements and the regularization parameter; see Table 7.9. However, due to the high cost of the ILU smoother, the computation is rather expensive, which needs further investigations on finding more robust and efficient smoothers for the solution of such an optimality system.

Table 7.5: The estimated order of convergence in $L^2(0,T;H_0^1(\Omega))$ for u, p and z, $\rho = 1$. $\|e_v\|_{0,1} := \|v - v_h\|_{L^2(0,T;H_0^1(\Omega))}$, $v = u,p,z$.

#Dofs	$\|e_u\|_{0,1}$	eoc	$\|e_p\|_{0,1}$	eoc	$\|e_z\|_{0,1}$	eoc
375	2.3×10^{-1}	–	7.7×10^{0}	–	7.7×10^{0}	–
2,187	1.4×10^{-1}	0.68	4.5×10^{0}	0.79	4.5×10^{0}	0.79
14,739	7.1×10^{-2}	1.00	2.4×10^{0}	0.92	2.4×10^{0}	0.92
107,811	3.2×10^{-2}	1.13	1.2×10^{0}	0.98	1.2×10^{0}	0.98
823,875	1.5×10^{-2}	1.09	6.0×10^{-1}	1.01	6.0×10^{-1}	1.01

Table 7.6: The estimated order of convergence for the optimal control problem in $L^2(0,T;H_0^1(\Omega))$ for u, p and z, $\rho = 0.001$. $\|e_v\|_{0,1} := \|v - v_h\|_{L^2(0,T;H_0^1(\Omega))}$, $v = u,p,z$.

#Dofs	$\|e_u\|_{0,1}$	eoc	$\|e_p\|_{0,1}$	eoc	$\|e_z\|_{0,1}$	eoc
375	2.2×10^{-1}	–	7.6×10^{-2}	–	7.6×10^{0}	–
2,187	1.3×10^{-1}	0.76	4.4×10^{-2}	0.79	4.4×10^{0}	0.79
14,739	6.6×10^{-2}	0.98	2.5×10^{-2}	0.82	2.5×10^{0}	0.82
107,811	3.1×10^{-2}	1.09	1.3×10^{-2}	0.94	1.3×10^{0}	0.94
823,875	1.5×10^{-2}	1.04	6.6×10^{-3}	1.01	6.6×10^{-1}	1.01

Table 7.7: The estimated order of convergence for the optimal control problem in $L^2(Q)$ for u, p and z, $\rho = 1$. $\|e_v\|_{0,0} := \|v - v_h\|_{L^2(Q)}$, $v = u,p,z$.

#Dofs	$\|e_u\|_{0,0}$	eoc	$\|e_p\|_{0,0}$	eoc	$\|e_z\|_{0,0}$	eoc
375	4.0×10^{-2}	–	1.1×10^{0}	–	1.1×10^{0}	–
2,187	2.5×10^{-2}	0.7	5.6×10^{-1}	0.9	5.6×10^{-1}	0.9
14,739	1.1×10^{-2}	1.2	2.3×10^{-1}	1.3	2.3×10^{-1}	1.3
107,811	3.5×10^{-3}	1.6	7.3×10^{-2}	1.6	7.3×10^{-2}	1.6
823,875	9.6×10^{-4}	1.9	2.0×10^{-2}	1.8	2.0×10^{-2}	1.8

Table 7.8: The estimated order of convergence for the optimal control problem in $L^2(Q)$ for u, p and z on five mesh levels, $\rho = 0.001$. $\|e_v\|_{0,0} := \|v - v_h\|_{L^2(Q)}$, $v = u,p,z$.

#Dofs	$\|e_u\|_{0,0}$	eoc	$\|e_p\|_{0,0}$	eoc	$\|e_z\|_{0,0}$	eoc
375	3.8×10^{-2}	–	1.0×10^{-2}	–	1.0×10^{0}	–
2,187	2.3×10^{-2}	0.7	5.0×10^{-3}	1.0	5.0×10^{-1}	1.0
14,739	9.9×10^{-3}	1.2	2.0×10^{-3}	1.3	2.0×10^{-1}	1.3
107,811	3.2×10^{-3}	1.6	6.4×10^{-4}	1.6	6.4×10^{-2}	1.6
823,875	8.7×10^{-4}	1.9	1.8×10^{-4}	1.9	1.8×10^{-2}	1.9

6 Conclusions

This work has reviewed space-time finite element methods for the approximate solution of related parabolic-type evolution equations. In particular, the following issues

Table 7.9: AMG iterations and computational costs for solving the optimality system.

#Dofs	$\rho = 10^2$		$\rho = 10^1$		$\rho = 10^0$		$\rho = 10^{-1}$		$\rho = 10^{-2}$	
	It	sec	It	sec	It	sec	It	sec	It	sec
375	2	0.03	2	0.03	2	0.02	2	0.03	2	0.03
2,187	3	0.09	3	0.10	3	0.09	3	0.10	3	0.10
14,739	3	0.98	3	1.21	3	1.03	3	1.04	3	1.01
107,811	3	19.6	3	18.9	4	24.3	4	23.3	4	25.3
823,875	4	509	4	522	4	500	4	518	4	585

have been addressed: space-time finite element discretization, a posteriori error estimates and corresponding space-time adaptive schemes, and modern parallel space-time solution methods.

Some numerical examples using the Petrov–Galerkin space-time finite element method proposed in [202] have been performed for both linear and nonlinear parabolic problems, which show advantages using simultaneous space-time adaptivity. We have provided numerical tests on algebraic multigrid methods for solving the large-scale linear systems of space-time finite element equations, which show the relative robustness of the solution methods with respect to the mesh refinements. Applicability of the proposed space-time finite element method [202] to the parabolic optimal control problems is confirmed by our numerical examples as well.

Bibliography

[1] Y. Achdou, F. Hecht, and D. Pommier. A posteriori error estimates for parabolic variational inequalities. *J. Sci. Comput.*, 37 (3): 336–366, 2008.

[2] M. Ainsworth. A synthesis of a posteriori error estimation techniques for conforming, nonconforming and discontinuous Galerkin finite element methods. In *Recent Advances in Adaptive Computation*. Volume 383 of Contemp. Math., pages 1–14, AMS, Providence, RI, 2005.

[3] M. Ainsworth and J. T. Oden. A posteriori error estimation in finite element analysis. *Comput. Methods Appl. Mech. Engrg.*, 142 (1): 1–88, 1997.

[4] M. Ainsworth and J. T. Oden. *A posteriori error estimation in finite element analysis*. Pure and applied mathematics. John Wiley & Sons, New York, 2000.

[5] G. Akrivis and P. Chatzipantelidis. A posteriori error estimates for the two-step backward differentiation formula method for parabolic equations. *SIAM J. Numer. Anal.*, 48 (1): 109–132, 2010.

[6] R. Andreev. *Stability of space-time Petrov–Galerkin discretizations for parabolic evolution problems*. PhD thesis, ETH Zürich, 2012.

[7] R. Andreev. Stability of sparse space–time finite element discretizations of linear parabolic evolution problems. *IMA J. Numer. Anal.*, 33: 242–260, 2013.

[8] R. Andreev. On long time integration of the heat equation. *Calcolo*, 53 (1): 19–34, 2016.

[9] H. Antil, E. Otárola, and A. J. Salgado. A space-time fractional optimal control problem: Analysis and discretization. *SIAM J. Control Optim.*, 54 (3): 1295–1328, 2016.

[10] T. Apel. Interpolation of non-smooth functions on anisotropic finite element meshes. *ESAIM: M2AN*, 33 (6): 1149–1185, 1999.

[11] D. N. Arnold and H. Chen. Finite element exterior calculus for parabolic problems. *ESAIM: M2AN*, 51 (1): 17–34, 2017.

[12] D. N. Arnold, A. Mukherjee, and L. Pouly. Locally adapted tetrahedral meshes using bisection. *SIAM J. Sci. Comput.*, 22 (2): 431–448, 2000.

[13] D. N. Arnold, R. S. Falk, and R. Winther. Finite element exterior calculus, homological techniques, and applications. *Acta Numerica.*, 15: 1–155, 2006.

[14] D. N. Arnold, R. S. Falk, and R. Winther. Finite element exterior calculus: from Hodge theory to numerical stability. *Bull. Amer. Math. Soc.*, 47: 281–354, 2010.

[15] D. G. Aronson. Non-negative solutions of linear parabolic equations. *Ann. Sc. Norm. Super. Pisa Cl. Sci.*, 22 (4): 607–694, 1968.

[16] A. K. Aziz and P. Monk. Continuous finite elements in space and time for the heat equation. *Math. Comput.*, 52 (186): 255–274, 1989.

[17] S. Badia and M. Olm. Space-time balancing domain decomposition. *SIAM J. Sci. Comput.*, 39 (2): C194–C213, 2017.

[18] W. Bangerth and R. Rannacher. *Adaptive finite element methods for differential equations.* Lectures in Mathematics, ETH Zürich. Springer, Berlin, 2006.

[19] R. E. Bank. *PLTMG: A software package for solving elliptic partial differential equations.* SIAM, Philadelphia, 1998.

[20] R. E. Bank and J. Xu. Asymptotically exact a posteriori error estimators. I. Grids with superconvergence. *SIAM J. Numer. Anal.*, 41 (6): 2294–2312, 2004.

[21] R. E. Bank, A. H. Sherman, and A. Weiser. Refinement algorithms and data structures for regular local mesh refinement. In R. Stepleman, editor, *Scientific Computing*. IMACS/North Holland, 1983.

[22] R. E. Bank, P. S. Vassilevski, and L. T. Zikatanov. Arbitrary dimension convection–diffusion schemes for space–time discretizations. *J. Comput. Appl. Math.*, 310: 19–31, 2017.

[23] E. Bänsch. Local mesh refinement in 2 and 3 dimensions. *IMPACT Comput. Sci. Eng.*, 3: 181–191, 1991.

[24] E. Bänsch, F. Karakatsani, and C. Makridakis. A posteriori error control for fully discrete Crank–Nicolson schemes. *SIAM J. Numer. Anal.*, 50 (6): 2845–2872, 2012.

[25] L. Banz and E. P. Stephan. hp-adaptive IPDG/TDG–FEM for parabolic obstacle problems. *Comput. Math. Appl.*, 67 (4): 712–731, 2014.

[26] S. Bartels and R. Müller. Quasi-optimal and robust a posteriori error estimates in $L^\infty(L^2)$ for the approximation of Allen–Cahn equations past singularities. *Math. Comput.*, 80 (274): 761–780, 2011.

[27] P. Bastian, K. Birken, K. Johannsen, S. Lang, N. Neuss, H. R. Reichert, and C. Wieners. UG–A flexible software toolbox for solving partial differential equations. *Comput. Visual. Sci.*, 1 (1): 27–40, 1997.

[28] R. Becker and R. R. Rannacher. An optimal control approach to a posteriori error estimation in finite element methods. *Acta Numer.*, 10: 1–102, 2001.

[29] R. Becker and B. Vexler. A posteriori error estimation for finite element discretization of parameter identification problems. *Numer. Math.*, 96 (3): 435–459, 2004.

[30] R. Becker and B. Vexler. Mesh refinement and numerical sensitivity analysis for parameter calibration of partial differential equations. *J. Comput. Phys.*, 206 (1): 95–110, 2005.

[31] R. Becker, H. Kapp, and R. Rannacher. Adaptive finite element methods for optimal control of partial differential equations: Basic concept. *SIAM J. Control Optim.*, 39 (1): 113–132, 2000.

[32] A. Bergam, C. Bernardi, and Z. Mghazli. A posteriori analysis of the finite element discretization of some parabolic equations. *Math. Comput.*, 74 (251): 1117–1138, 2005.

[33] C. Bernardi and R. Verfürth. A posteriori error analysis of the fully discretized time-dependent Stokes equations. *ESAIM: M2AN*, 38 (3): 437–455, 2004.

[34] S. Berrone. Adaptive discretization of stationary and incompressible Navier–Stokes equations by stabilized finite element methods. *Comput. Methods Appl. Mech. Engrg.*, 190 (34): 4435–4455, 2001.

[35] S. Berrone. Robust a posteriori error estimates for finite element discretizations of the heat equation with discontinuous coefficients. *ESAIM: M2AN*, 40 (6): 991–1021, 2006.

[36] S. Berrone. A local-in-space-timestep approach to a finite element discretization of the heat equation with a posteriori estimates. *SIAM J. Numer. Anal.*, 47 (4): 3109–3138, 2009.

[37] S. Berrone and M. Marro. Space-time adaptive simulations for unsteady Navier–Stokes problems. *Computers & Fluids*, 38 (6): 1132–1144, 2009.

[38] J. Bey. Tetrahedral grid refinement. *Computing*, 55 (4): 355–378, 1995.

[39] J. Bey. Simplicial grid refinement: on Freudenthal's algorithm and the optimal number of congruence classes. *Numer. Math.*, 85 (1): 1–29, 2000.

[40] P. Binev, W. Dahmen, and R. DeVore. Adaptive finite element methods with convergence rates. *Numer. Math.*, 97 (2): 219–268, 2004.

[41] M. Boman. *On a posteriori error analysis in the maximum norm*. PhD thesis, Chalmers University of Technology and Göthenburg University, 2000.

[42] F. Bornemann, B. Erdmann, and R. Kornhuber. Adaptive multivlevel methods in three space dimensions. *Int. J. Numer. Meth. Engng.*, 36 (18): 3187–3203, 1993.

[43] D. Braess and J. Schöberl. Equilibrated residual error estimator for edge elements. *Math. Comput.*, 77 (262): 651–672, 2008.

[44] D. Braess, V. Pillwein, and J. Schöberl. Equilibrated residual error estimates are p-robust. *Comput. Methods Appl. Mech. Engrg.*, 198 (13): 1189–1197, 2009.

[45] A. Brandt. Multi-level adaptive solutions to boundary-value problems. *Math. Comput.*, 31 (138): 333–390, 1977.

[46] A. Brandt. Algebraic multigrid theory: The symmetric case. *Appl. Math. Comput.*, 19 (1): 23–56, 1986.

[47] A. Brandt. Rigorous quantitative analysis of multigrid. I. Constant coefficients two-level cycle with L_2-norm. *SIAM J. Numer. Anal.*, 31 (6): 1695–1730, 1994.

[48] A. Brandt, S. F. McCormick, and J. W. Ruge. Algebraic multigrid (AMG) for sparse matrix equations. In D. J. Evans, editor, *Sparsity and its applications*, pages 257–284, Cambridge University Press Cambridge, 1984.

[49] J. J. Brannick and R. D. Falgout. Compatible relaxation and coarsening in algebraic multigrid. *SIAM J. Sci. Comput.*, 32 (3): 1393–1416, 2010.

[50] W. Briggs, V. Henson, and S. McCormick. *A multigrid tutorial*. SIAM, Philadelphia, second edition, 2000.

[51] A. N. Brooks and T. J. R. Hughes. Streamline upwind/Petrov–Galerkin formulations for convection dominated flows with particular emphasis on the incompressible Navier–Stokes equations. *Comput. Methods Appl. Mech. Engrg.*, 32 (1): 199–259, 1982.

[52] R. Buchholz, H. Engel, E. Kammann, and F. Tröltzsch. On the optimal control of the Schlögl–model. *Comput. Optim. Appl.*, 56 (1): 153–185, 2013.

[53] E. Burman and A. Ern. Continuous interior penalty hp-finite element methods for advection and advection–diffusion equations. *Math. Comput.*, 76 (259): 1119–1140, 2007.

[54] L. Caffarelli and L. Silvestre. An extension problem related to the fractional Laplacian. *Comm. Partial Differential Equations*, 32 (8): 1245–1260, 2007.

[55] C. Carstensen and S. Bartels. Each averaging technique yields reliable a posteriori error control in FEM on unstructured grids. I. Low order conforming, nonconforming, and mixed FEM. *Math. Comput.*, 71 (239): 945–969, 2002.

[56] C. Carstensen, W. Liu, and N. Yan. A posteriori error estimates for finite element approximation of parabolic p–Laplacian. *SIAM J. Numer. Anal.*, 43 (6): 2294–2319, 2006.
[57] E. Casas, C. Ryll, and F. Tröltzsch. Sparse optimal control of the Schlögl and FitzHugh–Nagumo systems. *Comput. Methods Appl. Math.*, 13: 551–566, 2013.
[58] M. Charalambos. Space and time reconstructions in a posteriori analysis of evolution problems. *ESAIM: Proc.*, 21: 31–44, 2007.
[59] Z. Chen and J. Feng. An adaptive finite element algorithm with reliable and efficient error control for linear parabolic problems. *Math. Comput.*, 73 (247): 1167–1193, 2004.
[60] Z. Chen, R. H. Nochetto, and A. Schmidt. Error control and adaptivity for a phase relaxation model. *ESAIM: M2AN*, 34 (4): 775–797, 2000.
[61] P. G. Ciarlet. *The finite element methods for elliptic problems*. North–Holland, Amsterdam, 1978.
[62] P. Clément. Approximation by finite element functions using local regularization. *ESAIM: M2NA*, 9 (R2): 77–84, 1975.
[63] J. de Frutos and J. Novo. A posteriori error estimation with the p-version of the finite element method for nonlinear parabolic differential equations. *Comput. Methods Appl. Mech. Engrg.*, 191 (43): 4893–4904, 2002.
[64] J. de Frutos and J. Novo. Element-wise a posteriori estimates based on hierarchical bases for non–linear parabolic problems. *Int. J. Numer. Meth. Engng.*, 63 (8): 1146–1173, 2005.
[65] L. Demkowicz and J. Gopalakrishnan. Analysis of the DPG method for the Poisson equation. *SIAM J. Numer. Anal.*, 49 (5): 1788–1809, 2011.
[66] L. Demkowicz and J. Gopalakrishnan. A class of discontinuous Petrov–Galerkin methods. Part II: Optimal test functions. *Numer. Methods Partial Differential Equations*, 27 (1): 70–105, 2011.
[67] L. F. Demkowicz and J. Gopalakrishnan. An overview of the discontinuous Petrov Galerkin method. In X. Feng, O. Karakashian, and Y. Xing, editors, *Recent developments in discontinuous Galerkin finite element methods for partial differential equations: 2012 John H Barrett memorial lectures*, pages 149–180, Springer, Cham, 2014.
[68] A. Demlow and C. Makridakis. Sharply local pointwise a posteriori error estimates for parabolic problems. *Math. Comput.*, 79 (271): 1233–1262, 2010.
[69] A. Demlow, O. Lakkis, and C. Makridakis. A posteriori error estimates in the maximum norm for parabolic problems. *SIAM J. Numer. Anal.*, 47 (3): 2157–2176, 2009.
[70] P. Destuynder and B. Métivet. Explicit error bounds in a conforming finite element method. *Math. Comput.*, 68 (228): 1379–1396, 1999.
[71] D. Devaud. *hp-approximation of linear parabolic evolution problems in* $H^{1/2}$. PhD thesis, ETH Zürich, 2017.
[72] D. Devaud and C. Schwab. Space-time hp-approximation of parabolic equations. *Calcolo*, 55, 2018.
[73] J. J. Douglas and T. Dupont. Galerkin methods for parabolic equations. *SIAM J. Numer. Anal.*, 7 (4): 575–626, 1970.
[74] S. D. Èidel'man and S. D. Ivasišen. Investigation of the Green's matrix of a homogeneous parabolic boundary value problem. *Tr. Mosk. Mat. Obs*, 23: 179–234, 1970.
[75] T. E. Ellis. *Space–time discontinuous Petrov–Galerkin finite elements for transient fluid mechanics*. PhD thesis, University of Texas at Austin, 2016.
[76] T. E. Ellis, L. Demkowicz, and J. Chan. Locally conservative discontinuous Petrov–Galerkin finite elements for fluid problems. *Comput. & Math. Appl.*, 68 (11): 1530–1549, 2014.
[77] K. Eriksson and C. Johnson. Error estimates and automatic time step control for nonlinear parabolic problems, I. *SIAM J. Numer. Anal.*, 24 (1): 12–23, 1987.
[78] K. Eriksson and C. Johnson. Adaptive finite element methods for parabolic problems I: A linear model problem. *SIAM J. Numer. Anal.*, 28 (1): 43–77, 1991.

[79] K. Eriksson and C. Johnson. Adaptive finite element methods for parabolic problems II: Optimal error estimates in $L_\infty(L_2)$ and $L_\infty(L_\infty)$. *SIAM J. Numer. Anal.*, 32 (3): 706–740, 1995.

[80] K. Eriksson and C. Johnson. Adaptive finite element methods for parabolic problems IV: Nonlinear problems. *SIAM J. Numer. Anal.*, 32 (6): 1729–1749, 1995.

[81] K. Eriksson and C. Johnson. Adaptive finite element methods for parabolic problems V: Long-time integration. *SIAM J. Numer. Anal.*, 32 (6): 1750–1763, 1995.

[82] K. Eriksson, C. Johnson, and V. Thomée. Time discretization of parabolic problems by the discontinuous Galerkin method. *ESAIM: M2AN*, 19 (4): 611–643, 1985.

[83] K. Eriksson, C. Johnson, and S. Larsson. Adaptive finite element methods for parabolic problems VI: Analytic semigroups. *SIAM J. Numer. Anal.*, 35 (4): 1315–1325, 1998.

[84] A. Ern and M. Vohralík. A posteriori error estimation based on potential and flux reconstruction for the heat equation. *SIAM J. Numer. Anal.*, 48 (1): 198–223, 2010.

[85] A. Ern and M. Vohralík. Polynomial-degree-robust a posteriori estimates in a unified setting for conforming, nonconforming, discontinuous Galerkin, and mixed discretizations. *SIAM J. Numer. Anal.*, 53 (2): 1058–1081, 2015.

[86] A. Ern, S. Nicaise, and M. Vohralík. An accurate H(div) flux reconstruction for discontinuous Galerkin approximations of elliptic problems. *C. R. Math. Acad. Sci. Paris*, 345 (12): 709–712, 2007.

[87] A. Ern, I. Smears, and M. Vohralík. Guaranteed, locally space-time efficient, and polynomial-degree robust a posteriori error estimates for high-order discretizations of parabolic problems. *SIAM J. Numer. Anal.*, 55 (6): 2811–2834, 2017.

[88] A. Ern, I. Smears, and M. Vohralík. Equilibrated flux a posteriori error estimates in $L^2(H^1)$-norms for high-order discretization of parabolic problems. *IMA J. Numer. Anal.*, published online, 2018. https://doi.org/10.1093/imanum/dry035.

[89] A. Ern, A. F. Stephansen, and M. Vohralík. Guaranteed and robust discontinuous Galerkin a posteriori error estimates for convection–diffusion–reaction problems. *J. Comput. Appl. Math.*, 234 (1): 114–130, 2010.

[90] R. D. Falgout and P. S. Vassilevski. On generalizing the algebraic multigrid framework. *SIAM J. Numer. Anal.*, 42 (4): 1669–1693, 2004.

[91] R. D. Falgout, S. Friedhoff, T. V. Kolev, S. P. MacLachlan, and J. B. Schroder. Parallel time integration with multigrid. *SIAM J. Sci. Comput.*, 36 (6): C635–C661, 2014.

[92] R. D. Falgout, S. Friedhoff, T. V. Kolev, S. P. MacLachlan, J. B. Schroder, and S. Vandewalle. Multigrid methods with space–time concurrency. *Comput. Visual. Sci.*, 18: 123–143, 2017.

[93] R. D. Falgout, T. A. Manteuffel, B. O'Neill, and J. B. Schroder. Multigrid reduction in time for nonlinear parabolic problems: A case study. *SIAM J. Sci. Comput.*, 39 (5): S298–S322, 2017.

[94] X. Feng and H. Wu. A posteriori error estimates and an adaptive finite element method for the Allen–Cahn equation and the mean curvature flow. *J. Sci. Comput.*, 24 (2): 121–146, 2005.

[95] F. Fierro and A. Veeser. A posteriori error estimators, gradient recovery by averaging, and superconvergence. *Numer. Math.*, 103 (2): 267–298, 2006.

[96] M. Fontes. *Parabolic equations with low regularity*. PhD thesis, Lund Institute of Technology, 1996.

[97] L. Formaggia and S. Perotto. New anisotropic a priori error estimates. *Numer. Math.*, 89 (4): 641–667, 2001.

[98] L. Formaggia and S. Perotto. Anisotropic error estimates for elliptic problems. *Numer. Math.*, 94 (1): 67–92, 2003.

[99] S. R. Franco, F. G. Francisco, M. A. V. Pinto, and C. Rodrigo. Multigrid method based on a space-time approach with standard coarsening for parabolic problems. *Appl. Math. Comput.*, 317: 25–34, 2018.

[100] H. Freudenthal. Simplizialzerlegungen von beschränkter Flachheit. *Annals. Math.*, 43 (3): 580–582, 1942.
[101] A. Friedman. *Variational principles and free boundary problems*. Wiley, New York, 1982.
[102] T. Führer, N. Heuer, and J. S. Gupta. A time-stepping DPG scheme for the heat equation. *Comput. Methods Appl. Math.*, 17: 237–252, 2016.
[103] A. Gaevskaya, R. H. W. Hoppe, and S. I. Repin. A posteriori estimates for cost functionals of optimal control problems. In *Numerical mathematics and advanced applications: Proceedings of ENUMATH: Santiago de Compostela, Spain, July 2005*, pages 308–316, Springer, Berlin, Heidelberg, 2006.
[104] M. J. Gander. 50 years of time parallel time integration. In T. Carraro, M. Geiger, S. Körkel, and R. Rannacher, editors, *Multiple shooting and time domain decomposition methods: MuS-TDD, Heidelberg, May 6–8, 2013*, pages 69–113, Springer, Cham, 2015.
[105] M. J. Gander and M. Neumüller. Analysis of a new space-time parallel multigrid algorithm for parabolic problems. *SIAM J. Sci. Comput.*, 38 (4): A2173–A2208, 2016.
[106] B. García–Archilla and E. S. Titi. Postprocessing the Galerkin method: The finite-element case. *SIAM J. Numer. Anal.*, 37 (2): 470–499, 2000.
[107] A. Gillette, M. Holst, and Y. Zhu. Finite element exterior calculus for evolution problems. *J. Comput. Math.*, 35 (2): 187–212, 2017.
[108] J. Grande. Red-green refinement of simplicial meshes in D dimensions. *Math. Comput.*, 88: 751–782, 2019.
[109] S. Gross and A. Reusken. Parallel multilevel tetrahedral grid refinement. *SIAM J. Sci. Comput.*, 26 (4): 1261–1288, 2005.
[110] C. Hesch, S. Schuss, M. Dittmann, S. R. Eugster, M. Favino, and R. Krause. Variational space-time elements for large-scale systems. *Comput. Methods Appl. Mech. Engrg.*, 326: 541–572, 2017.
[111] J. G. Heywood and R. Rannacher. Finite element approximation of the nonstationary Navier–Stokes problem, part II: Stability of solutions and error estimates uniform in time. *SIAM J. Numer. Anal.*, 23 (4): 750–777, 1986.
[112] J. G. Heywood and R. Rannacher. Finite element approximation of the nonstationary Navier–Stokes problem, part III: Smoothing property and higher order error estimates for spatial discretization. *SIAM J. Numer. Anal.*, 25 (3): 489–512, 1988.
[113] C. Hofer, U. Langer, M. Neumüller, and I. Toulopoulos. Time-multipatch discontinuous Galerkin space-time isogeometric analysis of parabolic evolution problems. RICAM Report 2017-26, Johann Radon Institute for Computational and Applied Mathematics, Linz, 2017.
[114] B. Holm and S. Matculevich. Fully reliable error control for first-order evolutionary problems. *Comput. Math. Appl.*, 75 (4): 1302–1329, 2018.
[115] M. Holst and A. Stern. Geometric variational crimes: Hilbert complexes, finite element exterior calculus, and problems on hypersurfaces. *Found. Comput. Math.*, 12 (3): 263–293, 2012.
[116] M. Holst and C. Tiee. Finite element exterior calculus for parabolic evolution problems on Riemannian hypersurfaces. *J. Comput. Math.*, 36: 792–832, 2018.
[117] G. Horton and S. Vandewalle. A space-time multigrid method for parabolic partial differential equations. *SIAM J. Sci. Comput.*, 16 (4): 848–864, 1995.
[118] T. J. R. Hughes and G. M. Hulbert. Space-time finite element methods for elastodynamics: Formulations and error estimates. *Comput. Methods Appl. Mech. Engrg.*, 66 (3): 339–363, 1988.
[119] T. J. R. Hughes, J. A. Cottrell, and Y. Bazilevs. Isogeometric analysis: CAD, finite elements, NURBS, exact geometry and mesh refinement. *Comput. Methods Appl. Mech. Engrg.*, 194 (39): 4135–4195, 2005.
[120] P. Jamet. Galerkin-type approximations which are discontinuous in time for parabolic

equations in a variable domain. *SIAM J. Numer. Anal.*, 15 (5): 912–928, 1978.
[121] S. Kaczmarz. Angenäherte Auflösung von Systemen linearer Gleichungen. *Bull. Acad. Polon. Sci. et Lett. A*, 35: 355–357, 1937.
[122] E. Karabelas. *Space-time discontinuous Galerkin methods for cardiac electro–mechanics*. Volume 28 of Monographic Series TU Graz: Computation in Engineering and Science, 2015.
[123] E. Karabelas and M. Neumüller. Generating admissible space-time meshes for moving domains in $d+1$-dimensions. In U. Langer and O. Steinbach, editors, *Space-time methods: Application to partial differential equations*, Volume 25 of Radon series on computational and applied mathematics, pages 185–206, de Gruyter, Berlin, 2019.
[124] E. Karabelas and O. Steinbach. Space-time DG methods for the coupled electro-mechanical activation of the human heart. *Proc. Appl. Math. Mech.*, 14 (1): 839–840, 2014.
[125] O. A. Karakashian and F. Pascal. A posteriori error estimates for a discontinuous Galerkin approximation of second-order elliptic problems. *SIAM J. Numer. Anal.*, 41 (6): 2374–2399, 2004.
[126] T. Katsaounis and I. Kyza. A posteriori error control and adaptivity for Crank–Nicolson finite element approximations for the linear Schrödinger equation. *Numer. Math.*, 129 (1): 55–90, 2015.
[127] D. Kessler, R. H. Nochetto, and A. Schmidt. A posteriori error control for the Allen–Cahn problem: Circumventing Gronwall's inequality. *ESAIM: M2AN*, 38 (1): 129–142, 2004.
[128] F. Kickinger. Algebraic multigrid for discrete elliptic second-order problems. In W. Hackbusch, editor, *Multigrid methods V. Proceedings of the 5th European Multigrid Conference, Lecture notes in computational sciences and engineering, vol. 3*, pages 157–172, Springer, Berlin, Heidelberg, 1998.
[129] N. Kopteva and T. Linss. Maximum norm a posteriori error estimation for parabolic problems using elliptic reconstructions. *SIAM J. Numer. Anal.*, 51 (3): 1494–1524, 2013.
[130] N. Kopteva and T. Linss. Improved maximum-norm a posteriori error estimates for linear and semilinear parabolic equations. *Adv. Comput. Math.*, 2017.
[131] I. Kossaczký. A recursive approach to local mesh refinement in two and three dimensions. *J. Comput. Appl. Math.*, 55: 275–288, 1994.
[132] C. Kreuzer. Reliable and efficient a posteriori error estimates for finite element approximations of the parabolic p-Laplacian. *Calcolo*, 50 (2): 79–110, 2013.
[133] C. Kreuzer, C. A. Möller, A. Schmidt, and K. G. Siebert. Design and convergence analysis for an adaptive discretization of the heat equation. *IMA J. Numer. Anal.*, 32 (4): 1375–1403, 2012.
[134] O. A. Ladyženskaya. *The boundary value problems of mathematical physics*. Springer, New York, 1985.
[135] O. A. Ladyženskaya, V. A. Solonnikov, and N. N. Ural'ceva. *Linear and quasi–linear equations of parabolic type*. AMS, Providence, 1968.
[136] O. Lakkis and C. Makridakis. Elliptic reconstruction and a posteriori error estimates for fully discrete linear parabolic problems. *Math. Comput.*, 75 (256): 1627–1658, 2006.
[137] O. Lakkis and T. Pryer. Gradient recovery in adaptive finite element methods for parabolic problems. *IMA J. Numer. Anal.*, 32 (1): 246–278, 2012.
[138] O. Lakkis, C. Makridakis, and T. Pryer. A comparison of duality and energy a posteriori estimates for $L_\infty(0,T;L_2(\Omega))$ in parabolic problems. *Math. Comput.*, 84 (294): 1537–1569, 2015.
[139] J. Lang. *Adaptive multilevel solution of nonlinear parabolic PDE systems*. Volume 16 of Lect. Notes Comput. Sci. Eng. Springer, Berlin, 2001.
[140] U. Langer, S. E. Moore, and M. Neumüller. Space–time isogeometric analysis of parabolic evolution problems. *Comput. Methods Appl. Mech. Engrg.*, 306: 342–363, 2016.
[141] U. Langer, S. I. Repin, and M. Wolfmayr. Functional a posteriori error estimates for

time–periodic parabolic optimal control problems. *Numer. Func. Anal. Opt.*, 37 (10): 1267–1294, 2016.

[142] U. Langer, S. I. Repin, and M. Wolfmayr. Functional a posteriori error estimates for parabolic time–periodic boundary value problems. *Comput. Methods Appl. Math.*, 15 (3): 353–372, 2016.

[143] U. Langer, S. V. Matculevich, and S. I. Repin. A posteriori error estimates for space–time IgA approximations to parabolic initial boundary value problems, 2017. arXiv:1612.08998 [math.NA].

[144] U. Langer, S. V. Matculevich, and S. I. Repin. Functional type error control for stabilised space–time IgA approximations to parabolic problems. In I. Lirkov and S. D. Margenov, editors, *Large-scale scientific computing: 11th International Conference, LSSC 2017, Sozopol, Bulgaria, June 5–9, 2017*, pages 57–66. Springer, 2017.

[145] U. Langer, S. V. Matculevich, and S. I. Repin. Adaptive space–time isogeometric analysis for parabolic evolution problems. In U. Langer and O. Steinbach, editors, *Space–time methods: Application to partial differential equations*. Volume 25 of Radon series on computational and applied mathematics, pages 141–183, de Gruyter, Berlin, 2019.

[146] U. Langer, M. Neumüller, and A. Schafelner. Space–time finite element methods for parabolic evolution problems with variable coefficients. In T. Apel, U. Langer, A. Meyer, and O. Steinbach, editors, *Advanced finite element methods with applications. Selected papers from the 30th Chemnitz FEM Symposium 2017*. Volume 128 of Lecture Notes in Computational Science and Engineering, pages 247–275, Springer, 2019.

[147] U. Langer, M. Neumüller, and I. Toulopoulos. Multipatch space–time isogeometric analysis of parabolic diffusion problems. In I. Lirkov and S. S. Margenov, editors, *Large–scale scientific computing*, pages 21–32. Springer, 2018.

[148] S. Larsson and C. Schwab. Compressive space-time Galerkin discretizations of parabolic partial differential equations. Technical report 2015-04, Seminar for Applied Mathematics, ETH Zürich, 2015.

[149] I. V. Lashuk and P. S. Vassilevski. The construction of coarse de Rham complexes with improved approximation properties. *Comput. Methods Appl. Math.*, 14: 257–303, 2014.

[150] D. Leykekhman and B. Vexler. Pointwise best approximation results for Galerkin finite element solutions of parabolic problems. *SIAM J. Numer. Anal.*, 54 (3): 1365–1384, 2016.

[151] D. Leykekhman and B. Vexler. Discrete maximal parabolic regularity for Galerkin finite element methods. *Numer. Math.*, 135 (3): 923–952, 2017.

[152] D. Leykekhman and L. B. Wahlbin. A posteriori error estimates by recovered gradients in parabolic finite element equations. *BIT Numer. Math.*, 48 (3): 585–605, 2008.

[153] B. Li and Z. Zhang. Analysis of a class of superconvergence patch recovery techniques for linear and bilinear finite elements. *Numer. Methods Partial Differential Equations*, 15 (2): 151–167, 1999.

[154] Y. Lin, X. Li, and C. Xu. Finite difference/spectral approximates for the fractional cable equation. *Math. Comput.*, 80 (275): 1369–1396, 2011.

[155] J. L. Lions. Sur les problémes mixtes pour certains systémes paraboliques dans des ouverts noncylindriques. *Ann. Inst. Fourier*, 7: 143–182, 1957.

[156] J. L. Lions and E. Magenes. *Non–homogeneous boundary value problems and applications. Vol. I–III*. Springer, New York, Heidelberg, 1972.

[157] A. Liu and B. Joe. Quality local refinement of tetrahedral meshes based on bisection. *SIAM J. Sci. Comput.*, 16 (6): 1269–1291, 1995.

[158] W. Liu and N. Yan. A posteriori error estimates for optimal control problems governed by parabolic equations. *Numer. Math.*, 93 (3): 497–521, 2003.

[159] S. MacLachlan and Y. Saad. A greedy strategy for coarse–grid selection. *SIAM J. Sci. Comput.*,

29 (5): 1825–1853, 2007.
[160] C. Makridakis and R. H. Nochetto. Elliptic reconstruction and a posteriori error estimates for parabolic problems. *SIAM J. Numer. Anal.*, 41 (4): 1585–1594, 2003.
[161] C. G. Makridakis and I. Babuška. On the stability of the discontinuous Galerkin method for the heat equation. *SIAM J. Numer. Anal.*, 34 (1): 389–401, 1997.
[162] O. Mali, P. Neittaanmäki, and S. I. Repin. *Accuracy verification methods*. Volume 32 of Comput. Methods Appl. Sci. Springer, Dordrecht, 2014.
[163] J. Mandel, B. Sousedík, and C. R. Dohrmann. Multispace and multilevel BDDC. *Computing*, 83 (2): 55–85, 2008.
[164] S. Matculevich and M. Wolfmayr. On the a posteriori error analysis for linear Fokker–Planck models in convection–dominated diffusion problems. *Appl. Math. Comput.*, 339: 779–804, 2018.
[165] S. V. Matculevich and S. I. Repin. Computable estimates of the distance to the exact solution of the evolutionary reaction–diffusion equation. *Appl. Math. Comput.*, 247: 329–347, 2014.
[166] J. M. Maubach. Local bisection refinement for N-simplicial grids generated by reflection. *SIAM J. Sci. Comput.*, 16 (1): 210–227, 1995.
[167] J. H. P. McKean. Nagumo's equation. *Advances in Math.*, 4 (3): 209–223, 1970.
[168] D. Meidner and B. Vexler. Adaptive space–time finite element methods for parabolic optimization problems. *SIAM J. Control Optim.*, 46 (1): 116–142, 2007.
[169] C. Mollet. Stability of Petrov–Galerkin discretizations: Application to the space–time weak formulation for parabolic evolution problems. *Comput. Meth. Appl. Math.*, 14: 231–255, 2014.
[170] C. Mollet. *Parabolic PDEs in space-time formulations: Stability for Petrov–Galerkin discretizations with B-splines and existence of moments for problems with random coefficients*. PhD thesis, Universität Köln, 2016.
[171] K. S. Moon, R. H. Nochetto, R. von Petersdorff, and C. Zhang. A posteriori error analysis for parabolic variational inequalities. *ESAIM: M2AN*, 41 (3): 485–511, 2007.
[172] M. Neumüller. Eine Finite Elemente Methode für optimale Kontrollprobleme mit parabolischen Randwertaufgaben. Master's thesis, TU Graz, 2010.
[173] M. Neumüller. *Space–Time Methods: Fast Solvers and Applications*. Volume 20 of Monographic Series TU Graz: Computation in Engineering and Science, 2013.
[174] M. Neumüller and O. Steinbach. Refinement of flexible space–time finite element meshes and discontinuous Galerkin methods. *Comput. Visual. Sci.*, 14 (5): 189–205, 2011.
[175] H. R. Nochetto, K. G. Siebert, and A. Veeser. Theory of adaptive finite element methods: An introduction. In R. DeVore and A. Kunoth, editors, *Multiscale, nonlinear and adaptive approximation*, pages 409–542. Springer, Berlin, Heidelberg, 2009.
[176] R. H. Nochetto. Error estimates for multidimensional singular parabolic problems. *Japan J. Appl. Math.*, 4 (1): 111–138, 1987.
[177] R. H. Nochetto, A. Schmidt, and C. Verdi. A posteriori error estimation and adaptivity for degenerate parabolic problems. *Math. Comput.*, 69 (229): 1–24, 2000.
[178] R. H. Nochetto, T. von Petersdorff, and C. Zhang. A posteriori error analysis for a class of integral equations and variational inequalities. *Numer. Math.*, 116 (3): 519–552, 2010.
[179] R. H. Nochetto, E. Otárola, and A. J. Salgado. A PDE approach to fractional diffusion in general domains: A priori error analysis. *Found. Comput. Math.*, 15 (3): 733–791, 2015.
[180] R. H. Nochetto, E. Otárola, and A. J. Salgado. A PDE approach to space-time fractional parabolic problems. *SIAM J. Numer. Anal.*, 54 (2): 848–873, 2016.
[181] M. Picasso. Adaptive finite elements for a linear parabolic problem. *Comput. Methods Appl. Mech. Engrg.*, 167 (3): 223–237, 1998.
[182] M. Picasso. Anisotropic a posteriori error estimate for an optimal control problem governed by the heat equation. *Numer. Methods Partial Differential Equations*, 22 (6): 1314–1336, 2006.

[183] C. Popa. Algebraic multigrid smoothing property of Kaczmarz's relaxation for general rectangular linear systems. *Electron. Trans. Numer. Anal.*, 29: 150–162, 2007.
[184] S. Prudhomme and J. T. Oden. A posteriori error estimation and error control for finite element approximations of the time-dependent Navier–Stokes equations. *Finite Elem. Anal. Des.*, 33 (4): 247–262, 1999.
[185] S. I. Repin. A posteriori error estimation for variational problems with uniformly convex functionals. *Math. Comput.*, 69 (230): 481–500, 2000.
[186] S. I. Repin. Estimates of deviations from exact solutions of initial-boundary value problem for the heat equation. *Rend. Mat. Acc. Lincei*, 13 (2): 121–133, 2002.
[187] S. I. Repin. *A posteriori estimates for partial differential equations*. Volume 4 of Radon Series on Computational and Applied Mathematics. Walter de Gruyter, Berlin, 2008.
[188] S. I. Repin and S. K. Tomar. A posteriori error estimates for approximations of evolutionary convection–diffusion problems. *J. Math. Sci.*, 170 (4): 554–566, 2010.
[189] M. Ries, U. Trottenberg, and G. Winter. A note on MGR methods. *Linear Algebra Appl.*, 49: 1–26, 1983.
[190] R. Rodríguez. Some remarks on Zienkiewicz–Zhu estimator. *Numer. Methods Partial Differential Equations*, 10 (5): 625–635, 1994.
[191] J. W. Ruge and K. Stüben. Algebraic multigrid (AMG). In S. F. McCormick, editor, *Multigrid Methods*, pages 73–130, 1987.
[192] Y. Saad. *Iterative methods for sparse linear systems*. SIAM, Philadelphia, second edition, 2003.
[193] A. Schafelner. Space-time finite element methods for parabolic initial-boundary problems. Master's thesis, Johannes Kepler Universität Linz, 2017.
[194] F. Schlögl. A characteristic critical quantity in nonequilibrium phase transitions. *Z. Phys. B - Condensed Matter*, 52 (1): 51–60, 1983.
[195] M. Schmich and B. Vexler. Adaptivity with dynamic meshes for space-time finite element discretizations of parabolic equations. *SIAM J. Sci. Comput.*, 30 (1): 369–393, 2008.
[196] C. Schwab and R. Stevenson. Space–time adaptive wavelet methods for parabolic evolution problems. *Math. Comput.*, 78 (267): 1293–1318, 2009.
[197] L. R. Scott and S. Zhang. Finite element interpolation of nonsmooth functions satisfying boundary conditions. *Math. Comput.*, 54 (190): 483–493, 1990.
[198] J. Sen Gupta, R. K. Sinha, G. M. M. Reddy, and J. Jain. A posteriori error analysis of two–step backward differentiation formula finite element approximation for parabolic interface problems. *J. Sci. Comput.*, 69 (1): 406–429, 2016.
[199] J. Sen Gupta, R. K. Sinha, G. M. M. Reddy, and J. Jain. New interpolation error estimates and a posteriori error analysis for linear parabolic interface problems. *Numer. Methods Partial Differential Equations*, 33 (2): 570–598, 2017.
[200] W. Shen, L. Ge, D. Yang, and W. Liu. Sharp a posteriori error estimates for optimal control governed by parabolic integro–differential equations. *J. Sci. Comput.*, 65 (1): 1–33, 2015.
[201] O. Steinbach. Adaptive finite element–boundary element solution of boundary value problems. *J. Comput. Appl. Math.*, 106 (2): 307–316, 1999.
[202] O. Steinbach. Space-time finite element methods for parabolic problems. *Comput. Meth. Appl. Math.*, 15: 551–566, 2015.
[203] O. Steinbach and H. Yang. An adaptive space–time finite element method for solving the heat equation. Technical report, TU Graz, 2019, in preparation.
[204] O. Steinbach and H. Yang. Comparison of algebraic multigrid methods for an adaptive space–time finite–element discretization of the heat equation in 3D and 4D. *Numer. Linear Algebra Appl.*, e2143, 2018.
[205] O. Steinbach and H. Yang. An algebraic multigrid method for an adaptive space–time

finite element discretization. In I. Lirkov and S. D. Margenov, editors, *Large-scale scientific computing: 11th International Conference, LSSC 2017, Sozopol, Bulgaria, June 5–9, 2017*, pages 66–73. Springer, 2018.
[206] O. Steinbach and M. Zank. Coercive space-time finite element methods for initial boundary value problems, 2018, submitted.
[207] R. Stevenson. The completion of locally refined simplicial partitions created by bisection. *Math. Comput.*, 77 (261): 227–241, 2008.
[208] K. Stüben. A review of algebraic multigrid. *J. Comput. Appl. Math.*, 128 (1–2): 281–309, 2001.
[209] K. Stüben and U. Trottenberg. Multigrid methods: Fundamental algorithms, model problem analysis and applications. In W. Hackbusch and U. Trottenberg, editors, *Multigrid methods: Proceedings of the conference held at Köln–Porz, November 23–27, 1981*, pages 1–176. Springer Berlin Heidelberg, Berlin, Heidelberg, 1982.
[210] V. Thomée. *Galerkin finite element methods for parabolic problems*. Springer Series in Computational Mathematics. Springer, Heidelberg, 2006.
[211] I. Toulopoulos. Stabilized space-time finite element methods of parabolic evolution problems. *Appl. Anal.*, published online, 2018. https://doi.org/10.1080/00036811.2018.1522630.
[212] C. Traxler. An algorithm for adaptive mesh refinement in n dimensions. *Computing*, 59 (2): 115–137, 1997.
[213] F. Tröltzsch. *Optimal control of partial differential equations: Theory, methods, and applications*. Graduate studies in mathematics. American Mathematical Society, 2010.
[214] U. Trottenberg, C. W. Oosterlee, and A. Schüller. *Multigrid*. Academic Press, New York, 2001.
[215] X. Tu. Three-level BDDC in three dimensions. *SIAM J. Sci. Comput.*, 29 (4): 1759–1780, 2007.
[216] K. Urban and A. T. Patera. An improved error bound for reduced basis approximation of linear parabolic problems. *Math. Comput.*, 83 (288): 1599–1615, 2014.
[217] P. Vaněk, J. Mandel, and M. Brezina. Algebraic multigrid by smoothed aggregation for second and fourth order elliptic problems. *Computing*, 56 (3): 179–196, 1996.
[218] R. Verfürth. A posteriori error estimates for nonlinear problems: Finite element discretizations of elliptic equations. *Math. Comput.*, 62 (206): 445–475, 1994.
[219] R. Verfürth. A posteriori error estimates for nonlinear problems. $L^r(0,T;L^\rho(\Omega))$-error estimates for finite element discretizations of parabolic equations. *Math. Comput.*, 67 (224): 1335–1360, 1998.
[220] R. Verfürth. A posteriori error estimates for nonlinear problems: $L^r(0,T;W^{1,\rho}(\Omega))$-error estimates for finite element discretizations of parabolic equations. *Numer. Methods Partial Differential Equations*, 14 (4): 487–518, 1998.
[221] R. Verfürth. A posteriori error estimates for finite element discretizations of the heat equation. *Calcolo*, 40 (3): 195–212, 2003.
[222] R. Verfürth. Robust a posteriori error estimates for nonstationary convection–diffusion equations. *SIAM J. Numer. Anal.*, 43 (4): 1783–1802, 2006.
[223] R. Verfürth. *A posteriori error estimation techniques for finite element methods*. Numerical mathematics and scientific computation. Oxford University Press, Oxford, 2013.
[224] M. Vohralík. A posteriori error estimates for lowest–order mixed finite element discretizations of convection–diffusion–reaction equations. *SIAM J. Numer. Anal.*, 45 (4): 1570–1599, 2007.
[225] M. Vohralík. Residual flux-based a posteriori error estimates for finite volume and related locally conservative methods. *Numer. Math.*, 111 (1): 121–158, 2008.
[226] P. Wesseling. *An introduction to multigrid methods*. John Wiley & Sons, Philadelphia, 2004.
[227] M. F. Wheeler. A priori L_2 error estimates for Galerkin approximations to parabolic partial differential equations. *SIAM J. Numer. Anal.*, 10 (4): 723–759, 1973.
[228] R. Wienands and W. Joppich. *Practical Fourier analysis for multigrid methods*. Chapman and Hall/CRC Press, 2005.

[229] J. Xu and L. Zikatanov. A monotone finite element scheme for convection–diffusion equations. *Math. Comput.*, 68 (228): 1429–1446, 1999.

[230] H. Yang. Partitioned solvers for the fluid-structure interaction problems with a nearly incompressible elasticity model. *Comput. Visual. Sci.*, 14 (5): 227–247, 2011.

[231] S. Zhang. *Multi-level iterative techniques*. PhD thesis, Pennstate University, 1988.

Index

Radon Series on Computational and Applied Mathematics

Volume 24
Maxwell's Equations. Analysis and Numerics
Ulrich Langer, Dirk Pauly, Sergey I. Repin (Eds.), 2019
ISBN: 978-3-11-054264-6, e-ISBN: 978-3-11-054361-2

Volume 23
Combinatorics and Finite Fields. Difference Sets, Polynomials, Pseudorandomness and Applications
Kai-Uwe Schmidt, Arne Winterhof (Eds.), 2019
ISBN: 978-3-11-064179-0, e-ISBN: 978-3-11-064209-4

Volume 22
The Radon Transform. The First 100 Years and Beyond
Ronny Ramlau, Otmar Scherzer (Eds.), 2019
ISBN: 978-3-11-055941-5, e-ISBN: 978-3-11-056085-5

Volume 21
Hamilton-Jacobi-Bellman Equations. Numerical Methods and Applications in Optimal Control
Dante Kalise, Karl Kunisch, Zhiping Rao (Eds.), 2018
ISBN: 978-3-11-054263-9, e-ISBN: 978-3-11-054359-9

Volume 20
Fluid-Structure Interaction. Modeling, Adaptive Discretisations and Solvers
Stefan Frei, Bärbel Holm, Thomas Richter, Thomas Wick, Huidong Yang (Eds.), 2017
ISBN: 978-3-11-049527-0, e-ISBN: 978-3-11-049425-9

Volume 19
Tensor Numerical Methods in Scientific Computing
Boris N. Khoromskij, 2018
ISBN: 978-3-11-037013-3, e-ISBN: 978-3-11-036591-7

Volume 18
Variational Methods. In Imaging and Geometric Control
Maitine Bergounioux, Gabriel Peyré, Christoph Schnörr, Jean-Baptiste Caillau, Thomas Haberkorn (Eds.), 2016
ISBN: 978-3-11-043923-6, e-ISBN: 978-3-11-043039-4

www.degruyter.com

Bei Fragen zur Produktsicherheit wenden Sie sich bitte an:
If you have any questions regarding product safety,
please contact:

Walter de Gruyter GmbH
Genthiner Straße 13
10785 Berlin
productsafety@degruyterbrill.com